Disruptive Technologies and Eco-Innovation for Sustainable Development

Ulas Akkucuk
Boğaziçi University, Turkey

IGI Global
PUBLISHER of TIMELY KNOWLEDGE

A volume in the Advances in Environmental Engineering and Green Technologies (AEEGT) Book Series

Published in the United States of America by
IGI Global
Engineering Science Reference (an imprint of IGI Global)
701 E. Chocolate Avenue
Hershey PA, USA 17033
Tel: 717-533-8845
Fax: 717-533-8661
E-mail: cust@igi-global.com
Web site: http://www.igi-global.com

Library of Congress Cataloging-in-Publication Data

Names: Akkucuk, Ulas, 1975- editor.
Title: Disruptive technologies and eco-innovation for sustainable development / Ulas Akkucuk.
Description: Hershey, PA : Engineering Science Reference, an imprint of IGI Global, [2022] | Includes bibliographical references and index. | Summary: "This book provides managers, academicians, scientists, and researchers in various government, public, and private sectors an in-depth look into new techniques, strategies, and technologies for achieving environmental sustainability through best business and technology practices"-- Provided by publisher.
Identifiers: LCCN 2021026045 (print) | LCCN 2021026046 (ebook) | ISBN 9781799889007 (h/c) | ISBN 9781799889014 (s/c) | ISBN 9781799889021 (ebook)
Subjects: LCSH: Environmental protection. | Disruptive technologies. | Sustainable development.
Classification: LCC TD170.3 .D58 2022 (print) | LCC TD170.3 (ebook) | DDC 363.7--dc23
LC record available at https://lccn.loc.gov/2021026045
LC ebook record available at https://lccn.loc.gov/2021026046

This book is published in the IGI Global book series Advances in Environmental Engineering and Green Technologies (AEEGT) (ISSN: 2326-9162; eISSN: 2326-9170)

British Cataloguing in Publication Data
A Cataloguing in Publication record for this book is available from the British Library.

Advances in Environmental Engineering and Green Technologies (AEEGT) Book Series

Sang-Bing Tsai
Zhongshan Institute, University of Electronic Science and Technology of China, China & Wuyi University, China
Ming-Lang Tseng
Lunghwa University of Science and Technology, Taiwan
Yuchi Wang
University of Electronic Science and Technology of China Zhongshan Institute, China

ISSN:2326-9162
EISSN:2326-9170

Mission

Growing awareness and an increased focus on environmental issues such as climate change, energy use, and loss of non-renewable resources have brought about a greater need for research that provides potential solutions to these problems. Research in environmental science and engineering continues to play a vital role in uncovering new opportunities for a "green" future.

The **Advances in Environmental Engineering and Green Technologies (AEEGT)** book series is a mouthpiece for research in all aspects of environmental science, earth science, and green initiatives. This series supports the ongoing research in this field through publishing books that discuss topics within environmental engineering or that deal with the interdisciplinary field of green technologies.

Coverage

- Renewable Energy
- Sustainable Communities
- Electric Vehicles
- Water Supply and Treatment
- Radioactive Waste Treatment
- Pollution Management
- Cleantech
- Biofilters and Biofiltration
- Green Transportation
- Waste Management

IGI Global is currently accepting manuscripts for publication within this series. To submit a proposal for a volume in this series, please contact our Acquisition Editors at Acquisitions@igi-global.com or visit: http://www.igi-global.com/publish/.

The Advances in Environmental Engineering and Green Technologies (AEEGT) Book Series (ISSN 2326-9162) is published by IGI Global, 701 E. Chocolate Avenue, Hershey, PA 17033-1240, USA, www.igi-global.com. This series is composed of titles available for purchase individually; each title is edited to be contextually exclusive from any other title within the series. For pricing and ordering information please visit http://www.igi-global.com/book-series/advances-environmental-engineering-green-technologies/73679. Postmaster: Send all address changes to above address.

Titles in this Series

For a list of additional titles in this series, please visit: www.igi-global.com/book-series

Handbook of Research on Microbial Remediation and Microbial Biotechnology for Sustainable Soil
Junaid Ahmad Malik (Government Degree College, Bijbehara, India)
Engineering Science Reference • ©2021 • 806pp • H/C (ISBN: 9781799870623) • US $295.00

Handbook of Research on Strategic Management for Current Energy Investments
Serhat Yüksel (Istanbul Medipol University, Turkey) and Hasan Dinçer (Istanbul Medipol University, Turkey)
Engineering Science Reference • ©2021 • 422pp • H/C (ISBN: 9781799883357) • US $295.00

Forest Fire Danger Prediction Using Deterministic-Probabilistic Approach
Nikolay Viktorovich Baranovskiy (National Research Tomsk Polytechnic University, Russia)
Engineering Science Reference • ©2021 • 297pp • H/C (ISBN: 9781799872504) • US $195.00

Role of IoT in Green Energy Systems
Vasaki Ponnusamy (Universiti Tunku Abdul Rahman, Malaysia) Noor Zaman (Taylor's University, Malaysia) Low Tang Jung (Universiti Teknologi PETRONAS, Malaysia) and Anang Hudaya Muhamad Amin (Higher Colleges of Technology, UAE)
Engineering Science Reference • ©2021 • 405pp • H/C (ISBN: 9781799867098) • US $195.00

Combating and Controlling Nagana and Tick-Borne Diseases in Livestock
Caleb Oburu Orenge (Department of Veterinary Anatomy and Physiology, Faculty of Veterinary Medicine and Surgery, Egerton University, Kenya)
Engineering Science Reference • ©2021 • 440pp • H/C (ISBN: 9781799864332) • US $185.00

Artificial Intelligence and IoT-Based Technologies for Sustainable Farming and Smart Agriculture
Pradeep Tomar (Gautam Buddha University, India) and Gurjit Kaur (Delhi Technological University, India)
Engineering Science Reference • ©2021 • 400pp • H/C (ISBN: 9781799817222) • US $215.00

Handbook of Research on Waste Diversion and Minimization Technologies for the Industrial Sector
Ashok K. Rathoure (M/s Akone Services, Lucknow, India)
Engineering Science Reference • ©2021 • 455pp • H/C (ISBN: 9781799849216) • US $295.00

IGI Global
PUBLISHER of TIMELY KNOWLEDGE

701 East Chocolate Avenue, Hershey, PA 17033, USA
Tel: 717-533-8845 x100 • Fax: 717-533-8661
E-Mail: cust@igi-global.com • www.igi-global.com

Table of Contents

Detailed Table of Contents

The Sustainable Development Goals (SDGs) can be seen as the critical goal for every country in the world. In this vein, a stable global financial system is needed these days to satisfy its duty to boost the private capital mobilisation to achieve sustainable development and steady economic growth. Nevertheless, several obstacles limiting such financial mobilisation have been identified by scholars, practitioners, and standard setters. Recently, digital transformation and advancement, specifically in the finance sector, include a wide range of technological developments, and applications such as blockchain, internet of things, big data, artificial intelligence are promised to enhance performance in the financial sector. The potential of digital applications in the finance sector to resolve critical obstacles in financing for inclusive and sustainable growth becomes evident. This chapter aims to provide a summary and a detailed discussion of the latest developments in financial technologies that both facilitate the SDGs and also contribute to future sustainable international business.

The governments, business firms, policymakers, advocacy groups, and even the public recently are hotly debating on the issues of environmentally friendly practices. In this context, being a part of 'going green', green banking, which plays an important role in environmental sustainability, has been a buzzword in the global baking industry. This study identifies how the customers perceive the emerging concept of green banking initiatives of banks and also analyzes the factors that influence such practices of the customers. Using a structured questionnaire, the primary data were collected from 403 commercial banks of Kathmandu valley, Nepal. The awareness index was prepared, and the binary logit model was applied for the econometric analysis. This study observed that the customers are positive towards the environmentally friendly practices of banks and ready to adopt the green banking practices. The research implies that in order to promote environmental sustainability, banks and financial institutions should be able to educate the customers about green banking practices and their benefits.

Renewable energy sources are clean energy sources that meet the energy needs in a sustainable way. Therefore, it is necessary to invest in renewable energy sources. However, there are some difficulties in renewable energy investments. It has problems such as high initial installation cost, underdeveloped technological infrastructure, and insufficient financial support. Several financial products are being developed in order to overcome the mentioned difficulties. In this context, the purpose of this study is to explain the role of green sukuk in the financing of renewable energy investments. Depending on this purpose, the study has been examined with a literature review. The production of renewable energy sources can be encouraged with green sukuk. However, green sukuk is an advantage for Islamic companies that want to realize environmental projects. On the other hand, it offers the opportunity to the environmentally friendly Islamic investors to evaluate their savings.

The purpose of this chapter is to review the Turkish private pension system as a key determinant of sustainable development of the country. The private pension system is of great significance in order to promote sustainable development. This is attributable to the fact that high level of total savings which are secured through individual savings are conducive to investments and in turn finances sustainable development and growth. Turkey, as an emerging economy, is considered to be one of those economies with low level of total savings. In an effort to deal with this issue, governments paid attention to an increasing level of savings of households. Individual retirement system (private pension system) is devised so as to increase savings and to fund investments with national sources. Therefore, it is safely argued that the very rationale behind the introduction of private pension system is associated directly with the aim of increasing total savings in the economy and ensuring sustainable development.

This chapter aims to analyse the concept and implementation strategies of Agriculture 4.0 within the framework of the study of disruptive technologies and eco-innovation, which allows facing the needs derived from a sustainable food system. To do this, it strategically reflects on the design requirements of a holistic model for the transformation of agricultural holdings, aimed at the implementation of sustainable agrotechnology. The Third Green Revolution, its antecedents, orientations, and purposes, as well as the concept and functional aspects of sustainable food systems are analysed. Finally, a model of transformation of agricultural holdings towards the implementation of Sustainable Agriculture 4.0 is proposed, as well as a generic methodology applicable to specific projects located in specific areas, through formula for change and cost-benefit analysis.

Sustainable production and consumption of food are vital for sustainable development. About one-third of all food produced for humans are either lost or wasted causing increased food insecurity and immense economic and social costs. In a world where famine has been an alarming issue, any action to reduce food loss and waste (FLW) is crucial. This chapter reviews, from a sustainable supply chain perspective, the extant literature on food supply chains and discusses FLW issues, especially within the context of sustainable consumption of fruits and vegetables. A framework for sustainable food supply chains (SFSCs) from both production and consumption ends are discussed. In doing so, such current disruptive intelligent technologies as blockchain and the internet of things are emphasized as potential enablers for SFSCs. Mainly driven by consumers' awareness of the pressing issues in the world and consumption behaviour, mitigating FLW in SFSCs would not only result in efficient land and water use but also positively impact climate change and livelihoods towards sustainable development.

The implementation of urban gardens, which are increasingly appearing in cities, aims to respond actively to the growing demand for urban spaces for the installation of urban gardens, creating conditions for the practice of sustainable agriculture in an urban context. Through these initiatives, it is intended to ensure that the needs of the population are met and to maximize the benefits arising from the practice of urban agriculture, both for the environment and for people's quality of life. Technology is a facilitating element in the process of acquiring and maintaining these urban gardens. This chapter presents an app that can be used by farmers to manage the production of consumer goods in this space, providing information about the status of crops, products to be grown, and types of required maintenance. This app simplifies the production process and also increases the sustainability of agriculture activities considering the economic, social, and logistical dimensions.

The objective of this chapter is to analyze the functioning of an integral model of entrepreneurship in green innovation business (GIB) that is currently emerging and in the process of internationalization. Therefore, this work aims to study the central perspectives of technology that are based on the phenomenon of entrepreneurship and thus develop a strategy that adapts to companies with an ecological basis to achieve internationalization. This chapter analyzes a particular company that specializes in ecological biomineral organic fertilizer, where no chemical product is used to produce the composition; everything that is marketed is made up of a base of organic minerals and other organic compounds.

The aim of the chapter was to model the impact of selected determinants (trade openness, human capital, entrepreneurship, and innovation) on regional competitiveness, as well as to propose future activities and measures required to be implemented to improve the competitive performance of the regions. The research was conducted on the sample of 18 regions in six European countries: Serbia, Croatia, Slovenia, Northern Macedonia, Montenegro, and Romania. The database was prepared, and the statistical processing was performed in SPSS. In this data analysis, the following methods were used: comparative analysis, correlation, and regression analysis. The results of the research showed that the impact of the determinants—import dependence, the number of pupils enrolled in secondary education, gross domestic expenditure on research and development, and the number of companies per 10,000 inhabitants on the competitiveness of the region—are (statistically) significant.

The aim of the chapter is to point the importance of eco-innovation and IT technologies for the sustainable development of health and recreational tourism in Serbia. The subject of the research is the rehabilitation center in Eastern Serbia. The main idea of the chapter is to show how those hospital institutions use artificial intelligence-IT technologies for improving recovery services to patients in the post-COVID condition. The chapter will discuss the most common types of support and measures to facilitate the functioning of eco-tourism in Serbia with the aim to adopt good practices of developed countries (Hungary). Ecological tourism takes place in areas of pure and preserved nature. The contribution of the chapter is to point to new strategies in spa tourism, to shorten the time and reduce business costs. This would contribute to the sustainability of tourism.

Building on the natural-resource-based view, and using a sample of 7,165 European SMEs, this chapter investigates the drivers of eco-design innovations among SMEs under three categories: (1) sustainability-oriented firm capabilities, (2) technological capabilities, and (3) access to finance. The findings reveal that sustainability-oriented capabilities achieved through investments into circular economy are the strongest driver of SMEs' eco-design innovations. Firms' technological capabilities are also found to boost their ability to adopt eco-design innovations. While equity finance increases the possibilities for SMEs to devote resources to eco-design, grant finance is interestingly observed to decrease such possibilities. The more traditionally used form of debt finance remains detached from eco-design implementations. The study contributes to a better understanding of how eco-design practices can be broadened within SMEs and highlights policy recommendations in this regard.

Green supplier selection has a crucial importance for businesses. In the past, the selection of suppliers was solely based on conventional criteria such as cost, quality, and flexibility whereas expectations of businesses transformed in today's world on grounds of raised environmental awareness, public pressure, and regulations. Alternatives called green suppliers sensitive to the environment, preserving the ecological balance, managing wastes, and preventing pollution increased in value. This study analyzes practices on the selection of green suppliers. The articles between 2014 and 2021 were analyzed from the perspective of green criteria. The green criteria in the 50 articles determined are divided into 28 groups. With the k-means algorithm, these criteria groups are divided into four clusters, which was aimed to analyze the usage frequency of green criteria. This study is intended to contribute to green supplier selection practices in academia and industry in the future.

Ranking and choosing research projects and analyzing experiments are usually difficult and complex responsibilities for professional research councils at universities and research centers. Its complexity stems from having more than one variable in each project, and the participation of many decision-makers in the

Preface

Disruptive technologies have dramatically altered the way traditional businesses operate and have been studied in the last two decades in every sector. Disruptive technologies are also very important from the point of view of sustainable development. These technologies can cause significant greenhouse gas reductions and other benefits in terms of logistics and smart cities. This book will explore this very important concept as it relates to all the relevant industries. This book offers research contributions, constructive debates, and investigations on new legislation on disruptive technologies, green (sustainable) IT and processes, Industry 4.0, healthcare informatics and applications in terms of environmental, defense and social issues for both the manufacturing and service industries.

Disruptive Technologies and Eco-Innovation for Sustainable Development provides managers, academicians, scientists, and researchers in various government, public, and private sectors an in-depth look into new techniques, strategies, and technologies for achieving environmental sustainability through best business and technology practices. The book covers topical issues including sustainable strategy and innovation, advanced technologies related to the 4th Industry Revolution, green digitalization, sustainable asset management (ILS, operation and maintenance, reliability, life cycle etc.), energy and defense efficient systems, green and sustainable transformation, as well as other topics like product usability, reverse and closed loop supply chain, environmental issues (carbon footprints, global warming, recycling and reuse systems, applied ergonomics, climate change), and all those topics related to logic, philosophy and history of science and technology from the green/sustainable point of view.

The many academic areas covered in this publication include:

- Disruptive logistics,
- Green IT,
- Reverse logistics,
- Industry 4.0,
- Green logistics,
- SCOR model,
- Sustainable farming,
- Information systems,
- Green energy,
- Sustainable energy,
- Carbon emissions,
- Robots,
- Sustainable cities,

- Smart cities
- Digital Transformation
- E-Commerce
- Green Supply Chains
- Life Cycle Costing
- Multi-Criteria Decision Making
- Quality Management
- Renewable Energy Sources

I am very happy to finalize the ninth book project I realized with IGI Global. I gave the first book proposal to IGI Global in the summer of 2013. This first book was finalized towards fall 2014 and published as hard copy in January 2015. This publication included 28 contributions. It has been nearly six years since the release. Later it also enjoyed Web of Science indexation. Sustainable development over the years has enjoyed ever more attention in the academic community. As a result I decided to pursue other book projects with similar titles. This last book contains 15 valuable contributions from eminent authors from nine different countries. The countries include Turkey, Canada, Pakistan, Nepal, Spain, Portugal, Republic of Serbia, Mexico and Vietnam.

Chapter 1 starts the discussion with an excellent analytical paper entitled "Developments in Financial Technologies for achieving Sustainable Development Goals (SDGs): FinTech and SDGs". The Sustainable Development Goals (SDGs) can be seen as the critical goal for every country in the world. In this vein, a stable global financial system is needed these days to satisfy its duty to boost the private capital mobilisation to achieve sustainable development and steady economic growth. Nevertheless, several obstacles limiting such financial mobilisation have been identified by scholars, practitioners and standard setters. Recently, digital transformation and advancement, specifically in the finance sector, include a wide range of technological developments, and applications such as blockchain, internet of things, big data, artificial intelligence are promised to enhance performance in the financial sector. The potential of digital applications in the finance sector to resolve critical obstacles in financing for inclusive and sustainable growth becomes evident. This chapter aims to provide a summary and a detailed discussion of the latest developments in financial technologies that both facilitate the SDGs and also contribute to future sustainable international business.

Chapter 2 is about "Customer Perception and Awareness on Green Banking Practices: An Alternative Strategy on Environmental Sustainability". The governments, business firms, policymakers, advocacy groups and even the public recently are hotly debating on the issues of environmentally friendly practices. In this context, being a part of 'going green', Green Banking which plays an important role in environmental sustainability has been a buzzword in the global baking industry. This study identifies how the customers perceive the emerging concept of green banking initiatives of banks and also analyzes the factors that influence such practices of the customers. Using a structured questionnaire, the primary data were collected from 403 commercial banks of Kathmandu valley, Nepal. The awareness index was prepared and the binary logit model is applied for the econometric analysis. This study observed that the customers are positive towards the environmentally friendly practices of banks and ready to adopt the green banking practices. The research implies that in order to promote environmental sustainability, banks and financial institutions should be able to educate the customers about green banking practices and their benefits.

Chapter 3 investigates "The Role of Green Sukuk for Sustainable Energy Production". Renewable energy sources are a clean energy source that meets the energy needs in a sustainable way. Therefore, it is necessary to invest in renewable energy sources. However, there are some difficulties in renewable energy investments. It has problems such as high initial installation cost, underdeveloped technological infrastructure, and insufficient financial support. Several financial products are being developed in order to overcome the mentioned difficulties. In this context, the purpose of this study is to explain the role of green sukuk in the financing of renewable energy investments. Depending on this purpose, the study has been examined with a literature review. The production of renewable energy sources can be encouraged with green sukuk. However, green sukuk is an advantage for Islamic companies that want to realize environmental projects. On the other hand, it offers the opportunity to the environment-friendly Islamic investors to evaluate their savings.

Chapter 4 is a discussion on "Financing Sustainable Development in an Emerging Economy: Private Pension System in Turkey". The purpose o this chapter is to review Turkish Private Pension System as a key determinant of sustainable development of the country. Private pension system is of great significance in order to promote sustainable development. This is attributable to the fact that high level of total savings which are secured through individual savings are conducive to investments and in turn finances sustainable development and growth. Turkey, as an emerging economy, is considered to be one of those economies with low level of total savings. In an effort to deal with this issue, governments paid attention to increasing level of savings of households. Individual retirement system (private pension system) is devised so as to increase savings and to fund investments with national sources. Therefore, it is safely argued that the very rationale behind the introduction of private pension system is associated directly with the aim of increasing total savings in the economy and ensuring sustainable development.

Chapter 5 focuses on "Agriculture 4.0 for a Sustainable Food System: A Holistic Model for the Transformation of Farms Towards a Sustainable Precision Agriculture". This chapter aims to analyse the concept and implementation strategies of Agriculture 4.0 within the framework of the study of disruptive technologies and eco-innovation, which allows facing the needs derived from a sustainable food system. To do this, it strategically reflects on the design requirements of a holistic model for the transformation of agricultural holdings, aimed at the implementation of sustainable agrotechnology's. The Third Green Revolution, its antecedents, orientations, and purposes, as well as the concept and functional aspects of sustainable food systems are analysed. Finally, a model of transformation of agricultural holdings towards the implementation of Sustainable Agriculture 4.0 is proposed, as well as a generic methodology applicable to specific projects located in specific areas, through Formula for Change and Cost-Benefit Analysis.

Chapter 6 is another look at the importance of Food Loss and Waste. Sustainable production and consumption of food are vital for sustainable development. About one-third of all food produced for humans is either lost or wasted causing increased food insecurity and immense economic and social costs. In a world where famine has been an alarming issue, any action to reduce food loss and waste (FLW) is crucial. This chapter reviews, from a sustainable supply chain perspective, the extant literature on food supply chains and discusses FLW issues, especially within the context of sustainable consumption of fruits and vegetables. A framework for sustainable food supply chains (SFSCs) from both production and consumption ends are discussed. In doing so, such current disruptive intelligent technologies as blockchain and the internet of things are emphasized as potential enablers for SFSCs. Mainly driven by consumers' awareness of the pressing issues in the world and consumption behaviour, mitigating FLW in SFSCs would not only result in efficient land and water use but also positively impacts climate change and livelihoods towards sustainable development.

Chapter 7 examines the concept of Managing a Municipal Urban Garden. The implementation of urban gardens, which are increasingly appearing in cities, aims to respond actively to the growing demand for urban spaces for the installation of urban gardens, creating conditions for the practice of sustainable agriculture in an urban context. Through these initiatives, it is intended to ensure that the needs of the population are met and to maximize the benefits arising from the practice of urban agriculture, both for the environment and for people's quality of life. Technology is a facilitating element in the process of acquiring and maintaining these urban gardens. This chapter presents an app that can be used by farmers to manage the production of consumer goods in this space, providing information about the status of crops, products to be grown, and types of required maintenance. This app simplifies the production process and also increases the sustainability of agriculture activities considering the economic, social, and logistical dimensions.

Chapter 8 focuses on a comprehensive entrepreneurship model for the internationalization of green innovation business. The objective of this paper is to analyze the functioning of an integral model of entrepreneurship in green innovation business (GIB) that is currently emerging and in the process of internationalization. Therefore, this work aims to study the central perspectives of technology that are based on the phenomenon of entrepreneurship and thus develop a strategy that adapts to companies with an ecological basis to achieve internationalization. This article analyzes a particular company, which specializes in ecological biomineral organic fertilizer, where no chemical product is used to produce the composition, everything that is marketed is made up of a base of organic minerals and other organic compounds.

Chapter 9 illustrates the notion of "Regional Competitiveness: Theoretical and Empirical Aspect". The aim of the paper was to model the impact of selected determinants (trade openness, human capital, entrepreneurship and innovation) on regional competitiveness, as well as to propose future activities and measures required to be implemented to improve the competitive performance of the regions. The research was conducted on the sample of 18 regions in six European countries: Serbia, Croatia, Slovenia, Northern Macedonia, Montenegro and Romania. The database was prepared and the statistical processing was performed in SPSS. In this data analysis, the following methods were used: comparative analysis, correlation and regression analysis. The results of the research showed that the impact of the determinants: Import Dependence, The Number of Pupils Enrolled in Secondary Education, Gross Domestic Expenditure on Research and Development and The Number of Companies per 10,000 inhabitants on the competitiveness of the region (statistically) significant.

Chapter 10 is a study on "Eco-Innovation and IT Technologies for Sustainable Development of Health and Recreational Tourism of Serbia". The aim of the chapter is to point the importance of eco-innovation and IT technologies for the sustainable development of health and recreational tourism in Serbia. The subject of the research is the rehabilitation center in Eastern Serbia. The main idea of the paper is to show how those hospital institutions, use artificial intelligence - IT technologies for improving recovery services to patients in the post-COVID condition. The chapter will discuss the most common types of support and measures to facilitate the functioning of eco-tourism in Serbia with the aim to with the aim to adopt good practices of developed countries (Hungary). Ecological tourism takes place in areas of pure and preserved nature. The contribution of the paper is to point to new strategies in spa tourism, to shorten the time and reduce business costs. This would contribute to the sustainability of tourism.

Chapter 11 provides examples on the drivers of eco-design innovations in European SMEs? Building on the natural-resource-based view, and using a sample of 7,165 European SMEs, this paper investigates the drivers of eco-design innovations among SMEs under three categories: (1) sustainability-oriented firm capabilities, (2) technological capabilities and (3) access to finance. The findings reveal that sustainability-oriented capabilities achieved through investments into circular economy are the strongest driver of SMEs' eco-design innovations. Firms' technological capabilities are also found to boost their ability to adopt eco-design innovations. While equity finance increases the possibilities for SMEs to devote resources to eco-design, grant finance is interestingly observed to decrease such possibilities. The more traditionally used form of debt finance remains detached from eco-design implementations. The study contributes to a better understanding of how eco-design practices can be broadened within SMEs and highlights policy recommendations in this regard.

Chapter 12 performs an analysis of investigation of green criteria with clustering analysis in green supplier selection. The selection of green suppliers is of crucial importance for businesses. In the past, the selection of suppliers was solely based on conventional criteria such as cost, quality, and flexibility whereas expectations of businesses transformed in today's world on grounds of raised environmental awareness, public pressure, and regulations. Alternatives called green suppliers sensitive to the environment, preserving the ecological balance, managing wastes, and preventing pollution increased in value. This study analyzes practices on the selection of green suppliers. The articles between 2014 and 2021 were analyzed from the perspective of green criteria. The green criteria in the 50 articles determined are divided into 28 groups. With the k-means algorithm, these criteria groups are divided into four clusters, which was aimed to analyze the usage frequency of green criteria. This study is intended to contribute to green supplier selection practices in academia and industry in the future.

Chapter 13 explores the "Application of Fuzzy Topsis and Taguchi Method for Optimization Problem With Disruptive Risk: A Systematic Review". Ranking and choosing research projects and analyzing experiments are usually difficult and complex responsibilities for professional research councils at universities and research centers. Its complexity stems from having more than one variable in each project, and the participation of many decision-makers in the ranking process and selection of research projects based on many variables. The Fuzzy set theory provides the required flexibility to show the uncertainty about the lack of knowledge, and also it can manage the uncertainty in the real world that the values of criteria are not defined properly. For this purpose, in the environment where the criteria of research projects are vaguely defined, the ranking methods such as Taguchi, which can reduce the number of experiments and making process more efficient, can be used for quality design in designing and processing product. In this work, first of all we will review fuzzy TOPSIS technique and the Taguchi method as well, then we can approach research efficiency and optimization of the level of effective parameters in an experiment.

Electric vehicles are gaining more popularity than ever. In this light Chapter 14 talks about "The Impact of Environmental Concern on Consumer Attitude and Intention Toward Electric Vehicles". This research aims to analyze the impact of environmental concerns on Turkish consumers' attitudes toward and intention to purchase EVs. It is also aimed to investigate the role of demographical characteristics –gender, age, income, and education- on those impacts. Given the purpose, a descriptive study was conducted through an online survey with a sample of 334 consumers. The research findings indicate that environmental concern plays a significantly influential role in the attitude toward and intention to purchase EVs. The study also confirms that environmental concern's impact on the attitude and intention toward EVs differs in terms of gender, age, and income. Only women show a positive influence of environmental concern on the attitude toward EVs. There is no difference in the direct or indirect link between EV purchase intention and environmental concern for age groups. Environmental concern's indirect influence on the intention to buy EVs is the highest and significantly different for the consumers with 7500-9999 TL income.

Wearable devices like watches coupled with smart phones are also gaining popularity in the health and sports sectors. The last chapter of the book will conclude with an up to date cross cultural study on this new technology. With rapid change in technology worldwide, innovative products such as wearable technology devices tend to have an uprising trend. Consumers, however, are not necessarily adaptive in their nature and their perception is shaped by many factors. The aim of this research is to investigate the consumer acceptance of wearable technology devices, specifically smartwatches. The study extends the widely used technology acceptance model with the introduction of new variables. For the purpose of the study, survey data was collected from German and Turkish university students. The overall results provide validation to previous literature while introducing new factors for consumer acceptance of technology products, wearable technology devices and smartwatches. Importance of this research comes from the innovative and promising nature of the wearable technology devices concept, the lack of work on smartwatches in literature as well as the cross-cultural nature of the study. The study also has managerial implications for technology companies who chase after growth in their businesses.

Ulaş Akküçük
Boğaziçi University, Turkey
July 2021

Chapter 1
Developments in Financial Technologies for Achieving the Sustainable Development Goals (SDGs):
FinTech and SDGs

Thinh Gia Hoang
https://orcid.org/0000-0001-5010-931X
School of Business and Management, RMIT University, Vietnam

Giang Ngo Tinh Nguyen
Faculty of Business and Law, The University of Newcastle, Callaghan Campus, Australia & School of Business and Management, RMIT University, Ho Chi Minh City, Vietnam

Dat Anh Le
Asia Commercial Bank, Vietnam & Centre for Applied Economics and Business Research, Hanoi, Vietnam

ABSTRACT

The Sustainable Development Goals (SDGs) can be seen as the critical goal for every country in the world. In this vein, a stable global financial system is needed these days to satisfy its duty to boost the private capital mobilisation to achieve sustainable development and steady economic growth. Nevertheless, several obstacles limiting such financial mobilisation have been identified by scholars, practitioners, and standard setters. Recently, digital transformation and advancement, specifically in the finance sector, include a wide range of technological developments, and applications such as blockchain, internet of things, big data, artificial intelligence are promised to enhance performance in the financial sector. The potential of digital applications in the finance sector to resolve critical obstacles in financing for inclusive and sustainable growth becomes evident. This chapter aims to provide a summary and a detailed discussion of the latest developments in financial technologies that both facilitate the SDGs and also contribute to future sustainable international business.

DOI: 10.4018/978-1-7998-8900-7.ch001

1. DIGITAL FINANCE AND SUSTAINABLE DEVELOPMENT GOALS (SDGS)

The Sustainable Development Goals (SDGs) were approved in September 2015 as a collective aim for the 194 participating countries in the United Nations to eradicate poverty, conserve the world and secure the prosperity for human beings. Scheyvens et al. (2016) stressed the long-term demand for driving sustainable and inclusive economic growth in the SDGs and the 2030 Agenda for Sustainable Development report. Furthermore, the G20 Sustainable Finance Study Group (SFSG) (2018) emphasises that, besides the environmental perspective, additional aspects of sustainable development must also be considered in order to gain benefits from private capital mobilisation to pursue sustainable development and the stability of the financial system. It is therefore proposing that the term sustainable finance should be adopted more broadly (Allen et al., 2016). The report indicates that, through direct and indirect support of the SDG framework, sustainable finance can be widely acknowledged as funding, related institutional and market arrangements which combine to lead to the achievement of solid, stable, healthy and inclusive growth. This movement was aimed at achieving the beneficial influences of investments for social and economic gains, such as job creation, technical innovation, poverty alleviation and social integration (Allen et al., 2018). A major problem in today's global financial environment is mobilising private capital to stimulate economic growth and stabilise the finance sector. Digitalisation, especially its nexus with finance (digital finance or financial technology), encompasses a wide variety of technology innovations, including big data, artificial intelligence (AI), mobile platforms, blockchain and the Internet of Things (IoT). Digital transformation is not a recent feature of the financial system. For the past twenty years, automated technologies and transformation have greatly increased performance in the financial industry (Collste et al., 2017; Hinson et al., 2019). Digital finance is increasingly demonstrating its potential to address obstacles relevant to the growth of finance for sustainable development.

In general, the term 'digital finance' represents the digitalisation or application of digital technologies in the financial sector. This trend covers digital advancement, services and applications in the financial services sector, such as online banking, chip cards, online payment, mobile wallets and payment apps (Zhou et al., 2018). Ozili (2018) describes digital finance enterprises as companies developing technologies for the incorporation of integrated digital banking, mobile applications and distribution networks, microfinance, payment solutions, peer-to-peer loans and crowd-funding. Although parts of the above-mentioned digital systems and services are well-developed (e.g. chip cards and online banking), there are several other services and business models that are innovative, yet not commonly employed, and also lead to the transformative potential for the financial sector. These creative services and business practices are focused on emerging technology, are at the forefront of recent debates and are also called "FinTech" solutions. Although there is no standard consensus on a definition, the term 'digital finance' is widely used in academia and the private sector, to represent a large number of the latest financial products, enterprises, applications and innovative ways of digital consumer engagement and interaction. Several definitions of digital finance have also been provided by international organisations and the private sector. George et al. (2020) defined digital financial services as any financial process involving information technology, namely digital money, mobile and online financial platforms, teller services and branchless banking. Meanwhile, according to the study by Przychodzen & Przychodzen (2015), it is characterised as the employment of the Internet and related innovative technology in the finance market.

In the same vein with "digital finance", the term "FinTech" is a terminology that derives from the terms "financial" and "technology" and generally explains the relationship between digital, in particular, the Internet-related technologies (including cloud technology, mobile technology, wireless technology, etc.) and existing business practices of the financial services sector (including loans, banking transactions and mortgages) (Puschmann, 2017). Traditionally, FinTech refers to pioneers and disruptors in the finance industry that employ the availability of instantaneous, flexible and protective financial communication channels, especially through the Internet and electronic data processing (Puschmann, 2017). Goldstein et al. (2019) distinguished between 'sustaining FinTech', i.e. existing financial services providers that aim to secure their market dominance through the use of information technology,' and "disruptive FinTech", i.e. new businesses and start-ups that threaten established providers by providing new products and services. Nicoletti et al., (2017) offered an insightful article on how differently the term "FinTech" is used and interpreted by numerous scholars and organisations of "FinTech", which stands for financial technology. It is generally defined as 'technically activated financial innovation that promotes new business models, applications, mechanisms and products, which could have a substantial influence on financial markets and financial services. Although there is no universal concept, the above definitions indicate that sustainable digital finance can be interpreted as the intentional employment of digital finance applications for funding, as well as support for related economic and market structures that contribute to sustainable growth.

There is a general consensus in research that FinTech can have a significant influence on financial growth. Nevertheless, whether FinTech encourages or restricts sustainable economic development has not yet been identified. The reason is that on the one hand, FinTech is related to networks, and data processing can improve the efficacy of financial services by enhancing financial products and service processes, and then encouraging technological advancement for future economic development. Furthermore, Puschmann (2017) confirms that FinTech can be incorporated into new businesses and social organisations, leading to an advanced economic paradigm that could boost economic development. On the other hand, Goldstein et al. (2019) claimed that neither an inadequate FinTech nor overregulation is beneficial to economic development. In addition, Nicoletti et al. (2017) suggested that the FinTech environment may have an impact on the transmission mechanism of monetary policy and the efficacy of macro-prudential government policies in shaping the financial cycle, thus constraining economic development.

Evidence also indicated that digital finance or FinTech has a great influence on ecological and social benefits (Deng et al., 2019). In regard to social development, scholars such as Zhang et al. (2021) have stated that the primary advantage of FinTech is its ability to build a more just and inclusive society. However, the potential consequences arising from the development of FinTech cannot be ignored (Ryu & Ko, 2020). As far as ecological sustainability is concerned, FinTech can stimulate the deployment of investments for energy security and environmental initiatives, encourage the construction of renewable energy and environmental facilities, and contribute to environmental development by providing cheap and adequate funding.

2. TECHNOLOGICAL APPLICATIONS IN FINANCIAL SECTOR

As stated earlier, FinTech can be described as emerging technologies that can expand the spectrum of finance and bring about a disruptive transformation of the way financial operations function in the world. Fintech is delivering many changes in the way business is operated in the traditional financial industry and turning them obsolete through disintermediation (Das, 2019). Although financial intermediaries such as commercial banks, investments banks and mutual funds still dominate the financial service and provide valuable financial service to customers and investors, their roles in the international market are being usurped by the fintech enterprises. For example, in the commercial banking area, the peer-to-peer lending platform allows individual borrowers or businesses to lend money directly from lenders at an affordable price rather than using commercial banking services (Duygun et al., 2020). Peer-to-peer lending is also widely known as social lending, which implies the facilitation of lending with disintermediation. The disintermediating financial service model offers benefits over traditional bank lending, such as lower capital requirements, flexible and favourable lending term, and lower operating expenses than banks (Boratyńska, 2019). Another example came from the increasing popularity of cryptocurrencies such as Bitcoin or Ethereum. Those currencies allow the transfer of value across the Internet without requiring intermediaries. Some applications and advantages came from the use of cryptocurrencies that have been reported by scholars such as early-stage start-ups crowdfunding, private transactions, a censorship-resistant alternative store of wealth (Vincent & Evans 2019). These changes result in serious threats to the traditional business model of any financial institutions and banks.

Detailed discussions of each technology - cellular or mobile technologies, blockchain, satellite imagery, big data, machine learning and artificial intelligence (MLAI), and Internet of things (IoT), are included in the following sections:

2.1. Mobile (or Cell) Technology

Mobile technological advances have allowed 'mobile money' and activated computer programs to work on smartphones through applications that provide access to a wide range of goods and services. These techno-financial transactions, variously known as mobile money (MM), m-banking, or mobile transfers, are gradually being incorporated into the global financial capital structure. Facilitating and promoting technology-aided resources for maximising financial inclusion is a recurring theme in the United Nations Sustainable Development Goals (Ochara & Mawela, 2015; Hoang et al., 2020). Electronic payment systems have converted mobile telephones into interfaces with the financial sector and are now trusted by over a billion people. According to Foth et al. (2011), 69 percent of adults had a financial institution account at the time, which increased by 7 percent in 2014. More than 60 percent of adults in many countries in sub-Saharan Africa have a digital wallet (Foth et al., 2011). Electronic payment systems in developing countries have also included new delivery mechanisms (through agent networks, secure communication and a power supply for them) and enhanced compatibility so that users of various applications and technologies can make seamless transactions that are as good as cash (Foth et al., 2011). Digital payment technology has contributed to socio-economic development through financial inclusion and security protection during crises and pandemics. Initially, mobile money platforms (MMPs) and associated mobile wallet technologies which allow electronic transactions through smartphones were developed for person-to-person (P2P) and money transfers, with the potential for broad accessibility (Ochara & Mawela, 2015).

2.2. Blockchain Technology

Blockchain is a collective database that is spread through vast peer-to-peer networks with secured exchanges. It is fundamentally recognised as a stable, permanent, provable and straightforward mechanism for storing transactions and documents, owing to the encrypted, distributed nature of the blockchain data and the consensus framework. Specifically, its technology enables users to 'tokenise' natural capital. It also allows pre-agreed programming conditions that are automatically fulfilled after certain procedures have been done and if certain conditions have been reached (Lund et al., 2019). Such 'smart contracts' (for example, the sale of assets at a certain price) may be negotiated without the need for a third party to control the release of assets. Blockchain technology is being used on a wide scale in the finance market, largely because of its ability to store transaction information and other confidential data safely. Each transaction is secured, and the likelihood of significant cyber-attacks is relatively minimal when blockchain technology is used (Giungato et al., 2017). Blockchain infrastructure is now the foundation of a range of cryptocurrencies. The idea of the blockchain originates from the invention of the blockchain bitcoin (Giungato et al., 2017). This "chain" comprises the entire history of legitimate and authenticated transactions among users of the network (Lund et al., 2019). Blockchain infrastructure was initially envisaged to be used mainly for digital currency transactions: 'BlockChain 1.0'. Nevertheless, beyond the framework, the platform has a wide scope, and it demonstrates capabilities that are also important to other fields. Blockchain infrastructure will also be used for contracting, crowdfunding and e-wallets, thus becoming "BlockChain 2.0" The third state, "BlockChain 3.0", focuses on applications in areas beyond business and finance, such as "government, health, literacy, culture and art" (Furfaro et al., 2019). It simplifies the internal management processes of institutional and non-state pension funds, accountability of financial transfers, assurances of purposeful usage, security and efficiency of administration and services, free and open transfer between funds, coordination of client and fund records, and the privacy of the data (Giungato et al., 2017).

2.3. Big Data

Big data is the utility of the large scale of structured and unstructured data to anticipate behaviours in finance, business and management of customers, that, overall, support the decision-making process. There are multiple trillion US dollars moving across international markets, and financial analysts who are responsible for working with the related data source aim to deliver precise and accurate predictions and decision-making (Del Vecchio et al., 2018a). Values behind this data fundamentally depend on how it is collected, synthesised, analysed and visualised. The current financial analysis system is unable to support the synthesis, analysis and interpretation of the huge source of structured and unstructured data without powerful ICT platforms; data analysts are increasingly implementing other supporting technologies such as cloud computing, big data analytic tools, machine learning and artificial intelligence (Garg et al., 2019). The integration of other advanced technologies such as cloud computing, big data analytic tools, machine learning and artificial intelligence helps to cut down the costs of human resource intensive, as well as enhancing scalability, accuracy, flexibility and security for targeting efficient ways of decision-making. One example of big data analytics is the accurate risk analysis, in which the decision-making based on big data analytics allow the unbiased analytic tools and procedures (Kudva & Ye, 2017). These analytical capabilities consider multiple aspects varied from the economic perspective, customer engagement and financial capital of each enterprise to identify potential risks and opportunities related to each business. Another example of big data analytics has resulted from the fraud detection and prevention powered by machine learning (Kudva & Ye, 2017). The employment of machine learning and big data allows the analytics of buying patterns, thus enabling the identification of fraud when a bank card is stolen and notifying the customer of threats. As a result, big data analytics aim to transform not only individual enterprise processes, but also the whole global financial service sector (Garg et al., 2019).

2.4. Satellite Imagery

Satellite images are earth images captured by imaging satellites operated by governments and corporations around the world. Satellite mapping providers sell images to governments and corporations, such as Apple Maps and Google Maps, by licensing them (Sutton, 2003). The spatially-collected astronomical images should not be confused. Big data is summed up with a large amount of increasingly complex data from a number of diverse internal (and external) channels, providing consumers with more possibilities to provide real-time market perspectives. Given the perspective of the financial sector, the analyst can access satellite images, which enable investigators to estimate historical and forecast future data such as timing and value of yields, farm crop yields, planting cycles, trends in production, the timing of the arrival of agricultural products, which can help to facilitate insights for credit assessment (Shen et al., 2013). In short, while the introduction and development of satellite imagery–based forecast is still in its infancy, emerging research in this area proves that satellite imagery forecasts will be able to reduce transaction expenses of accessing rural farmers, which makes financial capital more affordable for farmers, especially those in developing countries (Netzband et al., 2007).

2.5. Machine Learning & Artificial Intelligence (MLAI)

Machine Learning & Artificial Intelligence (MLAI) helps to evaluate massive data sets, develop actions and pricing patterns, automate and significantly improve the decision-making process by using sophisticated computer science and algorithms. Artificial Intelligence (AI) and Machine Learning (ML) are two of the most widely used technologies in FinTech, delivering the opportunity to contribute a greater role in the financial sector as developments continue (Nishant et al., 2020). Some of AI and ML's FinTech applications include automation, productivity enhancement, reduction of operational costs, security enhancement, customer engagement improvement, as well as personal marketing and selling strategy (Nilashi et al., 2019). First, regarding automation, AI-based working procedures are able to replace ineffective paperwork procedures, with the aim of providing more efficient approaches to manage, share, store and use the information that reduces expenses and time spent on document processing. Besides, in terms of productivity, MLAI can be used to replace the daily repetitive work of employees, hence reducing human resource-intensive and relevant costs. In addition, employees and managers can pay attention to more valuable duties such as decision-making tasks, thus generating additional value for a business (Khakurel et al., 2018). Moreover, MLAI allows efficient supervision of accomplished tasks through the relevant set of Key Performance Indicators (KPIs), which enhance security and compliance. MLAI can also help to maintain and improve continuous connection and communication with customers and other stakeholders through the automation capability. Another way in which MLAI can help is to decrease the time spent dealing with errors, consumer issues, contacting, searching for data and information from days to just a few minutes (Nilashi et al., 2019). Furthermore, other applications such as automated chatbot or MLAI-based virtual assistants are able to screen and sort customer complaints, which also helps to reduce long wait times and inefficient problem resolutions, thus, enhancing customer satisfaction (Nishant et al., 2020).

2.6. Internet of things (IoT)

Beier et al. (2018) explained that IoT could be determined as a global network of interconnected objects that is exclusively addressable, based on common communication protocols. In addition, Salam (2020) stated that IoT could be used as a complex and global network system based on common and integrated networking protocols and with self-configuring functionalities. In this particular network infrastructure, physical and virtual "objects" are embedded, have identities, physical attributes and virtual personalities, employ smart interfaces, and are seamlessly integrated into the information network. In the case of IoT, this new layer of "things" can be seen as an expansion of collective human-computer interactions and communications. More explicitly, IoT can be a mutualistic connection between the physical and virtual worlds, which enables people and "things" to be linked at any time, everywhere, with everything and everyone, preferably using any path/network and service via low-cost connected sensors, and AI contributes to machine learning that automates exploration, enables smart computers to perform non-routine tasks (Khatua et al., 2020). IoT can change the relationship between companies and customers by contributing to the fulfilment of communication and interconnectivity, then integrating intelligence into devices to process information and make real-time autonomous intelligent decisions without requiring any human involvement and/or intervention ((Del Vecchio et al., 2018b; Salam, 2020). There are diverse applications of IoT in the financial service sector such as immediate customer connection, smart customer navigation, customer satisfaction enhancement, and security and business efficiency improvements. First, through the capability of IoT and MLAI, a financial institution can develop an online customer support centre, which can provide customer support services 24 hours a day and reduce the physical contact between the customer and financial providers, thus saving costs, time and human resource-intensive for both the customer and the financial institution. In addition, the IoT allows the prediction and forecast of customer shopping behaviours based on their shopping history. The business users of IoT can thus navigate and introduce customers to new products and services related to historical shopping records (Hribar & DaSilva, 2019). Therefore, this simple approach enhances efficiency and customer satisfaction. In addition, IoT can also help to generate and send personalised emails and messages with personalised financial plans, hence, enhancing customer care and services. Another application of IoT comes from authentication and security, in which customer biometrics such as fingerprints can also be stored as a security approach in accessing online banking services. Finally, the most important application of IoT comes from the possibility of automation, in which IoT allows the deployment of daily automated financial services in the business; for instance, Citibank allows customers to access and withdraw money from the ATMs using their smartphone instead of a card, thus improving customer service and reducing the need for human resources (Perera & Zaslavsky, 2014).

This section has outlined the current development in financial technologies. In the next part, the importance of sustainable development goals is discussed.

3. THE SUSTAINABLE DEVELOPMENT GOALS (SDGS)

The UN global commitment is an important part of the SDGs. Following its introduction in September 2015, the paper "Transforming Our World: The 2030 Agenda for Sustainable Development" commits world leaders to poverty alleviation and sustainable development until 2030 (Allen et al., 2018). The current generation of priorities ensures both the urgency of progress and the case that this development must be equitable and foster equity. The priorities and objectives laid out an extraordinarily bold and optimistic vision, with the aim of freeing the world from poverty, hunger and disease, and also free from fear of violence, i.e. a world with universal literacy and equal and accessible access to quality education, health care and social security, and promising more prosperous and inclusive communities (Allen et al., 2016). Although taking into account various national conditions, growth would have to be uniformly applicable to all nations, and no one should be left behind.

The SDGs are a declaration of aspirations designed to direct sustainable development efforts. The goals are not part of a binding formal contract; they are part of a revitalised global alliance to function within a spirit of global unity. They also reflect voluntary recommendations that are expected to encourage policy and regulatory actions in areas of vital interest over the next 15 years. However, the targets are not set down in the Convention, and countries are not technically obligated to enforce them (Scheyvens et al., 2016). The adoption of the SDGs would therefore ultimately rely on civil society and people exerting influence on their respective governments to enforce the objectives. In addition, global commitment in promoting the execution of the priorities and objectives would be key to the accomplishment of the SDGs (Collste et al., 2017). While interconnected and indivisible, as well as being multinational in nature and widely applicable, the SDGs take into account various national realities, capacities and stages of growth. Each government would need to set its own national goals, led by the global degree of optimism while considering national circumstances. In addition, each government will determine how these aspirational and global goals can be integrated into national development systems, policies and strategies. While SDGs are not legally binding, it is expected that policymakers will create national structures to achieve the objectives (Purvis et al., 2019).

4. HOW CAN DIGITAL FINANCE CONTRIBUTE TO ACHIEVE 17 SDGS?

The financial service sector has an intrinsic relation to UN SDGs, although financial inclusion and FinTech are not impartial. Instead, both dimensions play a critical role in creating a prosperous world in the future. Capital markets have reached such a high degree of sophistication that they can offer payment services, long-term lending, insurance services and savings/investment options to support the financial service sector through FinTech that could indeed contribute to all of the 17 UN SDGs (Hinson et al., 2019). Financial inclusion is undoubtedly one of the most significant processes, if not *the* most significant, towards achieving the Sustainable Development Goals (SDGs), and FinTech is one of the best methods for these goals to be accomplished. National Economies should concentrate on implementing policies for digital financial transition while also maintaining financial inclusion (Ozili, 2018).

FinTech now includes diverse innovations, including the application of artificial intelligence and big data. This breakthrough is capable of promoting financial inclusion more efficiently. The most direct and clear solution is mobile money, although the long-term approach is more difficult. FinTech establishes a basis for a new financial environment (George et al., 2020). The incorporation degree of emerging modern technology is varying and it also needs significant improvement before being applied in practice. Big technology, machine learning, and to a lesser degree, blockchain have improved the efficiency of a significant number of financial institutions, which has led to a decrease in knowledge asymmetries and costs associated with developing funding instruments for long-term renewable assets, as well as providing transparency and enhancing stakeholders' expertise in recognising, evaluating and managing risk

(Puschmann, 2017). Of the new technologies, IoT has made some of the major differences in making investments more intricate and encouraging sustainable resource deployment.

Mobile money has been one of the fastest-growing inventions in the world (Bai et al., 2020). This massive, increasing popularity is supported by the spread of smartphones, as well as the speed, flexibility, convenience and affordability of mobile money (Al Amri & Almaiah, 2020). The main uses of mobile money are to send and receive cash transfers. Approximately a quarter of unbanked adults in sub-Saharan Africa are currently providing cash or over-the-counter (OTC) funds (Ramos-Soler, et al., 2019). Mobile money P2P transfers have emerged as an alternative to the alert that OTC services still outperform mobile money services in much of sub-Saharan Africa in terms of cost-effectiveness. However, the sending and receiving of remittances has become the main usage of mobile money in sub-Saharan Africa (Ma et al., 2017). Mobile money can also serve as a vital means of financing smallholder farmers and contributing to increased investment in agribusiness. However, the impacts on environmental protection are unclear since efficiency gains cannot be accomplished in environmentally friendly ways. As the reach of the MMPs grew, users were able to use mobile financial services to make deposits and withdrawals, make personal-to-business payments, pay taxes, collect payments from businesses and governments, save, lend, borrow, take out insurance and make investments contributing to the widespread inclusion of non-banking companies in the financial system (Zhao et al., 2019). The provision of these services may have a positive impact on health (SDG 3), jobs (SDG 8), education (SDG 4) and poverty alleviation (SDG 1) through improved productivity. However, as noted earlier, the effect on environmental sustainability may be unclear. However, initiatives that integrate mobile financial services with other (green) technologies, as well as interoperability between MMPs, monitoring technology and big data platforms (possibly integrating with digitised agricultural systems) have enormous potential to enhance synergies between the pillars of sustainability (Nguyen et al., 2021).

Financial technologies such as Machine Learning and AI are important for evaluating the investment decisions of local investors. Digital methods allow for the measurement of sustainability within financial investments. Digital finance provides a realistic solution for people to make more resource-conscious and equitable buying decisions. At the end of the day, buying decisions have a considerable impact on consumer actions. This information has been brought to our attention by emerging technology (Hinson et al., 2019). This is encouraged by the rapid development of the Internet of Things and provides much knowledge on the implications of sustainability. On one side, studies show that Blockchain and AI will transform monetary transaction information into carbon footprints, stressing real-time environmental impacts of financial transactions for the first time in history (Blakstad & Allen, 2018; Deng et al., 2019). On another side, the increase in the use of Blockchain technology and other fintech also deliver concerns over the enormous power consumption, heat generation, and additional costs for the operation of those technologies (Li et al., 2019; Sedlmeir et al., 2020). Online social media helps individuals to be more educated about the products which they have bought and ensures that these goods have been manufactured in an environmentally sound manner (Ozili, 2018). More environmentally conscious products and services would contribute to more purchases. People are inspired by digital investment, as institutional investors and as public pension policyholders, to balance environmental development goals with business behaviour in the best possible way (Nguyen et al., 2021). There is a plurality of customers in the European Union who are strongly geared toward the values of non-financial goals.

With the increased use of emerging technology, it is now possible to transform non-financial priorities into investments (Hoang et al., 2020). They will mobilise their assigned funds more efficiently and become even more interested in value-chain finance. Online networks tie together sustainable-tech firms

Table 1. Real example of how financial technologies can help Sustainable Development Goals (SDGs) achievement

Sustainable Development Goals (SDGs) and Indicators	Potential financial technologies able to help	Examples	References
Goal 1. End Poverty in All its Forms Everywhere			
Indicator 1.1.1. Proportion of the Population Below the International Poverty Line	mobile technology, ML, big data analytics	M-Pesa is a typical example for the use of mobile banking service in Kenya, which allows customers to store and transfer money through their cell phones. M-Pesa is proved to reduce poverty in Kenya by 2%. Those alternative financial services similar to M-Pesa have facilitated peer to peer (P2P) lending and crowdsourcing services in developing countries and contribute to the development of the financial sector, create more jobs and benefit small businesses and poor households.	Van Hove, L., & Dubus, A. (2019); Markus, M. L., & Nan, W. V. (2020); Ryu & Ko (2020).
Indicator 1.3.1. Proportion of population covered by social protection floors/ systems	mobile technology, IoT	In developing countries such as India and Bangladesh, the digitalization of cash transfers through online payments has been reported as efficiency improvement and corruption reduction.	Saini et al., (2017); Mohanty & Pawar (2019); Mishra (2019).
Goal 2. End hunger, achieve food security and improved nutrition and promote sustainable agriculture			
Indicator 2.3.2 Average income of small-scale food producers	Blockchain, mobile technology, AI	Blockchain technology and AI enable the application of smart contracts - a transaction protocol that automatically executes, control or document legally relevant events and actions. This application can help farmers from all over the world to securely connect and execute their contracts with the buyers.	Goldenfein & Leiter (2018); Cong & He (2019); Singh et al. (2020).
Goal 4. Ensure inclusive and equitable quality education and promote lifelong learning opportunities for all			
Indicator 4.4.1 Proportion of youth and adults with information and communications technology (ICT) skills	IoT, MLAI, satellite imagery	While the financial technologies have been embedded into the daily activities, the ICT skills of youth and adults will also increase. The ICT incubators and accelerators all over the world, such as F10 Incubator and Accelerator in Zurich, have developed multiple courses and programs that allow people to access and improve their digital business and financial innovation capability.	Jakobsen et al. (2017); Unterhalter (2019); Bondarenko et al. (2019).
Goal 7. Ensure access to affordable, reliable, sustainable and modern energy for all			
Indicator 7.1.2 Proportion of population with primary reliance on technology	Mobile technology, IoT, AI	The integration of IoT and mobile technology allows telecommunications companies to deliver their service for excluded communities by a pay-for-use model, such as Questsol (Guatemala) and Simpa Network (India).	Rastogi (2018); Yadav et al. (2019); Salvia & Brandli (2020).
Goal 8. Promote sustained, inclusive and sustainable economic growth, full and productive employment and decent work for all			
Indicator 8.4.1 Material footprint, material footprint per capita, and material footprint per GDP	Big data, IoT, MLAI, satellite imagery	The integration of advanced technologies such as big data, IoT, MLAI and satellite imagery can help to provide transparency about the environmental impact of consumption choices. One of the pioneers of this movement is the Ant Forest Programme, which interprets financial transaction data into individual carbon accounts, therefore encourage users to reduce carbon activities.	Lin et al. (2017); Zheng & Meng (2018); Zhang et al. (2021).

continued on following page

Table 1. Continued

Sustainable Development Goals (SDGs) and Indicators	Potential financial technologies able to help	Examples	References
Goal 9. Build resilient infrastructure, promote inclusive and sustainable industrialization and foster innovation			
Indicator 9.3.2 Proportion of small-scale industries with a loan or line of credit	Mobile technology, blockchain, big data	The AI, big data and blockchain technologies have enabled the expansion in the peer to peer lending and crowdfunding platforms. Thus, those advancements have offered new sources of funding for micro, small and medium enterprises (MSMEs).	Chatnani (2018); Piliyanti (2019); Candraningrat et al. (2021)
Goal 16. Promote peaceful and inclusive societies for sustainable development, provide access to justice for all and build effective, accountable and inclusive institutions at all levels			
Indicator 16.5.1 Proportion of persons who had at least one contact with a public official and who paid a bribe to a public official, or were asked for a bribe by those public officials, during the previous 12 months	Mobile technology, IoT, MLAI, big data	Online financial services through Mobile technology, IoT, MLAI, and big data can help to leverage transparency and security transactions and cut down corruption and money laundering. A recent study shows that online pension payments in India result in more than 40% reduction in bribe demands.	Klapper & Singer (2017); Blakstad & Allen (2018); Zarrouk et al. (2021)
Goal 17. Strengthen the means of implementation and revitalize the Global Partnership for Sustainable Development			
Indicator 17.17.1 Amount of United States dollars committed to (a) public- private partnerships and (b) civil society partnerships		Digital finance applications contribute to generating new partnerships between public-private and civil society; for instance, the Sustainable Digital Finance Alliance is developed by Ant Financial Services and UN Environment Programme to harness digital technologies for sustainability in the financial sector and global business context.	Babu et al. (2020); Adegbite & Machethe (2020); Mi et al. (2021).

and financiers so that they can cooperate. These technologies combine curated order flows, contract data and artificial intelligence data to construct artificial investments for investors (Nguyen et al., 2019). The mapping of the practice of digital finance through G20 Countries and the private sector illustrates how digital finance is helping to unlock multiple new possibilities.

Table 1 shows the potential contributions made by sustainable digital funding towards achieving consistent SDG targets in the global SDG indicator framework that was set and determined by the Inter-Agency/Expert Committee at the 48th UN Statistical Commission meeting in March 2017. It shows how to link sustainable financial technologies to simple SDG metrics, which are both highlighted and not comprehensive. Therefore, the issues which attribute to unique SDG metrics may not provide examples of the effects of digital funding on the inclusion by both active or passive investments of environmental and social factors in decision-making. Table 1 shows that sustainable digital finance already contributes to many SDGs. Further study is needed to determine the potential of current and evolving digital systems to achieve the SDGs and measure the effect of SDGs beyond the framework of financial inclusion practices.

Besides, digital finance can help to empower the future generation and support sustainable development. The most likely system to achieve the Sustainable Development Goals is by using business strategies, incentives, and legislation that shifts financial capital internationally. Due to the growing focus on environmental and social responsibility of companies in the EU, renewable energy and ESG-related risk capital have boomed in the region. In addition, this contributes to modernising the banking system and can assist in achieving sustainable development goals (SDGs). Financial inclusion and the service sector will be the main driving forces that can remove poverty from Malaysia (Mondal, 2015). China's development of digital transformation technology is a remarkable development example (Mi et al., 2021). Digitalisation will ensure that SDGs are responsive and appropriate to the context. This action will enable organisations to integrate SDGs into their current decision-making process. Many individuals feel that globalisation brought up enormous technological advancements. This will lead to an expanding global economy and an increased standard of living worldwide. Developments focus on data to sustain sustainable growth at the local level (Zheng & Meng, 2018; Zhang et al., 2021). This also applies to differences in levels of growth, levels of decentralisation and the extent of government e-services developed. There are countries with stronger centralisation of power, such as Colombia, that have an advantage over countries with less centralisation. There are several success stories in Europe and the rest of the world regarding e-government. As seen in Table 1, FinTech is directly or indirectly playing a role in some of the UN Sustainable Development Goals (SDGs).

5. CHALLENGES TO LEVERAGING FULL POTENTIAL OF SUSTAINABLE DIGITAL FINANCE

Besides potential contributions to the SDGs' accomplishments, there are multiple barriers to the influences of financial technologies to fulfil the 17 SDGs. The main obstacles in the industries for supporting SDGs are financial pressures and the impediments faced by faulty technical technology and expensive updates. Reinforcing such a platform is needed yet inadequate. One of the most critical and special issues that must be addressed relates to the minimal knowledge of sustainable business models and weak sustainability business models. In order to leverage the capacity of digital finance completely, many organisations are expected to be coordinated to meet this need (Puschmann, 2017). Reasonable costs and effective broadband services mean that the advantages of digital finance can be realised, particularly in cloud computing. This will ensure that a communication network will decrease the availability of gathering reliable economic data, resulting in inaccurate predictive analytics (Goldstein et al., 2019). Additionally, while financial institutions have significantly progressed the digital transformation of back-office operations, the larger financial institutions still run their traditional IT procedure, hindering the potential to integrate the data by the centrally-controlled dataset. Blockchain is not a well-suited platform to connect with most standard business work floors and networks. The expenses and operating challenges of this are considerably high.

Another point is that technologies such as blockchain and IoT are both in the early stages of deployment, with the high technical expense, uncertainty and minimal capabilities. Lacking these standards, it is impossible for countries to implement emerging innovations (Nicoletti et al., 2017). High costs of operations in Bitcoin networks will hinder mass consumer usage and acceptance. Blockchains have restricted bandwidth (between 10 and 100 transactions per second compared to 40,000 Visa transactions per second) and systems that use different consensus mechanisms. A greater focus is required on long-term impacts. This will also have a direct effect on investor morale as a result of cybersecurity concerns (Deng et al., 2019). Blockchain technology is far from homogeneous, particularly concerning its applications in not only business area or legitimacy, but also with regard to its technical characteristics and power consumption (Sedlmeir et al., 2020). This perspective is further supported by (Li et al., 2019), who argued that besides the technical expense, uncertainty and minimal capabilities, another concern for the use of Blockchain technology, especially in term of sustainable development is the enormous power consumption of this technology. Despite the huge consumption of energy of blockchain technology, the alternative solution for blockchain technology with lower power consumption have been developed and even could further decrease in particular the power consumption of blockchain systems and networks in the future (Greenberg & Bugden, 2019). The above studies critically highlight the negative perspective from the increase in the use of Blockchain technology that may create more cost involved from the transaction and energy consumption. One concern that needs to be asked, however, is whether the use of Blockchain technology, as well as other fintech, can create more heat as well as lead to the creation of additional solid or chemical waste. Thus, future studies on the current topic are therefore recommended. Moreover, another possible area of future research would be to investigate the potentials of fintech to eliminate, or significantly reduce financial contagion if so designed.

In spite of the current onset of sustainable electronic finance and increasingly emerging technology, there are still ongoing studies which provide the complete capacity and threats of digital finance. By combining sustainable finance and rapidly growing digital finance, there will be knowledge gaps between the two (Ryu & Ko, 2020). The market share of sustainable funding is still comparatively limited. The traditional financial participants did not regard digital finance to be a mechanism that can lift investment and improve the level of uncertainty to maintain sustainable investment.

There are variations between countries since government sectors typically believe that sustainable finance, integrated development and digital technology are separate fields. This makes it difficult to predict negative consequences because of the interconnectedness of problems, agendas and solutions (Zheng & Meng, 2018). The Financial Inclusion Agenda has been known to be focused on promoting financial inclusion by increasing access to quality financial services. It is very effective for financial inclusion in case of unsustainable financing.

Moreover, the pilot programs are more focused at the country level, while most digital funding focuses on the city or regional level. Digital finance is also still facing a few obstacles for international cooperation. Cross-border learning is another thing that can be done with more experience (Yadav et al., 2019). The success of digital finance technologies depends on the availability of vast quantities of material. With insights from the research, and knowledge and records on the environment, this is only one consideration used in making an investment decision. One issue with digital finance is that there are not yet any benchmarks for evaluating economic or behavioural results, or they are inconsistent from one country to another. Adoption costs and contingent personality also establish constraints (Jakobsen et al., 2017).

The final limitation is the nascent business models and markets. Many sustainable technology providers are start-ups with a higher risk of failure currently testing and creating solutions. Business models and markets are fresh, and it is unclear what works and what does not work (Piliyanti, 2019). As sustainable digital solutions are very recent, access to funds is difficult because they are not perceived or labelled as 'green' goods and cannot respond to key fund performance metrics (e.g. greenhouse gas reduction) (Nguyen et al., 2021). Sustainable implementation of IoT is extracted from the company's 'innovation' budgets for larger players, and it will take time to include these large-scale implementations into larger business budgets. Sustainable digital financing options may be increased by the incumbents in certain markets.

REFERENCES

G20 Sustainable Finance Study Group (SFSG). (2018). *Sustainable Finance Synthesis Report*. Technical report. Available at: http://www.gflp.org.cn/public/ueditor/php/upload/file/20181203/1543807391177171.pdf

Adegbite, O. O., & Machethe, C. L. (2020). Bridging the financial inclusion gender gap in smallholder agriculture in Nigeria: An untapped potential for sustainable development. *World Development, 127*, 104755.

Al Amri, M. M., & Almaiah, M. A. (2020). The Use of Mobile Gamification Technology for Sustainability Learning in Saudi Higher Education. *International Journal of Advanced Trends in Computer Science and Engineering*, *9*(5).

Alber, N., & Dabour, M. (2020). The Dynamic Relationship between FinTech and Social Distancing under COVID-19 Pandemic: Digital Payments Evidence. *International Journal of Economics and Finance*, *12*(11), 109–109.

Allen, C., Metternicht, G., & Wiedmann, T. (2016). National pathways to the Sustainable Development Goals (SDGs): A comparative review of scenario modelling tools. *Environmental Science & Policy*, *66*, 199–207.

Allen, C., Metternicht, G., & Wiedmann, T. (2018). Initial progress in implementing the Sustainable Development Goals (SDGs): A review of evidence from countries. *Sustainability Science*, *13*(5), 1453–1467.

Babu, M. M., Dey, B. L., Rahman, M., Roy, S. K., Alwi, S. F. S., & Kamal, M. M. (2020). Value co-creation through social innovation: A study of sustainable strategic alliance in telecommunication and financial services sectors in Bangladesh. *Industrial Marketing Management*, *89*, 13–27.

Bai, C., Dallasega, P., Orzes, G., & Sarkis, J. (2020). Industry 4.0 technologies assessment: A sustainability perspective. *International Journal of Production Economics*, *229*, 107776.

Beier, G., Niehoff, S., & Xue, B. (2018). More sustainability in industry through industrial internet of things? *Applied Sciences (Basel, Switzerland)*, *8*(2), 219.

Blakstad, S., & Allen, R. (2018). Green Fintech. In *FinTech Revolution* (pp. 183–199). Palgrave Macmillan.

Bondarenko, T. G., Zhdanova, O. A., & Klimova, N. (2019). Multi-Criteria Mechanism for Selecting Projects by Fintech Accelerators. In *International Science and Technology Conference" FarEastCon"(ISCFEC 2019)* (pp. 6-10). Atlantis Press.

Boratyńska, K. (2019). Impact of digital transformation on value creation in Fintech services: An innovative approach. *Journal of Promotion Management*, *25*(5), 631–639.

Candraningrat, I., Abundanti, N., Mujiati, N., & Erlangga, R. (2021). The role of financial technology on development of MSMEs. *Accounting*, *7*(1), 225–230.

Chatnani, N. N. (2018). Receivables management and supply chain finance for MSMEs: Analysis of TREDS. *Academy of Strategic Management Journal*, *17*(3), 1–8.

Collste, D., Pedercini, M., & Cornell, S. E. (2017). Policy coherence to achieve the SDGs: Using integrated simulation models to assess effective policies. *Sustainability Science*, *12*(6), 921–931. PMID:30147764

Cong, L. W., & He, Z. (2019). Blockchain disruption and smart contracts. *Review of Financial Studies*, *32*(5), 1754–1797.

Das, S. R. (2019). The future of fintech. *Financial Management*, *48*(4), 981–1007.

Del Vecchio, P., Mele, G., Ndou, V., & Secundo, G. (2018a). Open innovation and social big data for sustainability: Evidence from the tourism industry. *Sustainability*, *10*(9), 3215.

Del Vecchio, P., Secundo, G., & Passiante, G. (2018b). Analyzing big data through the lens of customer knowledge management: Evidence from a set of regional tourism experiences. *Kybernetes*, *47*(7), 1348–1362.

Deng, X., Huang, Z., & Cheng, X. (2019). FinTech and sustainable development: Evidence from China based on P2P data. *Sustainability*, *11*(22), 6434.

Duygun, M., Hashem, S. Q., & Tanda, A. (2020). Financial Intermediation Versus Disintermediation: Opportunities and Challenges in the FinTech Era. *Frontiers in Artificial Intelligence*, 3.

Foth, M., Forlano, L., Satchell, C., & Gibbs, M. (Eds.). (2011). *From social butterfly to engaged citizen: Urban informatics, social media, ubiquitous computing, and mobile technology to support citizen engagement*. MIT Press.

Furfaro, A., Argento, L., Saccá, D., Angiulli, F., & Fassetti, F. (2019). An infrastructure for service accountability based on digital identity and blockchain 3.0. In *IEEE INFOCOM 2019-IEEE Conference on Computer Communications Workshops (INFOCOM WKSHPS)* (pp. 632-637). IEEE.

Garg, R., Aggarwal, H., Centobelli, P., & Cerchione, R. (2019). Extracting knowledge from big data for sustainability: A comparison of machine learning techniques. *Sustainability*, *11*(23), 6669.

George, G., Merrill, R., & Schillebeeckx, S. (2020). Digital sustainability and entrepreneurship: How digital innovations are helping tackle climate change and sustainable development. *Entrepreneurship Theory and Practice*. Advance online publication. doi:10.1177/1042258719899425

Giungato, P., Rana, R., Tarabella, A., & Tricase, C. (2017). Current trends in sustainability of bitcoins and related blockchain technology. *Sustainability*, *9*(12), 2214.

Goldenfein, J., & Leiter, A. (2018). Legal engineering on the blockchain: 'Smart contracts' as legal conduct. *Law and Critique*, *29*(2), 141–149.

Goldstein, I., Jiang, W., & Karolyi, G. A. (2019). To FinTech and beyond. *Review of Financial Studies*, *32*(5), 1647–1661.

Greenberg, P., & Bugden, D. (2019). Energy consumption boomtowns in the United States: Community responses to a cryptocurrency boom. *Energy Research & Social Science*, *50*, 162–167.

Hinson, R., Lensink, R., & Mueller, A. (2019). Transforming agribusiness in developing countries: SDGs and the role of FinTech. *Current Opinion in Environmental Sustainability*, *41*, 1–9.

Hoang, C. V., Hoang, T. G., & Kane, V. (2020). Technological approaches to sustainability. In D. Crowther & S. Seifi (Eds.), *The Palgrave Handbook of Corporate Social Responsibility*. Palgrave Macmillan. doi:10.1007/978-3-030-22438-7_37-1

Hribar, J., & DaSilva, L. (2019). Utilising correlated information to improve the sustainability of internet of things devices. In *2019 IEEE 5th World Forum on Internet of Things (WF-IoT)* (pp. 805-808). IEEE.

Jakobsen, B. J., Tanev, S., & Jensen, S. (2017). The value of business incubation services for early stage start-ups. In *ISPIM Innovation Symposium* (pp. 1-23). The International Society for Professional Innovation Management (ISPIM).

Khakurel, J., Penzenstadler, B., Porras, J., Knutas, A., & Zhang, W. (2018). The rise of artificial intelligence under the lens of sustainability. *Technologies*, *6*(4), 100.

Khatua, P. K., Ramachandaramurthy, V. K., Kasinathan, P., Yong, J. Y., Pasupuleti, J., & Rajagopalan, A. (2020). Application and assessment of internet of things toward the sustainability of energy systems: Challenges and issues. *Sustainable Cities and Society*, *53*, 101957.

Klapper, L., & Singer, D. (2017). The opportunities and challenges of digitizing government-to-person payments. *The World Bank Research Observer*, *32*(2), 211–226.

Kudva, S., & Ye, X. (2017). Smart cities, big data, and sustainability union. *Big Data and Cognitive Computing*, *1*(1), 4.

Li, J., Li, N., Peng, J., Cui, H., & Wu, Z. (2019). Energy consumption of cryptocurrency mining: A study of electricity consumption in mining cryptocurrencies. *Energy*, *168*, 160–168.

Lin, J., Lobo, A., & Leckie, C. (2017). The role of benefits and transparency in shaping consumers' green perceived value, self-brand connection and brand loyalty. *Journal of Retailing and Consumer Services*, *35*, 133–141.

Lund, E. H., Jaccheri, L., Li, J., Cico, O., & Bai, X. (2019). Blockchain and sustainability: A systematic mapping study. In *2019 IEEE/ACM 2nd International Workshop on Emerging Trends in Software Engineering for Blockchain (WETSEB)* (pp. 16-23). IEEE.

Ma, Y. J., Gam, H. J., & Banning, J. (2017). Perceived ease of use and usefulness of sustainability labels on apparel products: Application of the technology acceptance model. *Fashion and Textiles*, *4*(1), 1–20.

Markus, M. L., & Nan, W. V. (2020). Theorizing the connections between digital innovations and societal transformation: learning from the case of M-Pesa in Kenya. In *Handbook of digital innovation*. Edward Elgar Publishing.

Mi, L., Xu, T., Sun, Y., Zhao, J., Lv, T., Gan, X., ... Qiao, L. (2021). Playing Ant Forest to promote online green behavior: A new perspective on uses and gratifications. *Journal of Environmental Management*, *278*, 111544. PMID:33129025

Mishra, A. (2019). Digitization in Indian Banking Sector. *ITIHAS The Journal of Indian Management*, *9*(4), 73–80.

Mohanty, B. R., & Pawar, B. R. (2019). Cashless (Less Cash) Economy & Digitization of Indian Economy. *Proceedings of 10th International Conference on Digital Strategies for Organizational Success.*

Mondal, S. (2015). Financial inclusion: A step towards eradicate poverty. *American Journal of Theoretical and Applied Business, 1*(1), 21–26.

Najaf, K., Subramaniam, R. K., & Atayah, O. F. (2021). Understanding the implications of FinTech Peer-to-Peer (P2P) lending during the COVID-19 pandemic. *Journal of Sustainable Finance & Investment*, 1-16.

Netzband, M., Stefanov, W. L., & Redman, C. (Eds.). (2007). *Applied remote sensing for urban planning, governance and sustainability*. Springer Science & Business Media.

Nguyen, H., Hoang, T. G., & Luu, H. (2019). Corporate Social Responsibility in Vietnam: Opportunities and Innovation Experienced by Multinational Corporation Subsidiaries. *Social Responsibility Journal, 16*(6), 771–792. doi:10.1108/SRJ-02-2019-0082

Nguyen, H. T., Hoang, T. G., Nguyen, L. Q. T., Le, H. P., & Mai, H. X. V. (2021). Green technology transfer in developing country: Mainstream practitioner views. *The International Journal of Organizational Analysis*. Advance online publication. doi:10.1108/IJOA-11-2019-1941

Nicoletti, B. (2017). Fintech innovation. In *The Future of FinTech* (pp. 81–159). Palgrave Macmillan.

Nicoletti, B., Nicoletti, W., & Weis. (2017). *Future of FinTech*. Basingstoke, UK: Palgrave Macmillan.

Nilashi, M., Rupani, P. F., Rupani, M. M., Kamyab, H., Shao, W., Ahmadi, H., ... Aljojo, N. (2019). Measuring sustainability through ecological sustainability and human sustainability: A machine learning approach. *Journal of Cleaner Production, 240*, 118162.

Nishant, R., Kennedy, M., & Corbett, J. (2020). Artificial intelligence for sustainability: Challenges, opportunities, and a research agenda. *International Journal of Information Management*, *53*, 102104.

Ochara, N. M., & Mawela, T. (2015). Enabling social sustainability of e-participation through mobile technology. *Information Technology for Development*, *21*(2), 205–228.

Ozili, P. K. (2018). Impact of digital finance on financial inclusion and stability. *Borsa Istanbul Review*, *18*(4), 329–340.

Perera, C., & Zaslavsky, A. (2014). Improve the sustainability of internet of things through trading-based value creation. In *2014 IEEE World Forum on Internet of Things (WF-IoT)* (pp. 135-140). IEEE.

Piliyanti, I. (2019). Fintech achieving sustainable development: The side perspective of crowdfunding platform. *Shirkah: Journal of Economics and Business*, *3*(2).

Przychodzen, J., & Przychodzen, W. (2015). Relationships between eco-innovation and financial performance–evidence from publicly traded companies in Poland and Hungary. *Journal of Cleaner Production*, *90*, 253–263.

Purvis, B., Mao, Y., & Robinson, D. (2019). Three pillars of sustainability: In search of conceptual origins. *Sustainability Science*, *14*(3), 681–695.

Puschmann, T. (2017). Fintech. *Business & Information Systems Engineering*, *59*(1), 69–76.

Ramos-Soler, I., Martínez-Sala, A. M., & Campillo-Alhama, C. (2019). ICT and the sustainability of World Heritage Sites. Analysis of senior citizens' use of tourism Apps. *Sustainability*, *11*(11), 3203.

Rastogi, C. (2018). M-Kopa solar: Lighting up the dark continent. *South Asian Journal of Business and Management Cases*, *7*(2), 93–103.

Ryu, H. S., & Ko, K. S. (2020). Sustainable Development of Fintech: Focused on Uncertainty and Perceived Quality Issues. *Sustainability*, *12*(18), 7669.

Sahay, M. R., von Allmen, M. U. E., Lahreche, M. A., Khera, P., Ogawa, M. S., Bazarbash, M., & Beaton, M. K. (2020). *The promise of fintech: Financial inclusion in the post COVID-19 era*. International Monetary Fund.

Saini, S., Sharma, S., Gulati, A., Hussain, S., & von Braun, J. (2017). Indian food and welfare schemes: Scope for digitization towards cash transfers. *ZEF Discussion Papers on Development Policy*, 241.

Salam, A. (2020). Internet of things for sustainability: perspectives in privacy, cybersecurity, and future trends. In *Internet of Things for Sustainable Community Development* (pp. 299–327). Springer.

Salvia, A. L., & Brandli, L. L. (2020). Energy sustainability at universities and its contribution to SDG 7: a systematic literature review. *Universities as Living Labs for Sustainable Development*, 29-45.

Scheyvens, R., Banks, G., & Hughes, E. (2016). The private sector and the SDGs: The need to move beyond 'business as usual'. *Sustainable Development*, *24*(6), 371–382.

Sedlmeir, J., Buhl, H. U., Fridgen, G., & Keller, R. (2020). The energy consumption of blockchain technology: Beyond myth. *Business & Information Systems Engineering*, *62*(6), 599–608.

Shen, L., Kyllo, J. M., & Guo, X. (2013). An integrated model based on a hierarchical indices system for monitoring and evaluating urban sustainability. *Sustainability*, *5*(2), 524–559.

Singh, A., Parizi, R. M., Zhang, Q., Choo, K. K. R., & Dehghantanha, A. (2020). Blockchain smart contracts formalization: Approaches and challenges to address vulnerabilities. *Computers & Security*, *88*, 101654.

Sutton, P. C. (2003). An empirical environmental sustainability index derived solely from nighttime satellite imagery and ecosystem service valuation. *Population and Environment*, *24*(4), 293–311.

Unterhalter, E. (2019). The many meanings of quality education: Politics of targets and indicators in SDG 4. *Global Policy*, *10*, 39–51.

Van Hove, L., & Dubus, A. (2019). M-PESA and financial inclusion in Kenya: Of paying comes saving? *Sustainability*, *11*(3), 568.

Vincent, O., & Evans, O. (2019). Can cryptocurrency, mobile phones, and internet herald sustainable financial sector development in emerging markets? *Journal of Transnational Management*, *24*(3), 259–279.

Yadav, P., Heynen, A. P., & Palit, D. (2019). Pay-As-You-Go financing: A model for viable and widespread deployment of solar home systems in rural India. *Energy for Sustainable Development*, *48*, 139–153.

Zarrouk, H., Ghak, T. E., & Bakhouche, A. (2021). Exploring Economic and Technological Determinants of FinTech Startups' Success and Growth in the United Arab Emirates. *Journal of Open Innovation*, *7*(1), 50.

Zhang, Y., Chen, J., Han, Y., Qian, M., Guo, X., Chen, R., ... Chen, Y. (2021). The contribution of Fintech to sustainable development in the digital age: Ant forest and land restoration in China. *Land Use Policy*, *103*, 105306.

Zhao, H., Zhao, Q. H., & Ślusarczyk, B. (2019). Sustainability and digitalization of corporate management based on augmented/virtual reality tools usage: China and other world IT companies' experience. *Sustainability*, *11*(17), 4717.

Zheng, X., & Meng, Q. (2018). Business Model of Green Finance under the Background of "Internet Plus": A Case Study of Ant Forest. *Open Journal of Social Sciences*, *6*(8), 127–134.

Zhou, Q., Chen, X., & Li, S. (2018). Innovative financial approach for agricultural sustainability: A case study of Alibaba. *Sustainability*, *10*(3), 891.

Chapter 2
Customer Perception and Awareness of Green Banking Practices:
An Alternative Strategy of Environmental Sustainability

Niranjan Devkota
https://orcid.org/0000-0001-9989-0397
Quest International College, Pokhara University, Lalitpur, Nepal

Rekha Rai
Quest International College, Pokhara University, Lalitpur, Nepal

Ghanashyam Khanal
Tribhuvan University, Nepal

Ihtsham Ul Haq Padda
Department of Economics, Federal Urdu University of Arts, Science, and Technology, Islamabad, Pakistan

Udaya Raj Paudel
Principal-Quest International College, Pokhara University, Lalitpur, Nepal

Seeprata Parajuli
Quest International College, Pokhara University, Lalitpur, Nepal

Udbodh Bhandari
Quest International College, Pokhara University, Lalitpur, Nepal

ABSTRACT

The governments, business firms, policymakers, advocacy groups, and even the public recently are hotly debating on the issues of environmentally friendly practices. In this context, being a part of 'going green', green banking, which plays an important role in environmental sustainability, has been a buzzword in the global baking industry. This study identifies how the customers perceive the emerging concept of green banking initiatives of banks and also analyzes the factors that influence such practices of the customers. Using a structured questionnaire, the primary data were collected from 403 commercial banks of Kathmandu valley, Nepal. The awareness index was prepared, and the binary logit model

DOI: 10.4018/978-1-7998-8900-7.ch002

was applied for the econometric analysis. This study observed that the customers are positive towards the environmentally friendly practices of banks and ready to adopt the green banking practices. The research implies that in order to promote environmental sustainability, banks and financial institutions should be able to educate the customers about green banking practices and their benefits.

1. INTRODUCTION

All over the world in the last few decades, the governments, business firms, policy makers, advocacy groups and even the public are providing prime concerns on the issues of environmentally friendly practices (Shaumya, & Arulrajah, 2017). To mitigate environmental degradation, banking sector has introduced the concept of green banking recently (Tara et al., 2015). Green banking refers to the environmental- friendly practices that reduce the carbon footprint from both internally as well as externally by using online practices for banking activities to contribute environment protection with green banking products (Deepa & Karpagam, 2018). Green banking adopts the modern technology, change client habits in banking activities and improve operation of bank by considering environmental, economic and social factors to endorse environmental-friendly practices and reducing carbon foot print from banking activities (Singh and Singh 2012; Bihari, 2015; Masukujjaman & Aktar, 2013; Thombre, 2011; Jha and Bhoome, 2013; Mishra, 2013; Biswas, 2011). Therefore, greening the banking sector refers to ethical and environmental banking that promots corporate social responsibility in financial aspects (Ibe-enwo et al., 2019).

In the view of Shaumya & Arulrajah (2016) green banking saves the environment in two ways: technological innovation and behavioral management. Technological innovation encourage banking system to deplete their negative environmental effect using online banking instead of traditional banking system; whereas behavioral management helps to reduce negative environmental effect of the banks using environmental-riendly initiatives of bank employees, waste reduction efforts of bank employees, energy saving behavior of bank staff in their respective branches, providing loans to the environmental friendly project and etc. In regard to banks themselves, green banking offers the substantial benefits include increased goodwill and reputation, customers' loyalty, positive effects on the environment on the environment and simplicity of bank process which is more than monetary benefits (Vijay& Natarajan 2015). It also helps to aware bank's corporate and social responsibility & environmental activities along with maintaining their ethical standards (Manzano et al. 2009).

To the customers, green banking helps to reduce wastage from their lives by eliminating paper wastage and also makes their life easier and more secure with online banking by providing facilities like balance enquiry, check balance statement, fund transfer and deposit, opening and closing account and easy-to-access location to prevent identity theft (Wessel & Drennan, 2010). Similarly, it saves time and money through mobile and electronic banking by reducing amount of fuel and time consumed on those numerous trips to bank for the banking activities (Sahni & Dhamija 2018). Likewise, it also helps to online payment services avoiding late payments and save the fines. Customers' satisfaction and expectation will be different as per customers comparatively analysis between the perception and services performance because the success of any organization depends on it (Parasuraman, 1991). Customers expect that banks should provide the required information about the green banking financial products (Shampa and Jobaid 2017). It shows customer's interest on green banking practices.

At present, both the banks and customers encourage each other to "go green" (Ko et al., 2012). It is also found that customers show their prime concerns with environmentally friendly products and the organizations that take care of both the society and customer. So, customers have higher expectations towards the financial institution with the aim that they should take environmental protection responsibilities and this is the same reason behind going green for financial institution (Sharma, Sarika & Gopal, 2014). Understanding such type of costumer's psychology, banks are encouraging the environmental friendly business (Thombre, 2011).

Nepalese banks have adopted green banking concept in their banking activities by converting the manual activities into computerized, providing services like balance enquiry, checking balance statement, fund transfer and deposit, opening and closing account via using online and providing loan at low level interest rate for green projects (Risal & Joshi, 2018). Laxmi bank is the initiator of green banking strategies in Nepal followed by Standard Charter Bank (Mehta & Sharma, 2016; Risal, 2018). Then after, today, almost all the commercial banks have adopted green banking practices in one or other forms. Mobile banking and internet banking are the most common practice to avoid customer-counter delay and provide access to easy finance. Nepal Rastra Bank (2019) in its monetary policy indicates that in Nepal the number of ATMS reached 3188, internet banking users 893 thousand, mobile banking users 7 million, debit card issuer 6.28 million and credit card issuer 113 thousand in mid-April 2019 (Nepal Rastra Bank, 2019). It shows use of the green banking products by the customers is increasing day by day.

Despite the growing importance of green banking concept worldwide and customer's keen interest on online transaction with time and cost utility, they are not aware about the operational part of it (Jani, 2012). Studies on green banking from different countries including Nepal also revealed that customers are less aware about the green banking practices. For example, Prasad (2017) finds out that the customers are less aware with the term Green banking even if they are enjoying the benefits and features of green banking. Similarly, Sharma et al. (2014) analyzes that most of the respondents think online transaction, online bill payment and point of sales systems as a green banking but they were not aware about the Green CDs, solar powered ATM, and bonds for environment protection.

In the context of Nepal research conducted about the green banking practices are scant. Mehta & Sharma (2016) examined the persistence of customer toward the green banking in Nepal and found that the customers are unaware about the green banking concept but they agree on the point that green banking initiatives are essential for the environment protection and sustainable growth in future. Similarly, Risal & Joshi (2018) have investigated the green banking practices on banking environmental performance in Kathmandu valley and revealed that there is positive and statistically significant relationship between green banking practices and their environmental performance in Nepal.

Existing literature shows that the majority of studies conducted on green banking are on the awareness and perception of customers towards the adoption of green banking practices. Paucity of research scholars sought to understand the various factors that influence customer's attitude or behavioral intention to adopt green banking practices. So, the study tries to fill this gap by analyzing the various factors that affect the adoption of green banking practices. This study aims to measure customers' perception on green banking practices in Nepalese commercial banks. Likewise, this study provides the information of factors that influence customer's inclination towards the green banking products. This study will benefit greatly to the banking institutions, employees of the banks, customers of the banks, and various stakeholders associated to banks. Also, this research aims to make it easier for policy makers to identified and implement related policies in all the banking and financial institution.

The remaining part of the study is structured as follows; section 2 discusses the methodology with study area, data collection and method of data analysis. Section 3 incorporates data and results. Section 4 presents discussion and the final section concludes.

2. METHODOLOGY

2.1. Conceptual Framework

There are significant research studies that determine the key forces behind the adoption of green banking practices such as Islam et al. (2014), Biswas (2011) and Ahmad et al. (2013). The customer satisfaction and expectation will be different as per their experience perceived from the organization and attitudes towards the products (Ankit 2011). Parasuraman,(1991) states that understanding customer expectation is required for every organization because customers always make a comparative analysis between the perception and services performance.

To understand the customer expectation, the organization needs to know about what customers really want from the product and services they delivered. If the service is below the minimum expectation level of customer, it will not meet their anticipation and if the service exceeds than desired service, they are satisfied (Zeithaml, 2013). Nowadays, customers expect cashless services and 24 hour access to electronic banking. In this context, green banking offers these facilities like the e-cards, internet banking and ATMs (Alinvi, 2009). As a solution of global warming customers like the idea of saving energy, for this bank make their infrastructure environment –friendly by using laptops, desktops computers, resource efficient environmentally responsible green building to meet customers expectation (IDRBT, 2013).

Customers expect their banks & financial institution will use eco-friendly technology. The bank adopts eco-friendly technology for energy efficiency & resources utilization by using solar powered branches, eco-friendly ATMs & use of less power consuming device which will reduce carbon foot prints & render a clear image about banks green (Sabharwal, 2013). Every bank, which have environment directed guidelines which they can increase green concern among stakeholders, customers anticipate that the financial institution will endorse by the guidelines (Islam et al., 2014). The factor that influence the customers expectation are green financial product and services, internet banking, commitment to energy saving, economic factor, compliance with environmental & ethical and eco- friendly technology. These factors are taken by study the different review and from the different customers view point. Similarly, customers perceived green banking as a promote social responsibility, advocates cleanliness, reduces, resources wastage, supported by government law, upholds ethics in business. However, green banking determines by the organizational pressure, environmental policy, operationa wealth of bank, green policy by bank, related parties instruction.

In the context of Nepal only few people research on green banking practices. Still to identify the factors that influence in customer expectation and usage of green banking products is unique. A comprehensive study on the area is worth for policy perspectives as well.

2.2. The Empirical Model

For the empirical analysis of the study, a logit regression model was selected to identify the significant variables that determine the customer perception on green banking practices in study area. The logistic regression equation that is used to ascertain variables influencing green banking practices is given as:

$$Y= f(S, P_e, D, I, E_l, B, P_r, E_c) \tag{1}$$

Where, Y= Dependent Variables, S = Matrix of Socio-demographic variables that contains age, education, gender, work experience. P_e = Matrix of Perception that contains green banking promotion on social responsibility, advocate on cleanliness, reduces resources wastage, whether green banking is supported by government law, upholds ethics in business, D = Matrix of Determinants that contains organizational pressure, environmental policy, operational wealth of bank, green policy by bank. Related parties instruction on green banking, I = Matrix of Importance that contains whether saving paper, solar use, power save equipment, attract customers is important for green banking, E_l = Matrix of Effectiveness level that contains green mortgage, green loan, green credit card, green saving account, effectiveness of green checking account, green money market account, remote deposit, mobile banking, online banking, ethical banking and effectiveness of ATM, B = Matrix of Benefits that contains reduce stationary cost, raise profit, accelerate service delivery, covers CSR, P_r = Matrix of Problems that contains data security and privacy, lack of education, technical issues, traditional approach, lack of infrastructure, technical procedure problem, administration problem, E_c = Matrix of Environmental conservation that contains reduce resource wastage and protect environment.

Based on the functional form and variables undertaken for the study, the logistic regression equation that is used to ascertain variables influencing green banking practices is given as:

$$Y = \beta_1 + \beta_2 x_1 + \beta_3 x_2 + \ldots + \beta_{43} x_{43} + \varepsilon \tag{2}$$

Where, Y= Dependent Variables[1], β_0 = Regression Constant, x_1–x_{43}= Independent variables as given in table 1 and β_1– β_{43} =is respective coefficient of independent variables (i.e. x). The details of variables are presented in result section.

2.3. Study Area and Data

The study is conducted in Kathmandu Valley of Nepal. Among 77 districts of Nepal, Kathmandu Valley includes 3 districts; Kathmandu, Bhaktapur and Lalitpur situated in Bagmati Province were taken for the study area of this research. The total population of Kathmandu is 1,442,271 which has the largest population than other district (CBS, 2020). There are many financial institutions but we select only 28 commercial banks located in Kathmandu Valley. The Kathmandu valley was suitable area for research because it has a largest population of Nepal and important in analyzing the impact of proximity to urban areas in terms of access to ICTs and climate change awareness in the banking sectors.

For the study, data were collected from both the primary and secondary sources. This study was descriptive based on sample survey method. Basically, this study depends on the primary data. The primary data was collected by using structured questionnaire administrated to the 403 customers of

28 commercial banks in Kathmandu valley to elicit their responses. The target populations were banks customers who have an account in banks and use the banking services. A non-probability sampling was used for the survey. The sampling unit in the study was the banks customers. The secondary data were collected from survey, journals, document, research, National Planning Commission (NPC), Ministry of Finance (MOF), Banker Association, Sustainable development goals (SDGs) and Nepal Rastriya Bank policy which is used for strengthening research and its finding.

3. RESULTS

3.1. Descriptive Analysis

Characteristic of Respondents

The number of male and female respondents taken in the sample are almost equal; 56% male, 44% female (as in table 1). Predominant numbers of respondents are youngsters and the middle aged people. But in terms of their qualification and work experience male are more educated and experienced in comparison to female respondents. The study observed that most of the female respondents are housewife and they are not involved in any organization. But on the other hand, it was found that male respondents work even after their retirement from one organization and involved in other organizations.

Table 1. Respondents personal characteristics

Field		**Number (N)**	**Percentage (%)**
Gender	Male	228	56
	Female	175	44
Age	Below 20 Years	58	14.39
	21-30 Years	204	50.62
	31-40 Years	63	15.63
	41-50 Years	59	14.64
	Above 50	19	4.71
Education	Below SLC/SEE	66	16.37
	Upto SLC/SEE	56	13.89
	Higher Secondary Level	87	21.58
	Bachelor	111	27.54
	Masters and Above	83	20.59
Work Experience	No Experience	150	37.22
	Less than 10 Years	190	47.14
	10-20 Years	32	7.94
	21-30 Years	27	6.69
	Above 30 Years	4	0.99

Source: Filed Survey data, 2020

Customer Awareness Level on Green Banking Practices

The study expects to assess the awareness level of the respondents by asking 7 yes/no questions which helps to determine the individual respondents' awareness level towards the green banking practices adopted by their banks. Each 'yes' answer will get one point. If overall awareness score exceeds 75, it is considered as highly aware; if it remains between 51 to 75, it is medium aware and the score below 50 means less aware about green banking. In this study, the awareness index contains 4 independent variables; gender, age, education and work experience (See figure 1). The awareness index was prepared to understand age, gender, work experience and education. In summary, majority of respondent 385(≈96%) were less aware about the green banking practices whereas only 17 of them and 1 of them are moderately aware and highly aware respectively.

Figure 1. Overall awareness on Green Banking

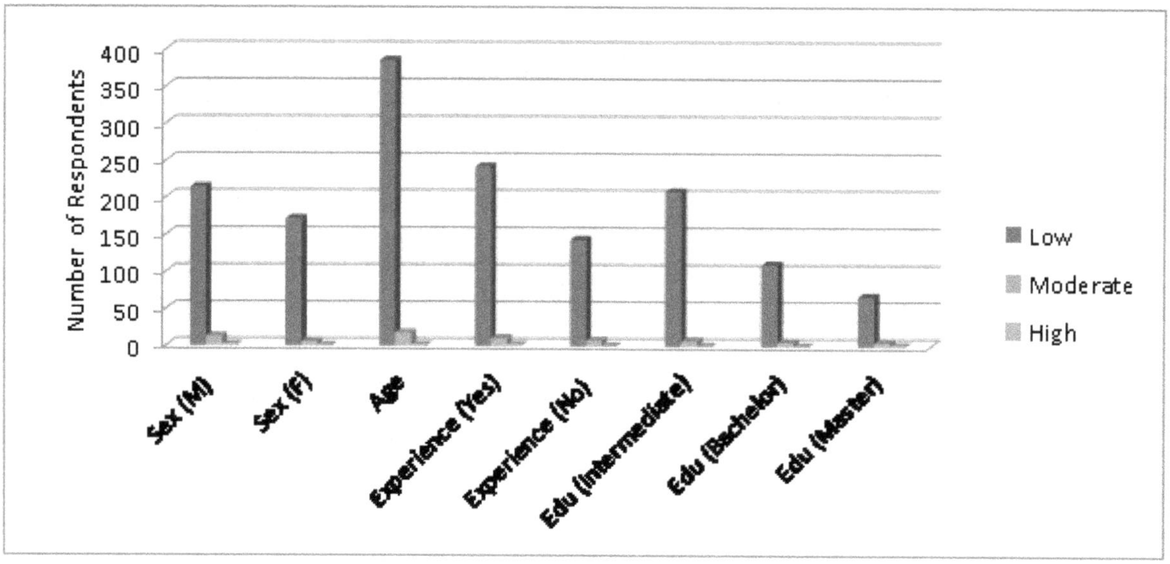

Customers Perception on Green Banking Concept

The result indicates that the most of the people are not aware about the green banking concept (See table 2). Only 109 respondents (27%) are aware about the green banking practices by their bank and 294 are not aware about such practices adopted in their bank. Most of respondents (41%) thought green banking is the environmental banking whereas 24 percent opined green banking as social responsibility banking. Those who are aware about the green banking practices of their banks, most of them responded that the level of awareness is moderate indicates that only few people aware about the green banking practices adopted by their banks. From the analysis, it is observed that 88% respondents thought that the green banking practices are important.

Majority of the respondents know about the ethical banking (367), half of respondents aware about mobile banking (263), and online banking (232). Similarly, save paper (114) and use solar energy also known by the respondents and other components are known moderately. But the very few respondents were known about the components of green banking, green mortgage (20), green money markets account (25), and green checking accounts (30). Five most popular practices adopted by banks are mobile banking and online banking. Likewise, ethical banking, save paper and use solar energy is the 3rd, 4th and 5th components of green banking respectively. The components adopted by bank are also helpful to promote green banking concept with the respect to reduce paper consumption, recycle waste properly, sponsor tree plantation, spend in green publicity, launch green product and services and others. According to the respondents, most of them thought it helps to reduces paper consumption (29%) and recycle

Table 2. Customers perception on Green Banking

Variable		Number (N)	Percentage (%)
Awareness on Green Banking Practices	Yes	92	22.82
	No	196	48.63
	May Be	37	9.18
	No Idea	78	19.35
Green Banking Practices	Ethical Banking	80	19.85
	Sharing Based Banking	139	34.49
	Environmental Banking	89	22.08
	Social Responsibility Banking	224	55.58
	Sustainable Banking	20	4.96
	Others	11	2.72
Importance of Green Banking	Very Important	103	25.55
	Important	138	34.24
	Moderate	113	28.03
	Less Important	13	3.22
	Not Important	36	8.93
Five Popular Strategies Adopted by Banks	Mobile Banking	168	41.68
	Online Banking	158	39.20
	Ethical Banking	69	17.12
	Saves Paper	65	16.12
	Use Solar Energy	61	15.13
Strategies to Promote Green Banking	Reduce Paper Consumption	92	22.82
	Recycle Waste Properly	73	18.11
	Sponsor Tree Plantation	61	15.13
	Green Publicity	45	11.16
	Launch Green Products & Services	43	10.66
	Others	6	1.48
Bank Provides Green Banking Trainings	Yes	7	1.73
	No	113	28.03
	May Be	40	9.92
	No Idea	243	60.29

wastage properly (22%). Training is an essential factor in order to initiate new concept or technology and to know how it works and what are the benefits and consequences of it.

Majority of the respondents perceived that green banking reduce resources wastage (27%) and promotes social responsibility (26%). Respondents thought the factors that determine the green banking practices are environmental policy (38%) and green banking by bank (25%) (see figure 2). It indicates that major factor that determines green banking is environmental policy and adoption of green banking practices by banks.

For the levels of effectiveness about the green banking compared to other mode of banking, majority of respondents (84%) have positive response. It indicates that there is a high level of effectiveness about the green banking compared to other mode of banking in Nepal. But 75% of the respondents don't get the opportunity of significant benefits of green banking compared to other mode of banking.

Figure 2. Determinants of Green Banking

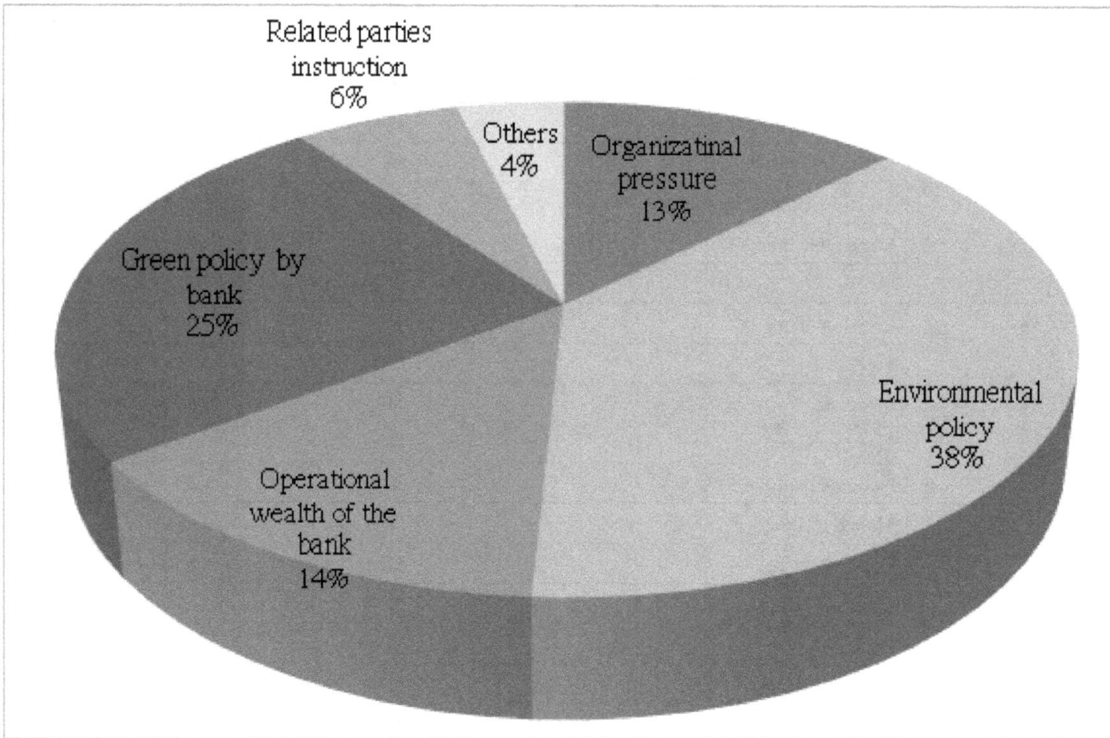

Managerial Suggestion for Promotion of Green Banking

This section includes the customers' response towards the existing technology sufficient to promote green banking practices or not. 70% respondents positively thought the existing technology is sufficient to promote green banking practices. One third customers thought bank should organize the awareness programs by coordinating with different media channels, pamphlets and so on. Other 18%, 13% and 15% respondents opined that update the new technology of the banks, training to employees and the customers, and electronic transaction and marketing are major tools to use to promote green banking practices respectively. In addition to this, quarter of the respondents (23%) argued different things like international payment system, increase green loan and green saving account and so on.

The major things are awareness program as 51% customers support on it (see figure 3). Another thing needed to be done is promote the green banking practices. The awareness level of customers towards the concept depends on the promotion level and it is supported by 14% customers. To adopt the new concept and use it both the employees and customers need training. Training helps to develop the concept by making understandable how it works and what are the benefits of it. 12% customers' support the training is needed to improve the scope of the green banking practices. 13% customers indicated the different views like banks should make mobile banking & online banking compulsory, translate the language into Nepali words, bank should separate some amount of money for green banking activities and so on.

Figure 3. Managerial implication to promote Green Banking

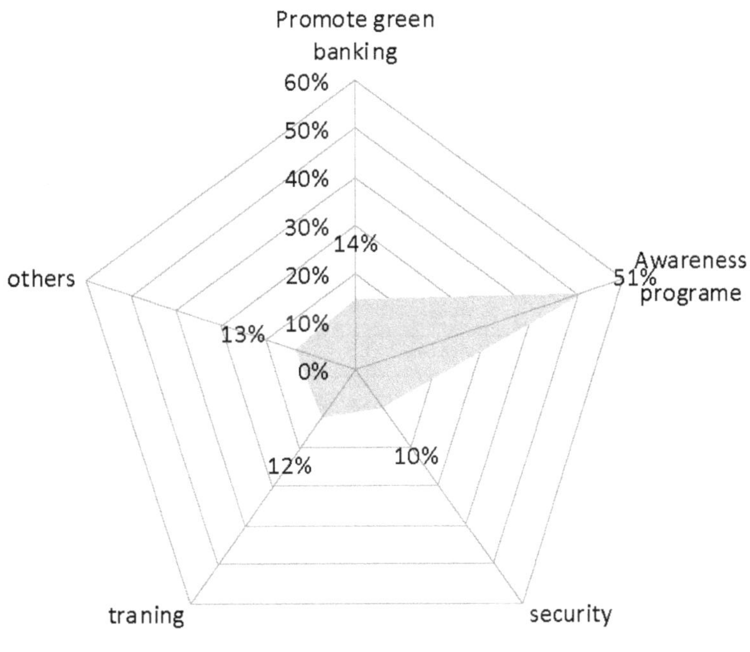

3.2. Econometrics Estimation

Before running the Binary Logit regression, necessary post estimation test is done. Multicollinearity test using VIF and heteroscedasticity test using Breusch- Pagan/ Cook- Weisberg test were performed to find out about any existing repetitions or similarities between various data sets and ascertain whether the data. The post estimation result indicates the heteroscedasticity problem in the existing data set. Similarly, $\chi 2$ value was significant and the value of the Pseudo-R2 ranged from 0.22 to 0.65. The VIF value is less than 1.79 in all models. Hence, we performed robust standard error test to correct the heteroscedasticity problem. The final output result for this study is presented in table 3.

Table 3. Logistic Regression Result of Green Banking

Variables		Awareness	Importance	Effectiveness	Benefits	Problems	Environment conservation
Gen		0.133	0.855*	-0.253	-0.122	0.211	-0.284
		(0.294)	(0.464)	(0.347)	(0.350)	(0.471)	(0.332)
Age		-0.0172	0.00335	-0.0143	-0.0506**	-0.0125	-0.0290
		(0.0187)	(0.0318)	(0.0217)	(0.0231)	(0.0248)	(0.0216)
Edu		-0.0604	1.600***	0.627	1.377**	0.539	-0.274
		(0.434)	(0.550)	(0.442)	(0.627)	(0.654)	(0.450)
work_exp		-0.0149	-0.0318	0.0239	0.0740**	0.0422	-0.0150
		(0.0292)	(0.0432)	(0.0293)	(0.0323)	(0.0335)	(0.0300)
received_any_train		2.882**	NA	NA	2.780**	-2.841	NA
		(1.120)			(1.173)	(2.415)	
GB_promo_SR		0.289	1.814***	0.862**	0.199	-0.201	0.649*
		(0.334)	(0.655)	(0.436)	(0.394)	(0.511)	(0.380)
GB_advocates_clean		0.788**	0.966	0.732*	-0.620	-0.109	-0.231
		(0.329)	(0.651)	(0.437)	(0.412)	(0.508)	(0.367)
GB_reduce_resource_wastage		0.342	0.378	0.0691	1.343***	-0.656	-0.776*
		(0.370)	(0.651)	(0.472)	(0.445)	(0.596)	(0.427)
GB_support_ govt		-0.215	1.027	-0.00679	-0.114	0.376	-0.110
		(0.352)	(0.754)	(0.453)	(0.440)	(0.599)	(0.413)
GB_upholds_ethics_business		-0.215	1.680**	0.663	-0.377	0.348	0.0845
		(0.370)	(0.850)	(0.452)	(0.437)	(0.572)	(0.419)
organizational_pressure		-0.438	-1.590**	-0.850	-0.0992	-2.000***	0.371
		(0.419)	(0.714)	(0.529)	(0.482)	(0.752)	(0.491)
environmental_policy		0.237	-0.212	-0.489	1.030**	0.906	1.487***
		(0.393)	(0.702)	(0.510)	(0.452)	(0.629)	(0.453)
operational_wealth_bank		-0.897**	0.888	-1.137**	-0.690	-0.292	-0.629
		(0.423)	(0.765)	(0.478)	(0.460)	(0.662)	(0.466)
green_policy_by_bank		-0.202	-0.133	-1.009**	0.188	-0.881	0.285
		(0.372)	(0.707)	(0.480)	(0.406)	(0.588)	(0.417)
related_partie_instruction		0.0397	-3.39***	-0.934	0.252	1.665*	-0.131
		(0.513)	(0.936)	(0.618)	(0.559)	(0.947)	(0.620)
save_paper		0.614	0.522	0.125	-0.637	-1.094*	-0.141
		(0.408)	(0.934)	(0.494)	(0.448)	(0.606)	(0.470)
use_solar_energy		-0.441	-1.428	0.585	-0.525	-0.0109	0.898*
		(0.392)	(0.885)	(0.499)	(0.444)	(0.562)	(0.460)
power_supply_equipment		-0.333	3.850**	0.153	0.00751	-0.134	-0.231
		(0.497)	(1.689)	(0.618)	(0.543)	(0.743)	(0.552)

continued on following page

Table 3. Continued

Variables	Awareness	Importance	Effectiveness	Benefits	Problems	Environment conservation
attract_customers	0.621*	-0.395	0.914*	-0.347	-2.564***	-0.340
	(0.357)	(0.601)	(0.493)	(0.423)	(0.717)	(0.423)
green_mortgage	1.327*	0.712	0.774	-0.285	-0.658	-2.295**
	(0.703)	(1.470)	(1.181)	(0.821)	(1.220)	(0.997)
green_loan	-0.0404	1.466	0.981*	0.893*	2.378***	0.447
	(0.446)	(1.099)	(0.585)	(0.536)	(0.756)	(0.576)
green_credit_card	-0.134	0.551	0.544	0.566	1.190*	1.074**
	(0.424)	(0.851)	(0.551)	(0.510)	(0.636)	(0.504)
green_saving_acc	-0.00963	-2.526**	0.716	0.333	-1.043	0.472
	(0.477)	(1.039)	(0.662)	(0.566)	(0.766)	(0.567)
green_checking_acc	0.988	0.123	-1.390*	0.193	-2.430***	0.396
	(0.633)	(1.543)	(0.835)	(0.717)	(0.940)	(0.817)
green_money_market_acc	0.160	1.379	2.000	-0.987	2.047**	3.166**
	(0.616)	(1.682)	(1.374)	(0.706)	(0.919)	(1.277)
mobile_banking	-0.444	0.738	-1.480**	-0.545	0.532	-0.0200
	(0.565)	(1.041)	(0.653)	(0.669)	(0.761)	(0.611)
online_banking	0.445	0.273	-0.0590	0.910	0.176	0.662
	(0.542)	(1.014)	(0.616)	(0.633)	(0.668)	(0.568)
Remot_deposit	-0.111	-0.337	-0.100	-0.0679	1.177*	1.016*
	(0.434)	(0.991)	(0.615)	(0.503)	(0.714)	(0.586)
ethical_banking	-0.310	-0.311	-0.979	1.474**	-1.196	0.0150
	(0.513)	(0.909)	(0.744)	(0.731)	(0.773)	(0.680)
Other (ATM)	-0.812	-4.24***	-0.302	-2.039**	0.923	0.125
	(0.687)	(0.891)	(0.543)	(0.976)	(0.840)	(0.557)
reduce_stationary_cost	0.602*	0.697	-0.568	0.376	-0.247	0.876**
	(0.315)	(0.578)	(0.363)	(0.381)	(0.518)	(0.347)
raise_profit	-0.738*	-1.903**	1.351**	1.459***	-1.716**	1.147**
	(0.405)	(0.756)	(0.630)	(0.427)	(0.762)	(0.492)
accelerate_service_delivery	0.770**	0.502	-1.117**	1.166**	-0.132	-0.567
	(0.388)	(0.703)	(0.444)	(0.458)	(0.605)	(0.450)
covers_CSR	-0.473	-0.128	1.523***	-0.478	1.109**	-0.246
	(0.371)	(0.643)	(0.469)	(0.431)	(0.561)	(0.418)
data_security_and_privacy	-0.731*	-0.206	-0.383	1.272***	3.185***	-0.654
	(0.422)	(0.974)	(0.611)	(0.457)	(0.674)	(0.589)
lack_of_education	0.368	2.481**	0.479	1.160***	2.319***	0.629
	(0.391)	(1.225)	(0.582)	(0.421)	(0.571)	(0.522)
technical_issue	0.789*	0.195	-1.095**	0.541	2.450***	1.524***
	(0.416)	(0.896)	(0.540)	(0.436)	(0.603)	(0.541)

continued on following page

Table 3. Continued

Variables	Awareness	Importance	Effectiveness	Benefits	Problems	Environment conservation
traditional_approach	0.527	-1.225	0.756	-0.367	1.543**	1.993***
	(0.481)	(1.125)	(0.721)	(0.516)	(0.678)	(0.716)
lack_of_infrastructure	-1.54***	-1.930*	-0.979	0.793	2.603***	0.0135
	(0.547)	(1.017)	(0.685)	(0.528)	(0.781)	(0.655)
technical_procedure_problem	0.177	1.445	2.639**	0.0683	1.978**	2.771**
	(0.497)	(1.431)	(1.160)	(0.557)	(0.849)	(1.103)
admistration_problem	0.162	-1.345	-0.899	-0.682	1.008	1.327**
	(0.453)	(1.114)	(0.644)	(0.567)	(0.783)	(0.650)
reduc_resources_wastage	1.030***	0.618	1.551***	1.414***	1.574***	1.202***
	(0.367)	(0.554)	(0.442)	(0.472)	(0.593)	(0.381)
protect_environment	0.346	1.004*	-0.215	-0.392	0.813	0.550
	(0.333)	(0.586)	(0.398)	(0.410)	(0.511)	(0.364)
Constant	-2.27***	-0.335	2.730***	-5.98***	-4.294***	-2.710***
	(0.862)	(1.354)	(1.055)	(1.202)	(1.310)	(0.986)
Pseudo R^2	0.2255	0.4608	0.2704	0.4176	0.6475	0.4833
VIF	1.66	1.67	1.66	1.67	1.71	1.68
Breusch- Pagan/ Cook- Weisberg test	56.66 (0.0000)	93.85 (0.0000)	84.54 (0.0000)	57.69 (0.0000)	75.81 (0.0000)	2.13 (0.0000)
Observations	403	397	397	403	403	397

Standard errors in parentheses
*** p<0.01, ** p<0.05, * p<0.1

The first model result presented on the table depicts seven independent variables were found to have significant relationship with the green banking awareness. Four of these variables namely, received training, green banking advocate's cleanliness, technical issue and reduce stationary cost were significant at 5% level while green mortgage and green checking account significant at 1% level and reduce resource wastage were significant at 10% level. This result indicates the probability of increase in green banking awareness with the increase the above significant independent variables. However, operational wealth of bank had a significant negative relationship (at 5% level), lack of infrastructure (at 10% level), raise profit (at 1%level). Looking at their odds ratio, our result indicates that for one unit change in received training, green mortgage, green banking advocate cleanliness, green checking account, technical issue and reduce resource wastage, the odds of being awareness on green banking increase by a factor of 17 times, 3.8 times, 2 times, 2.95 times, 2.24 times, 1.9 times and 2.6 times respectively. This study indicates that training is the most significant factors to promote green banking among the customers. The first hypothesis is rejected as there is no significant relationship between the green banking awareness and the given explanatory variables.

The result presented in **the model 2** in the table shows the probability of increase in importance of green banking increase in gender, education, power supply equipment, green banking promotion social responsibility, green banking upholds ethics in business, lack of education, protect environment but decrease with green saving account, ATM, organizational pressure, related parties instruction, lack of infrastructure, raise profit. The odds ratio of gender, education, power save equipment, green banking promotion social responsibility, green banking upholds ethics in business, lack of education, protect environment are 2.3 times, 4.9 times, 47 times, 6 times, 5 times, 12 times, 2 times respectively. The second hypothesis is also rejected since the model found significant relationship between importance of green banking with seven of the explanatory variables.

In the third model green banking effectiveness was captured where variables green banking promotion social responsibility, attract customers, green money market significant at (1% level) while, technical procedure problem significant at (5% level) and cover CSR, raise profit and reduce resources wastage significant at (10% level). This result indicates that the probability of increasing in effectiveness level depends on the increase in these independent variables. The odds ratio of these seven variables are 2.37 times, 2.49 times, 7.4 times, 13.97 times, 4.58 times, 3.8 times, and 4.7 times respectively which signifies effectiveness level among customers with these variables. The study indicates that technical procedure problem is the most significant factors to the effectiveness of green banking. However, lack of infrastructure, green checking account negative significant at (1% level) and operational wealth of the bank, green policy by bank, accelerate service delivery, technical issues, mobile banking at (5% level). It shows that increase in any independent variables decreases effectiveness level. It mirrors that the third hypothesis is also rejected in case of effectiveness level indicating significant relationship.

In **the fourth model** benefits have been assessed. Here too, out of the thirteen, ten variables i.e. received training, green loan positively significant at (1% level), while work experience, lack of education, environmental policy, accelerate service delivery, ethical banking significant at (5%level) and data security and privacy, raise profit and reduce resource wastage at (10% level) are taken into account. Further, odds ratio of these variables are 16 times, 2.44 times, 1 times, 3.9 times, 2.8 times, 3.2 times, 4.3 times, 3.5 times, 4.3 times and 4 times. Similarly, two variables age and other (ATM) are negatively significant at (5% level).This result indicates that probability of age and other (ATM) affects negatively to the benefits of green banking. Similar to previous three hypotheses the fourth hypothesis faced the same fate. It shows that there is significant association between the benefits and the given eleven explanatory variables among customers.

The model fifth result depicts that work experience, green credit card significant at (1%level) while traditional approach, lack of infrastructure, reduce resource wastage significant at (5%level) and data security, lack of education, technical issue, green loan and green money market significant (10%level) with odds ratio of 1 times, 3.29 times, 4.67 times, 13.5 times, 4.8 times, 24 times, 10 times, 11.5 times, 10.7 times and 7.7 times respectively. The result indicates that data security, lack of infrastructure, lack of education and technical issues are major problem of green banking. However, out of fifteen significant variables five variables are negatively significant. Received training and green policy are significant at 1%level, organizational pressure at 5%level and attract customer, green checking account significant at 10%level.The fifth hypothesis is also rejected.

The sixth model of the study explains the relation of environment conservation where green banking promotion social responsibility, remote banking, use solar energy are positively significant at (1%level). Similarly, technical issue, green credit card, green money market, reduce stationary cost, raise profit and technical process procedure (5% level) and environment policy, traditional approach and reduce resources wastage (10%). It entails that if there is increase in these variables, there is decrease in environment conservation. The odds ratio are 1.9 times, 2.7 times, 2.4 times 4.5 times,2.9 times, 23.7 times and 2,4 times,3 times,15 times, 4.4 times, 7.3 times, 3.3 times respectively. Only green banking reduces resources wastage and green mortgage negatively significant at 1 and 5% level. As all other hypothesis, the hypothesis is also rejected since the model found significant association between the environment conservation and seven variables.

4. DISCUSSION

The study tries to identify the customers' perceptions on green banking practices by focusing on the awareness level, importance of green banking, opinion on green banking, effectiveness level, benefits of green banking, determinants of green banking, problems of green banking and environmental conservation. From the table 1, we see that 96% of the respondents are less aware about the green banking practices and 4% moderately and only 1 respondent is found highly aware. It refers that bank should focus on encouraging the promotion of the concept green banking towards their customers (Dixit & Datta 1970; Sharma et al., 2014;Vijai & Natarajan 2015; Deka 2015; Mehta & Sharma, 2016; Malliga & Revathy,2016; Pariag-maraye et al.,2017; Risal & Joshi,2018). Similarly, very few 12% respondents think that green banking concept is not important because they are unknown about the benefits of green banking practices but the majority of respondents 88% think that it is important, and will benefit to customers, banks and society. It indicates that green banking concept in financial institution is necessary in current context to change the traditional way of banking transaction into technological banking (Sathey 1999; Mehta &Sharma, 2016; Malliga & Revathy, 2016).

Green banking concept has various components like mobile banking, online banking, green mortgage, green loan, ethical banking, green credit card, green saving account, green checking account, green money market account, remote deposit, power save equipment, save paper, use solar energy and others (Faruque et al., 2016; Ganesan & Bhuvaneswari,2016; Deepa & Karpagam,2018 Masukujjaman et al., 2017; Mehta & Sharma, 2016) . Comparatively the customers ranked the five most adopted practices by their banks like this 1st mobile banking, 2nd online banking, 3rd ethical banking, 4th save paper and 5th use solar energy respectively. From the analysis, it is found out that respondents are still unaware about whether their bank adopt such practices on their banking practices or not. Whereas respondents are positive about mobile banking, online banking, ethical banking, save paper and use solar energy as a green banking practices. To adopt the green banking practices by customers, the bank should not only aware the customers about various components of the green banking practices adopted by their banks but also the advantages of using green banking practices (Deka, 2018).

41% of respondents agree on green banking is the environmental banking which means customers perceived the green banking as to solve the environmental problem. It means customers are known about green banking practices is environmental friendly concept that promotes and contributes to protect environment. So, bank should adopt the environmental friendly practices by focusing on clean and healthy environment through the reduction of carbon foot print (Sharma, Sarika & Gopal, 2014; Raj & Rajan, 2017).

To solve the problem of environmental degradation and upgrade the banking services with technology, Nepal needs to work hard. In this context, bank and financial institutions play a vital role to maintain sustainable growth by financing the environmental friendly activities and change on the customer's habits by introducing green products and services. Recently, the green banking practices adopted by Nepalese banks and respondents thought that the success and improve the concept will be determined by the environmental policy (38%) and green policy by bank (25%) more than other factors. Green banking has benefits to both the banks and customers, on that context respondents said that green banking concept helps to attract customers (18%), cover CSR (18%) to the banks and for customers it helps to protect environment (20%) and accelerate service delivery (9%) that saves both time and money, and facilitates to do banking transaction easily from anywhere. Moreover, it helps to reduce the stationary cost and raise profit of both banks and customers. As green banking is a new concept and most of the people are unfamiliar with it, the reason of this problem is rooted in the issues like lack of education, data security, comfortable with traditional method, lack of infrastructures.

The foremost contribution of this research is to find out the customers expectation towards the new concept green banking practices. The previous researcher find out that the practices have a huge potential in Nepal if banks are able to introduced the concept towards the customers and benefit both the customers and banks which is significant for the development of nation (Mehta &Sharma, 2016;Risal & Joshi, 2018). This study helps to know the customers perceptions towards the green banking practices. Customers are ready to adopt the green banking practices but the problem is they are not familiar with it. This means bank and government should focus on the increasing the awareness of the concept and make a policy that supports the green banking practices.

There are some limitations in this study. First, this study is conducted based on the data collected only from the 28 commercial banks, there are also other development banks and financial institution in Kathmandu Valley which are not included in the study. Second, the information was collected only from the commercial banks of Kathmandu valley Third, targeted sample size of the study is limited. Despite these limitations, this study makes significant contribution to future research in Nepal by updating the recent information and data. The current data and information are important for future studies as a reference for their studies and find out the new research gap. In addition to that, future studies have the further opportunities to consider the predecessor variables related to this study. Hence, it is suggested that there is possibility to conduct the study in customers' perception on green banking practices in all other districts of Nepal. To overcome from these above mentioned limitations, further studies are suggested.

5. CONCLUSION

The governments, business firms, policy makers, advocacy groups and even the public are hotly debating on the issues of environmentally friendly practices around the world in the last few decades. In this scenario, being a part of 'going green', Green Banking which plays an important role in environmental sustainability has been a buzzword in global baking industry. The research concludes that in the context of Nepal, green banking is still the new concept and issues in banking and financial institutions are also in incipient form in comparison to other countries. There is positive response of the customers towards the green banking practices and they are ready to adopt such practices. Also, customers feel that green banking practices is important to them. But the major problems are less awareness about the practices adopted by their banks and benefits of it. The result from 403 banking customers suggest that to adopt the practices in the context of Nepal, bank and financial institutions should focus on awareness programs to promote the new concept initiated by their banks, and on the other hand training is also the most influencing factor in regard to such practices. This study observed that the customers are positive towards the environmental friendly practices of banks and ready to adopt the green banking practices and such practices will help to maintain sustainable environment. It means that before adoption of the new concept in banks and financial institutions, they should introduce such new concepts to customers who are the users and their perceptions towards the practices means a lot. The research implies that the banks and financial institutions should be able to educate the customers about green banking practices and their benefits. Furthermore, when such practices and benefits are realized, this will lead the environment towards sustainability from the part of banking and financial sector.

REFERENCES

Ahmad, F., Zayed, N. M., & Harun, M. (2013). Factors behind the adoption of green banking by Bangladeshi commercial banks. *ASA University Review*, *7*(2), 241–255.

Aldás-Manzano, J., Lassala-Navarré, C., Ruiz-Mafé, C., & Sanz-Blas, S. (2009). The role of consumer innovativeness and perceived risk in online banking usage. *International Journal of Bank Marketing*, *27*(1), 53–75. doi:10.1108/02652320910928245

Alinvi, F. (2009). *Customers' Expectations of Banks Becoming Cashless-How could Banks meet customer's expectations when changing from cash services to deeper customer relationship?* Retrieved in July 20, 2016 from www.divaportal.org/smash/get/diva2:220556/FULLTEXT01.pdf

Ankit, S. (2011). Factors Influencing Online Banking Customer Satisfaction and Their Importance in Improving Overall Retention Levels: An Indian Banking Perspective. *Information and Knowledge Management*, *1*(1), 45–54.

Arumugam.D.,& Chirute.T.(2018). Factors determining the adoption of green banking amongst commercial banks in Malaysia. *Electronic Journal of Business & Management,* 50 - 62.

Bihari, S. C., & Pandey, B. (2015). Green banking in India. *Journal of Economics and International Finance*, *7*(1), 1–17. doi:10.5897/JEIF2014.0599

Biswas, N. (2011). Sustainable Green Banking Approach: The Need of the Hour. *Business Spectrum*, *1*, 32–38.

Bouma, J. J., Jeucken, M., & Klinkers, L. (2001). *Sustainable Banking: The Greening of Finance*. Greenleaf Publishing Limited.

Deepa, P., & Karpagam, C. R. (2018). A study on customer's awareness on green banking in selected public and private sector banks with reference to Tirupur. *International Journal of Advanced Research and Development*, *3*(1), 58–63.

Deka, G. (2015). Green Banking Practices: A Study on environmental strategies of banks with special reference to State bank of India. *Indian Journal of Commerce and Management Studies*, *6*(3), 11–19.

Deka, G. (2018). Customers' perception on green banking practices of state bank of India: A SWOC Analysis. *International Journal of Management Studies*, *5*(1), 20–42. doi:10.18843/ijms/v5iS1/04

Devkota, N., & Phuyal, R. K. (2017). An analysis of Nepalese youth understanding level on climate change. *Asian Journal of Economic Modelling*, *5*(3), 342–353. doi:10.18488/journal.8.2017.53.342.353

Dixit, N., & Saroj, K. (1970). Acceptance of e-banking among adult customers: An empirical investigation in India. *Journal of Internet Banking and Commerce*, *15*(2), 1–17.

Dixit, N., & Saroj, K. (1970). Acceptance of e-banking among adult customers: An empirical investigation in India. *Journal of Internet Banking and Commerce*, *15*(2), 1–17.

Faruque, M. O., Biplob, M. N. K., Al-Amin, M., & Patwary, M. S. H. (2016). Green banking and its potentiality & practice in Bangladesh. *Science Journal of Business and Management*, *4*(2), 28–33. doi:10.11648/j.sjbm.20160402.12

Ganesan, D. R., & Bhuvaneswari, A. (2016). customer perception towards green banking. *IOSR Journal of Economics and Finance*, *7*(5), 5–17. doi:10.9790/5933-0705010517

Institute of Development and Research in Banking Technology. (2013). *Publications: IDRBT.* Retrieved in July 21, 2016 from http://www.idrbt.ac.in/publications/Frameworks/Green%20Banking%20Framework%20

Islam, M. A., Yousuf, S., Hossain, K. F., & Islam, M. R., (2014). Green financing in Bangaladesh: Challenges and opportunities- a descriptive approach. *International Journal of Green Economics*, *8*(1), 74-91.

Jahangir, N., & Begum, N. (2008). The role of perceived usefulness, perceived ease of use, security and privacy, and customer attitude to engender customer adaption in the context of electronic banking. *African Journal of Business Management*, *2*(1), 32–40.

Jani, A. S. (2012). A study of consumer perception on the use of e-technology in the retail banking sector: A comparative study of public sector and private sector banks. *Indian Journal of Marketing*, *42*(1), 46–58.

Jayabal & Saundarya. (2016). Green Banking: As Banks Initiative for Sustainable Development. *International Journal of Management*, *7*(7), 276–280.

Jha, N., & Bhome, S. (2013). A study of green banking trends in India. *International Monthly Referred Journal of Research in Management and Technology*, *2*, 127–132.

Klein, A. G., Gerhard, C., Büchner, R. D., Diestel, S., & Schermelleh-Engel, K. (2016). The detection of heteroscedasticity in regression models for psychological data. *Psychological Test and Assessment Modeling*, *58*(4), 542–568.

Ko, M., Mancha, R., Beebe, N., & Yoon, H. S. (2012). Customers' personality, their perceptions, and green concern on internet banking use. *Journal of Information Technology Management*, *23*(4), 21–32.

Mahfuzur, SM, R., & Barua, S. (2016). The design and adoption of green banking framework for environment protection: Lessons from Bangladesh. *Australian Journal of Sustainable Business and Society*, *2*(1), 1–19.

Malliga, A.L., & Revathy, K. (2016). *Customers awareness on green banking- an initiative by private sector bank in Theni district*. Academic Press.

Masukujjaman, M., & Aktar, S. (2013). Green banking in Bangladesh: A commitment towards the global initiatives. *Journal of Business and Technology (Dhaka)*, *8*(1-2), 17–40. doi:10.3329/jbt.v8i1-2.18284

Masukujjaman, M., Siwar, C., Mahmud, M. R., & Alam, S. S. (2017). Bankers' perception of Green Banking: Learning from the experience of Islamic banks in Bangladesh. *Geografia: Malaysian Journal of Society and Space*, *12*(2), 144–153.

Mehta, K., & Sharma, R. (2016). Customers" Persistence for Green Banking in Nepal. *Asian Journal of Research in Banking and Finance*, *6*(10), 30–44. doi:10.5958/2249-7323.2016.00050.X

Mishra, D. K. (2013). Green strategies: Response of Indian banks to climate change. *An International Quarterly Journal of Environmental Sciences*, *3*, 345 – 348.

Pallant, J. (2011). *SPSS survival manual: a step by step guide to data analysis using SPSS* (4th ed.). Allen & Unwin.

Pant, B. (2016). *Financial Inclusion in Nepal: Policy Review and Prescriptions* (No. 34). NRB Working Paper.

Parasuraman, A., Berry, L. L., & Zeithaml, V. A. (1991). Understanding customer expectations of service. *Sloan Management Review*, *32*(3), 39–48.

Pariag-Maraye, N., Munusami, N., & Ansaram, K. (2017). A customer's perspective of green banking: A case study of commercial banks in Mauritius. *Theoretical Economics Letters*, *7*(07), 19–75. doi:10.4236/tel.2017.77134

Prasad, V. M. (2017). Green Banking- Is The Concept Known By The Bank Customers? *Journal of Advances in Business Management.*, *3*(1), 23–27.

Raj, D. G. P., & Rajan, D. A. P. (2017). A Study on the customer awareness on green banking initiatives. *Intercontinental Journal of Finance Research Review*, *5*(7), 54–65.

Risal, N., & Joshi, S. K. (2018). Measuring Green Banking Practices on Bank's Environmental Performance: Empirical Evidence from Kathmandu valley. *Journal of Business and Social Sciences*, *2*(1), 44–56. doi:10.3126/jbss.v2i1.22827

Sabharwal, M. (2013). The Use of Eco-Friendly Technology and Gree Methods to Bring Down the Carbon Footprint by Indian Banks. *Masters International Journal of Management Research and Development*, *1*(1), 76–85.

Sahni, D., & Dhamija, A. (2018). Green Banking: Perception and Willingness of Customer to Adapt Green Banking. *International Journal of Financial Management*, *7*(2), 1–8.

Sathye, M. (1999). Adoption of Internet banking by Australian consumers: An empirical investigation. *International Journal of Bank Marketing*, *17*(7), 324–334. doi:10.1108/02652329910305689

Shampa, T. S., & Jobaid, M. I. (2017). Factors Influencing Customers' Expectation Towards Green Banking Practices in Bangladesh. *European Journal of Business and Management*, *9*(12), 140–152.

Sharma, N., Sarika, K., & Gopal, R. (2014). A Study on customer awareness on green banking initiatives in selected public and private sector banks with special reference to Mumbai. *IOSR Journal of Economics and Finance.*, *3*(7), 28–35.

Shaumya, K., & Anton Arulrajah, A. (2017). The Impact of Green Banking Practices on Bank's Environmental Performance: Evidence from Sri Lanka. *Journal of Finance and Bank Management*, *5*(1), 77–90.

Shaumya, K., & Arulrajah, A. (2016). Measuring green banking practices: evidence from Sri Lanka. *University of Sri Jayewardenepura, Sri Lanka, 13th International Conference on Business Management (ICBM)*, 999-1023.

Singh, H., & Singh, B. P. (2012). An Effective and Resourceful Contribution of Green Banking towards Sustainability. *International Journal of Advances in Engineering Science and Technology*, *1*(2), 3.

Tara, K., Singh, S., & Kumar, R. (2015). Green banking for environmental management: A paradigm shift. *Current World Environment*, *10*(3), 1029–1038. doi:10.12944/CWE.10.3.36

Thombre, K. A. (2011). The new face of banking: Green banking. *Indian Streams Research Journal*, *1*(2), 1–4.

Vijay & Natarajan. (2015). Customers' Awareness towards Green Banking Products of the Select Commercial Banks in Cuddalore District. *An Empirical Assessment*, *3*(10), 1–5.

Wessels, L., & Drennan, J. (2010). An investigation of consumer acceptance of M-banking. *International Journal of Bank Marketing*, *28*(7), 547–568. doi:10.1108/02652321011085194

Wong, C. C., & Hiew, P. L. (2005). Diffusion of mobile entertainment in Malaysia: Drivers and barriers. *Proceedings of World Academy of Science. Engineering and Technology*, *5*, 263–266.

Yadav, R., & Pathak, G. (2013). Environmental sustainability through green banking: A study on private and public sector banks in India. *OIDA International Journal of Sustainable Development*, *6*(08), 37–48.

Zeithaml, V. A., Bitner, M. J., Gremler, D. D., & Pandit, A. (2013). *Services Marketing: Integrating customer focus across the firms* (6th ed.). Tata McGraw Hill Education Private Limited.

ENDNOTE

1 This study has set 6 different dependent variables with the same independent variables to check the independent variables in different situations.

APPENDIX

Table 4. Variables and the expected signs undertaken for the study

Variables	Description	Value	Expected signs
	Socio-demographic		
Gen	Customers gender	1= male, 0= otherwise	+
Age	Customers age	in years	+
Edu	Formal schooling year	Year	+
Work _Exp	Total customers experience year	In years	±
Awareness			
Received_any_train	Whether received any green banking related training	1=yes, 0= otherwise	+
Perception			
GB_promo_SR	Whether green banking promotes social responsibility	1=yes, 0= otherwise	±
GB_advocates_clean	Whether green banking advocates cleanliness	1=yes, 0= otherwise	±
GB_reduce_resource_wastage	Whether green banking reduces resources wastage	1=yes, 0= otherwise	±
GB_support_ govt	Whether green banking is supported by government law	1=yes, 0= otherwise	±
GB_upholds_ethics_business	Whether green banking upholds ethics in business	1=yes, 0= otherwise	±
Determinants			
Organizational_ pressure	Green banking determine by organizational pressure	1=yes, 0= otherwise	±
Environmental_policy	Green banking determine by Environmental Policy	1=yes, 0= otherwise	±
operational_wealth_bank	Green banking determine by Operational Wealth of the bank	1=yes, 0= otherwise	±
green_policy_by_bank	Green banking determine by Green Policy by Bank	1=yes, 0= otherwise	±
related_partie_ instruction	Green banking determine by Related parties instruction	1=yes, 0= otherwise	±
Importance			
Save_ paper	Whether save paper is important for green banking	1=yes, 0= otherwise	±
Use_ solar_ energy	Whether use solar is important for green banking	1=yes, 0= otherwise	±
power_supply_equipment	Whether power save equipment is important for green banking	1=yes, 0= otherwise	±
Attract_customers	Whether attract customers is importance of green banking	1=yes, 0= otherwise	±
Effectiveness level			
Green_ mortgage	Whether green mortgage is effective	1=yes, 0= otherwise	±

continued on following page

Table 4. Continued

Variables	Description	Value	Expected signs
Socio-demographic			
Green_ loan	Whether green loan is effective	1=yes, 0= otherwise	±
Green _credit _card	Whether green credit card is effective	1=yes, 0= otherwise	±
Green _saving_ acc	Whether green saving account is effective	1=yes, 0= otherwise	±
Green_ checking _acc	Whether green checking account is effective	1=yes, 0= otherwise	±
Green _money_ market _acc	Whether green money market account is effective	1=yes, 0= otherwise	±
Mobile_ banking	Whether mobile banking is internet banking	1=yes, 0= otherwise	±
Online_ banking	Whether online banking is internet banking	1=yes, 0= otherwise	±
Remote_ deposit	Whether remote deposit is green products	1=yes, 0= otherwise	±
Ethical_ banking	Whether ethical banking is effective	1=yes, 0= otherwise	±
Other (ATM)	Whether ATM is effective		±
Benefits			
Reduce_stationary_cost	Whether reduce stationary cost is benefit of green banking	1=yes, 0= otherwise	±
Raise _profit	Whether raise profit is benefit of green banking	1=yes, 0= otherwise	±
Accelerate_service_delivery	Whether accelerate service delivery is benefit of green banking	1=yes, 0= otherwise	±
Covers_CSR	Whether cover CSR is benefit of green banking	1=yes, 0= otherwise	±
Problems			
Data_security_and_privacy	Data security and privacy is challenges to implement gb	1=yes, 0= otherwise	±
Lack_of_education	Lack of education can provide problem in adopting gb	1=yes, 0= otherwise	±
Technical_issue	Technical issue is problem of gb	1=yes, 0= otherwise	±
Traditional_approach	Traditional approach impact on gb	1=yes, 0= otherwise	±
Lack_ of_ infrastructure	Lack of infrastructure is problem	1=yes, 0= otherwise	±
technical_procedure_problem	Technical procedure problem	1=yes, 0= otherwise	±
admistration_problem	Administration problem	1=yes, 0= otherwise	±
Environmental Conservation			
Reduc_ resources_wastage	Reduce resources wastage is helpful to environment conservation	1=yes, 0= otherwise	±
Protect_environment	Protect environments helpful to environment conservation	1=yes, 0= otherwise	±

Chapter 3
The Role of Green Sukuk for Sustainable Energy Production

Hasan Dinçer
Istanbul Medipol University, Turkey

Serhat Yüksel
Istanbul Medipol University, Turkey

Hüsne Karakuş
https://orcid.org/0000-0003-1354-9580
Istanbul Medipol University, Turkey

ABSTRACT

Renewable energy sources are clean energy sources that meet the energy needs in a sustainable way. Therefore, it is necessary to invest in renewable energy sources. However, there are some difficulties in renewable energy investments. It has problems such as high initial installation cost, underdeveloped technological infrastructure, and insufficient financial support. Several financial products are being developed in order to overcome the mentioned difficulties. In this context, the purpose of this study is to explain the role of green sukuk in the financing of renewable energy investments. Depending on this purpose, the study has been examined with a literature review. The production of renewable energy sources can be encouraged with green sukuk. However, green sukuk is an advantage for Islamic companies that want to realize environmental projects. On the other hand, it offers the opportunity to the environmentally friendly Islamic investors to evaluate their savings.

INTRODUCTION

Renewable energy sources are among the energy types that are inexhaustible and renew themselves in nature. In this context, it meets the energy demand uninterruptedly and ensures energy supply security. On the other hand, renewable energy sources are clean energy. It reduces the carbon emissions caused by fossil resources (Zhong et al., 2020). Therefore, it has a positive effect on the climate. However, it

DOI: 10.4018/978-1-7998-8900-7.ch003

lowers the import dependency of countries that are dependent on foreign countries in energy. It promotes domestic energy production in countries. This situation positively affects the current account balance of countries (Ubay and Karakuş, 2020). In this context, renewable energy resources are important in terms of ensuring the social and economic welfare of the countries. Because it increases the social welfare as it provides the energy demand and energy supply security in an uninterrupted manner. At the same time, it increases economic prosperity in countries as it nationalizes energy production (Zhu et al., 2020; Zhao et al., 2021). Therefore, renewable energy investments should be encouraged.

Renewable energy is an environmentally friendly energy source. However, some systems need to be developed in order to benefit from renewable energy sources. In this context, there are many difficulties in renewable energy investments (Li et al., 2020; Xie et al., 2021). The initial setup costs of renewable energy investments are high. At the same time, the technological infrastructure used in renewable energies is not very developed. This situation cannot provide the targeted efficiency from energy. However, renewable energy sources are affected by seasonal changes (Li et al., 2021; Zhou et al., 2021). This situation increases the cost of the systems. Therefore, the funds provided are important in promoting renewable energy investments. Government incentives fund renewable energy investments. However, tax cuts and low interest loans by banks encourage renewable energy investments (Dixon et al., 2016). On the other hand, it is important to consider financial instruments in the financing of renewable energy investments.

Sukuk is an Islamic instrument that firms or entrepreneurs export to provide financing. The instrument in question is issued based on the asset or the project. In this context, the asset or project constitutes a guarantee for investors. The issuing party is a special purpose organization. The resource institution transfers a certain part of its income from its assets to a special purpose institution until a certain maturity in order to obtain financing (Halim et al., 2017). The special purpose organization, on the other hand, structures the income as a certificate. It sells the structured certificates to investors and transfers the funds obtained to the source institution. The resource institution meets the funding needs in the short term. In addition, the resource institution pays the lease to the special purpose company until the agreed maturity. Income from the rent payment is distributed to investors in proportion to their shares. At the end of maturity, the special purpose organization sells the revenues it holds back to the originating organization. In this way, the principal capital obtained from the investors at the beginning is paid back to the investors in proportion to their shares (Smaoui and Khawaja, 2017).

There are some types of sukuk that have been developed based on assets or projects. However, sukuk types such as ijara, mudaraba, musharaka, murabaha and selem are often used. There is a new instrument that is similar to the operating structure of the mentioned types of sukuk but attaches importance to the financing of environmentally friendly and renewable energy projects. Green sukuk is an Islamic instrument created to finance environmental and renewable energy investments. Green sukuk, which is the same as the functioning of the sukuk market, focuses only on investors and resource institutions that are Islamic and environmentally sensitive. Thanks to green sukuk, renewable energy production is encouraged. However, it provides funds to Islamic investors who want to make environmental projects. On the other hand, considering environmentally sensitive non-Islamic investors increases the investor portfolio (Abubakar and Handayani, 2020). Considering all these issues, green sukuk is important in financing renewable energy investments. The aim of this study is to explain the role of green sukuk in the financing of renewable energy investments.

This study consists of 5 parts. This section includes basic information on renewable energy, sukuk and green sukuk. In the second part of the study, the literature will be reviewed and the studies on green sukuk will be mentioned. In the third part of the study, theoretical information about sukuk will be given. The definition of the concept of sukuk will be made and the functioning of the sukuk market will be explained. Later, the sukuk types developed according to their usage areas and models will be mentioned. In the fourth part of the study, the green sukuk issue that arises in the financing of renewable energy investments will be evaluated. In the last part of the study, the obtained information will be shared.

LITERATURE REVIEW

Sukuk is a certificate issued by states, companies, or entrepreneurs to provide financing. The mentioned certificate is issued to investors by the state, company, or entrepreneurs through special purpose organizations. In this way, cash needs are met in the short term (Halim et al., 2017). However, sukuk has a structure in accordance with Islamic principles. This issue has been emphasized by many researchers in the literature. For example, Razak et al. (2019) investigated the structures of sukuk contracts. As a result, it has been determined that investors earn dividends or rental income by issuing sukuk. Since there is no interest-based income, it is stated that it is suitable for the Islamic structure. A similar study was conducted by Hossain et al. (2020). They dealt with sukuk and the bond market in their study. It has been determined that there is a lender and borrower relationship in the bond market. In addition, it was stated that there is a relationship between tenant and lessor in the sukuk market. It has been emphasized that the relationship is in accordance with Islamic conditions. Sukuk is an important tool in financing investment projects. Because it meets the funds required for large projects in the short term. But sukuk is a newly developing market. Therefore, there are some deficiencies in the sukuk market. There is no regulation regarding the measurement of the suitability of investment projects to religious structure. However, the lack of sufficient knowledge of the investors about the sukuk market decreases the transaction volume. This situation cannot provide the financial support required for investment projects. Therefore, the regulations to be made affect the existence of the sukuk market (Duqi and Al-Tamimi, 2019; Samoui and Khawaja, 2017).

Combating climate change, which is one of the sustainable development goals, is one of the important issues because climate change affects human life. Carbon emissions are increasing with climate change. Increasing carbon emissions negatively affect human health and social life. Therefore, measures should be taken against the factors that cause carbon emissions. The use of non-renewable energy sources causes carbon emissions. In this context, as one of the issues that cause climate change. Therefore, it is necessary to invest in renewable energy sources. However, since renewable energy projects are costly, financing tools are needed. Green sukuk contributes to the financing of renewable energy projects (Zain et al., 2021). There are many studies on this subject in the literature. For example, Suroso et al. (2020) focused on the green sukuk issue. Indonesia was included in the study. As a result, it was stated that due importance should be given to green projects due to climate change. It was emphasized that one of the green project financing methods is green sukuk. In parallel with these studies, Azhgaliyeva et al. (2020) examined green sukuk and green sukuk policies. The Southeast Asia region was included in the study. As a result, it was stated that importance should be attached to renewable energy projects in reducing climate change and meeting energy demand. It has been stated that the green sukuk is an important financing tool in renewable energy investments.

Green Sukuk is an instrument developed by considering investors who are sensitive to the environment. This issue has been emphasized by many researchers in the literature. Alam et al. (2016) evaluated the potential of green sukuk. France was included in the study in the study. The relevant study was supported by a literature review. It has been stated that the green sukuk is an important tool that finances projects such as solar and wind energy. However, it has been stated that the investors in Islamic markets have been developed by taking into account their environmental awareness (Eti et al., 2020). Morea and Poggi (2017) researched the green sukuk issue in wind energy investments. The study, in which Italy was included in the study, was examined with a case study. As a result, it was stated that importance should be attached to wind energy to reduce climate change and greenhouse gas emissions. It was stated that green sukuk should be used in order to harmonize wind energy investments with Sharia. Green sukuk is important for socially responsible people who attach importance to renewable energy projects in order to reduce their carbon emissions. Green sukuk is a financing model that places emphasis on climate change. In this context, it is important in financing green projects (Boutti and El Mosaid, 2015; Suherman, 2019). In addition to the studies, Abubakar and Handayani (2020) focused on the green sukuk issue, which is one of the financing instruments in his study. Indonesia was included in the study. In the study, which was examined through a literature review, it was stated that green sukuk is important in financing environmental projects.

Green sukuk is a tool that takes both environmentalist and Islamic investors into account. There are many studies on this subject in the literature. Anugrahaeni (2017) focused on the issue of green bond and green sukuk issuance in his study. Indonesia was included in the study. On the other hand, the study was examined with a literature review. As a result, it was stated that green sukuk attracted Islamic and environmentalist investors. However, it was emphasized that the support of the banks is important in ensuring the participation of the environmentalist Islamic investors in the market. In addition to these studies, Ramadhan (2020) investigated the issue of green sukuk issuance. In the study in which Indonesia was included in the scope of investigation, it was stated that green sukuk is important in terms of attracting both Islamic and environmentalist investors. There is a relationship between Islamic finance and social responsibility. Green sukuk is an opportunity for Islamic investors who aim to fulfill their social responsibilities. In this context, green sukuk is important for Islamic investors who want to invest in renewable energy projects (Güçlü, 2019).

The sustainability of green sukuk is important because it provides an alternative for Islamic investors who want to invest in renewable energy projects and Islamic companies who want to realize environmental projects. The issue price of green sukuk is also important in terms of sustainability. The main reason for this is that profit returns affect investors' demand for green sukuk (Munir et al., 2020; Siswantoro, 2018). However, problems arising in green sukuk issuance affect sustainability. There are many studies on this subject in the literature. There are three factors that affect green sukuk issuance. Competitiveness affects green sukuk issuance. Especially the competitive action of companies in sukuk issuance in order to increase their profitability affects sukuk issuance. Another factor affecting the issuance of Sukuk is legitimacy. Firms should take into account the values, norms and beliefs of the countries in which they operate in order to sustain their existence. Finally, the concept of ecological responsibility also affects the issuance of sukuk (Abdullah and Keshminder, 2020). Similar to the purpose of this study, Keshminder et al. (2019) evaluated the green sukuk issue. Malaysia in the period between 2014-2018 was included in the study. As a result, it has been determined that green sukuk is a slow developing market. However, issues such as companies' asset and debt level, credit rating, and tax regulations are found to be important in green sukuk issuance. Sukuk that emerged in Islamic financial markets is a new financial instrument and needs to be developed. For this purpose, the green sukuk that has been developed has the potential to attract Islamic investors to this field. Therefore, it is necessary to determine the deficiencies in green sukuk (Muhmad and Muhmad, 2018).

There are many studies in the literature aimed at solving the problems arising in green sukuk issuance. Moghul and Safar-Aly (2014) investigated green sukuk, which is seen as a way to create environmental awareness in Islamic markets. As a result, it was stated that financiers should take part in order to strengthen the environmental perceptions of Islamic markets. It is stated that financiers' disclosure of green sukuk to Islamic investors will increase environmental awareness. Parallel to this study, Rahim and Mohamad (2018) discussed the green sukuk issue in the financing of renewable energy projects. Malaysia was examined in the study. The relevant study has been tested with case analysis. As a result, it has been determined that green sukuk is important in the financing of renewable energy projects. However, it has been stated that the existence of financial institutions is important in creating awareness and demand for green sukuk. In addition to these studies, Abdullah and Nayan (2020) examined the issue of green sukuk, which is one of the financing methods. Malaysia was included in the study in the relevant study. It was stated that strong investment policies should be established to encourage the issuance of green sukuk and attract investors to this field. Green sukuk is important in financing environmental projects. Green sukuk replaces green bonds and appeals to the Islamic market. However, certain regulations need to be made in order to increase green sukuk export. In this process, responsibilities fall on the government (Anggraini, 2018).

According to the results of the literature review, the issue of green sukuk has been discussed by many researchers. In some of the studies, it was stated that investments should be made in renewable energy projects in order to reduce climate change and greenhouse gas emissions, and green sukuk is important in financing renewable energy projects. In others of the studies, it has been emphasized that green sukuk are based on Islamic and environmentalist investors. It was stated that green sukuk should be used for companies or investors that are sensitive to the environment. It has been stated that green sukuk is a guide in creating environmental awareness for Islamic investors. In some of the studies, the problems arising in green sukuk issuance have been discussed. Issues such as competitiveness, legitimacy and ecological responsibility were found to be important in green sukuk issuance. However, it has been determined that the asset and debt levels and credit ratings of companies needing funds affect green sukuk issuance. On the other hand, Indonesia and Malaysia were generally included in the study. Most of the related studies have been examined with a literature review. Therefore, the purpose of this study is to explain the role of green sukuk in the financing of renewable energy investments.

THE CONCEPT OF SUKUK

General Information About Sukuk

Sukuk, which is described as the plural of the word "sak", means certificate. Sukuk is the issuance of the revenues from a real asset of companies, governments or entrepreneurs within certain rates and periods to investors through special-purpose organizations or asset leasing companies. In this context, the issuer meets its cash needs and investors are entitled to the certificate within the ratios they have (Klein and Weill, 2016; Klein et al., 2018). On the other hand, the Accounting and Auditing Organization for Islamic Financial Institutions defines sukuk. According to this organization, sukuk can be defined as certificates representing common rights on real assets, investing with the exported income in a pre-planned manner, providing equal values that provide partnership rights in a project or private investment activities.

Considering all these definitions, sukuk is an asset-based financial instrument and there is no interest yield (Smaoui and Khawaja, 2017; Smaoui and Nechi, 2017).

There are three parties in the sukuk market, namely fund establishment, special purpose organization and sukuk investors. The state, company or entrepreneurs defined as resource institutions are the parties who want to meet their fund needs. For this purpose, the originating organization has to establish or agree to a special purpose company because the originating organization has no right to issue green sukuk. The special purpose institution, on the other hand, divides the transferred revenues into certain numbers and rates and turns them into certificates. In this process, the special purpose company has the ownership right over the asset transferred by the originating organization. In this way, the entity owned by the special purpose organization constitutes a guarantee. Divided by a certain number and proportions, the certificate is sold to sukuk investors and the funds obtained are transferred to the source institution. The resource institution pays rent to a special purpose institution at certain rates until the end of maturity. The special purpose organization distributes lease payments to sukuk investors. At the end of maturity, the resource institution makes the payment of the initial fund it obtained from the sukuk investors to the special purpose institution. The special purpose entity makes the payment of the initial fund by considering the rates the investors have on the certificate (Halim et al., 2017; Smaoui and Khawaja, 2017).

Types of Sukuk

Sukuk is a financial instrument that states or companies issue to provide financing. With Sukuk issuance, the party in question meets its cash needs, while investors use the excess funds they have and earn returns (Klein et al., 2018). However, sukuk are divided according to their usage areas and models. In this section, sukuk types are explained according to their usage areas and models.

Sukuk are issued based on any underlying asset. However, it is issued in order to obtain financing in the projects planned to be made. In this context, sukuk are exported and used based on the asset, the project. Asset-based sukuk is one of the types of sukuk that companies and entrepreneurs use to provide financing in a short time. According to this type of sukuk, the income generated from existing assets is sold to some investors as certificates at certain rates. In this way, the segment in need of funds meets their funding needs in a shorter time. While meeting the funding needs, companies or entrepreneurs are required to enter into agreements with a special purpose organization or asset leasing companies (Borhan and Ahmad, 2018; Zolfaghari, 2017). For example, assume that you have one plane as a company, and you want to buy a second plane in the short term. In this case, you must transfer some of your revenues from the current aircraft to special purpose organizations, taking into account certain terms. The special purpose organization, on the other hand, structures it as a certificate, taking into account the timing of the investors' demand for the product, and sells it to the investors. It gives the funds it has obtained to the company. In this way, the company can buy airplanes in the short term. In this type of sukuk, it must be a real asset. In this way, funds are obtained in the short term by selling the earnings targeted to be made on real assets.

One of the sukuk types according to their usage areas is project-based sukuk. States or companies need financing in order to make new projects. In this case, companies, or governments issue sukuk based on projects to provide financing. As with asset-based sukuk, special purpose organizations need to be established in order to realize project-based sukuk issuance. The special purpose organization, on the other hand, sells the certificates to the segment with surplus funds in order to finance new projects and transfers the funds required for the project to companies or states. According to this type of sukuk, investors share the profit and loss of the project. Because the project has risks such as not being completed. In this case, the risk is shared according to the agreement (Klein and Weill, 2016).

There are a number of sukuk types created according to their models. However, sukuk types such as ijara, mudaraba, musharaka, murabaha, and selem are generally used. Ijara sukuk is a type of sukuk that is similar to the structure of a lease certificate. In this type of sukuk, there are three parties: resource establishment, special purpose company and sukuk investors. In ijara sukuk, the resource institution goes to special purpose companies due to insufficient funds. In this type of sukuk, a special purpose company is usually a participation bank. The resource institution goes to the participation bank. The participation bank, on the other hand, issues sukuk to sukuk investors in order to purchase the asset. It buys the asset with the sukuk income it has obtained and leases it to the source institution for a certain lease until a certain maturity. The participation bank shares its rental income among sukuk investors. At the end of the agreed maturity, the asset purchased by the participation bank is sold to the source institution. The participation bank distributes its revenues to sukuk investors in proportion to their shares. In this context, the functioning structure of ijara sukuk is similar to leasing transactions (Smaoui and Khawaja, 2017).

Mudarebe sukuk, which means labor-capital partnership, is a common type of sukuk that is mostly used in the financing of large projects. In this type of sukuk, there are a labor-holding company or state, which is defined as a resource institution, as well as a private-purpose organization, and investors who want to offer their funds (Razak et al., 2019). For example, let's assume that the labor company, which is described as a resource institution, wants to implement a project worth $ 1 million. But assume that he does not have enough funds to carry out the project. The retired company goes to the special purpose organization to provide this fund. The special purpose institution distributes the value of this project to various certificates and exports it to investors with surplus funds. In this process, the special purpose company becomes the capitalist of this project. However, the source institution can also be a kind of financier. Because it has the right to establish a special purpose company in order to realize such an export. The special purpose company transfers the funds obtained by the sukuk issuance to the source institution and in this process, the resource organization has signed a mudarabe agreement with the special purpose company called the capitalist. In this context, the mudaraba agreement is a labor-capital partnership agreement. Generally, according to this agreement, liabilities in case of loss from the project belong to the capitalist. Unless the labor is an activity that will be violated by the company, the responsibility belongs to the capital. Assume that a project of 1 million TL is completed and a profitability of 1.5 million dollars has been achieved. In this case, the special purpose company distributes a profit of 500,000 dollars to investors in proportion to their shares. However, it also repays the capital that they initially raised. In the event that the project is damaged, or the project is not completed, the loss is shared between the investors and the company. In this context, mudarebe sukuk is a type of sukuk in which risk sharing is made (Smaoui and Khawaja, 2017).

One of the types of sukuk that is similar to mudaraba sukuk but has some differences is musharaka sukuk. This type of sukuk is also carried out by a partnership agreement between the special purpose organization and the source firm in Musharaka. Sukuk issues through special purpose organizations, assets or projects in order to fund the source firm and to implement projects in the short term. Investors who want to use their funds invest in sukuk. After this process, unlike the Mudaraba Sukuk, the special purpose establishment capital, the source company, signs its own capital and signs the Musharaka agreement. In this way, they realize the partnership agreement. According to this contract, periodic profit or loss is distributed at certain rates from an asset or a project. The special purpose institution distributes its profits to investors at certain rates. At the end of the partnership agreement, the assets and projects that are held are sold. The distribution of the revenues is made to the investors in proportion to their shares. The difference of this type of sukuk from mudaraba sukuk is that the liability belongs to the shareholder in case of loss unless otherwise stated in the mudaraba sukuk; In musharaka sukuk, on the other hand, loss sharing is made as much as the capital rate revealed (İsabet and Kazak, 2019).

Murabaha sukuk, which means profit or loss in Arabic, refers to the special purpose company purchasing the equipment needed by the source firm and selling it to the source firm in certain terms by putting a profit on it. For example, let's say a company wants to buy a machine worth $ 2,000,000. However, the company does not have sufficient funds to purchase this machine. For this, it goes to a special purpose organization and makes an agreement. The special purpose institution, on the other hand, issues 1000 sukuk worth $ 2000 each to investors and buys the machine with the funds from the issue. In the following process, the special purpose organization increases the value of the machine, which is worth $ 2,000,000, to $ 2,100,000, and sells it to the source firm in installments up to a certain maturity. The special purpose institution, on the other hand, distributes the installments it obtains to the sukuk investors in proportion to their shares (Shaikh and Saeed, 2010).

Another type of sukuk developed according to its models is selem sukuk. Selem sukuk is a type of sukuk where funds and assets are exchanged between a special purpose organization and the source company. One of the other definitions made in the literature is a purchase and sale contract in which an asset that the source firm has bought in advance is guaranteed to give the property to the special purpose organization at a later date (Smaoui and Khawaja, 2017). For example, suppose a farmer wants to produce 500 tons of potatoes. Suppose the special purpose company made a sukuk issuance worth $ 500,000 to raise the funds for this production. With the 500.000 TL provided by the special purpose company, the source company produces potatoes and transfers these potatoes to special purpose organizations. The special purpose organization aims to sell potatoes worth $ 500,000 with a profit of $ 200,000. The profit in question is shared at the rate agreed between sukuk investors and special purpose companies. This sukuk issuance is mostly carried out on assets such as metal, petroleum, agricultural products.

Istisna contracts with a special purpose company to finance projects of a sukuk type of source company. The special purpose company, on the other hand, issues sukuk based on the value of the projects. Sukuk investors, on the other hand, invest in sukuk that are issued on a project basis. However, the source firm and the special purpose company are signing an agreement between them. With this agreement, the source company undertakes to transfer the ownership of the project, which it will realize with the funds it has obtained, to special purpose organizations at a later date. The source company realizes the project with the funds obtained from the issued sukuk and the ownership is transferred to special purpose organizations. The special purpose company sells the purchased project by putting a profit on it and shares its revenues with the investors (Chik, 2012).

GREEN SUKUK IN RENEWABLE ENERGY INVESTMENTS

Energy is an integral part of human life. Because needs are met with energy and life is continued. Throughout history, people meet their energy needs by using fossil resources. However, fossil resources are scarce in nature and have negative effects on the environment and climate. The use of fossil resources increases carbon emissions. Increase in carbon emissions causes climate change. Experiencing climate change negatively affects human and natural life (Qiu et al., 2020). However, fossil resources are not evenly distributed all over the world. Therefore, countries that are inefficient in terms of fossil resources have to import energy. This situation makes energy-importing countries economically fragile. Therefore, it is necessary to invest in energy resources that do not have a negative impact on the environment, do not cause climate change, and do not make countries economically fragile (Bekun et al., 2019; Zhu et al., 2020).

One of the sustainable development goals is to ensure the use of accessible and clean energy in the world by 2030. Because the increase in the world population increases the energy demand. Fossil resources are insufficient to meet the energy demand. However, the use of fossil resources causes climate change and negatively affects human life (Du et al., 2020). Therefore, it is necessary to invest in renewable energy sources. Renewable energy is a clean energy that constantly renews itself in nature (Haiyun et al., 2021). Renewable energy sources meet the energy demand without interruption. However, it does not cause carbon emissions. This situation enables the reduction of climate change (Yüksel et al, 2020). On the other hand, it encourages domestic energy production in countries that are dependent on foreign sources in terms of fossil resources. This situation positively affects the foreign trade balance. Because countries that import most of the energy are faced with a foreign trade deficit. Increasing foreign trade deficit in the long-term causes countries to experience current account deficit (Ubay and Karakuş, 2020).

Renewable energy investments need to be made in order to reduce climate change, to ensure uninterrupted energy demand and to reduce the current account deficit by encouraging domestic energy production (Cheng et al., 2020). However, some problems arise in renewable energy investments. The initial setup costs of investments in renewable energy systems are high. However, its insufficiency in terms of technological infrastructure leads to the inability to obtain the necessary efficiency from energy. Technological costs are encountered in systems depending on seasonal changes. However, the insufficient financial support provided for renewable energy investments also creates a problem. Therefore, green sukuk is important in financing renewable energy investments (Wang et al., 2018; Zhu et al., 2020).

Green sukuk is an Islamic instrument that focuses solely on the financing of environmental and renewable energy projects. In green sukuk, there are Islamic resource organizations that want to carry out renewable energy projects, a special purpose institution that mediates the funding and Islamic sukuk investors who will invest in environmental sukuk. A special purpose company is established for the environmental projects to be made. The special purpose company issues green sukuk in proportion to the value of the project. Islamic and environmentalist sukuk investors purchase sukuk and fund the resource institution. The resource institution has to use the funds it has obtained only in environmental and renewable energy projects. However, the resource institution pays rent to the special purpose institution until the agreed maturity. The special purpose organization shares its rental income among sukuk investors. After the resource organization completes the project, it transfers the ownership to the special purpose organization. The special purpose organization increases the value of the project and sells it. It distributes its revenue to sukuk investors in proportion to their shares. In this context, the operating structure is similar to the exceptional sukuk. However, the issuance of green sukuk is done by considering the types of sukuk. In this way, necessary financial support is provided for Islamic investors who are environmentally friendly (Keshminder et al., 2019; Alam et al., 2016).

Green Sukuk provides support to companies and entrepreneurs in financing renewable energy investments. Because it helps companies, entrepreneurs who are Islamic and sensitive to the environment, to obtain funds. On the other hand, it offers the opportunity to evaluate the savings of the environmentally conscious Islamic investors. Funds obtained through green sukuk are used only in environmentally friendly and renewable energy projects. In this way, renewable energy production is encouraged. However, environmentalist and traditional (non-Islamic) investors can invest in green sukuk. In this context, the investor portfolio is wide. Green sukuk is important in achieving sustainable development goals, reducing climate change and ensuring access to clean energy (Boutti and El Mosaid, 2015).

SOLUTIONS AND RECOMMENDATIONS

In this context, green sukuk should be taken into account when financing renewable energy investments. At the same time, Islamic investors who are sensitive to the environment should also consider green sukuk. In this way, Islamic investors who are sensitive to the environment have the opportunity to make use of their savings. It is stated in the literature that green sukuk have developed in countries such as Malaysia and Indonesia. Therefore, other countries with high renewable energy potential need to make regulations for green sukuk. In this way, the process progresses faster in the financing of renewable energy investments.

FUTURE RESEARCH DIRECTIONS

The most important aspect of this study is the focus on green sukuk in the financing of renewable energy investments. Therefore, other financial instruments considered in the financing of renewable energy investments were excluded from the study. This situation shows that green sukuk alone will not be sufficient in financing renewable energy investments. Therefore, it is important to be included in the scope of other financial instruments in new studies. Another limitation of this study is that it does not focus on any particular region or country. Therefore, in new studies, it is important to consider the countries and regions with high renewable energy potential and where the green sukuk market is developed. Another limitation of the study is that the study is not analyzed using any econometric analysis method. Therefore, it will be useful to examine new studies to be conducted using econometric analysis method.

CONCLUSION

Renewable energy is one of the energy resources that do not cause carbon emissions and climate change and ensure energy supply security by meeting energy demand without interruption. However, renewable energy sources reduce dependency on foreign sources. This issue positively affects the current account balance of the countries. Because countries that are not rich in fossil resources import energy. This situation causes countries to experience foreign trade deficit. However, foreign currency imports also trigger the current account deficit. This issue makes countries that are dependent on foreign energy economically fragile. Therefore, investments in renewable energies are important. There are some difficulties in renewable energy investments. The high initial installation cost, being technologically low, and the inability to provide sufficient financial support adversely affect renewable energy investments. Therefore, financial instruments developed in financing renewable energy investments become important. Based on this issue, the purpose of this study is to explain the role of green sukuk in the financing of renewable energy investments. The study has been examined with a literature review. In the literature, it has been emphasized that green sukuk is important in renewable energy investments. However, it has been stated that the green sukuk takes environmental and Islamic investors and firms into consideration.

On the other hand, it is stated that green sukuk encourages renewable energy production and contributes to climate change mitigation.

REFERENCES

Abdullah, M. S., & Keshminder, J. S. (2020). What drives green sukuk? A leader's perspective. *Journal of Sustainable Finance & Investment*, 1-21.

Abdullah, N., & Nayan, M. A. (2020). *Green Sukuk: Financing The Future To Sustainable Environment.* Academic Press.

Abubakar, L., & Handayani, T. (2020, May). Green Sukuk: Sustainable Financing Instruments for Infrastructure Development in Indonesia. In *1st Borobudur International Symposium on Humanities, Economics and Social Sciences (BIS-HESS 2019)* (pp. 983-987). Atlantis Press.

Alam, N., Duygun, M., & Ariss, R. T. (2016). Green Sukuk: An Innovation in Islamic Capital Markets. In *Energy and Finance* (pp. 167–185). Springer. doi:10.1007/978-3-319-32268-1_10

Anggraini, Y. (2018). Peran green sukuk dalam memperkokoh posisi Indonesia di pasar keuangan syariah global. *El-Barka: Journal of Islamic Economics and Business*, *1*(2), 251–268. doi:10.21154/elbarka.v1i2.1453

Anugrahaeni, P. (2017). Analysis of Indonesian Sovereign Green Bond and Green Sukuk Initiatives. *Kajian Ekonomi dan Keuangan, 1*(1), 1-24.

Azhgaliyeva, D., Kapoor, A., & Liu, Y. (2020). Green bonds for financing renewable energy and energy efficiency in South-East Asia: A review of policies. *Journal of Sustainable Finance & Investment, 10*(2), 113–140. doi:10.1080/20430795.2019.1704160

Bekun, F. V., Alola, A. A., & Sarkodie, S. A. (2019). Toward a sustainable environment: Nexus between CO2 emissions, resource rent, renewable and nonrenewable energy in 16-EU countries. *The Science of the Total Environment*, *657*, 1023–1029. doi:10.1016/j.scitotenv.2018.12.104 PMID:30677870

Borhan, N. A., & Ahmad, N. (2018). Identifying the determinants of Malaysian corporate Sukuk rating. *International Journal of Islamic and Middle Eastern Finance and Management.*

Boutti, R., & El Mosaid, F. (2015). Sukuk or The Green Sukuk: A Promising Opportunity For The Financing of Renewable Energies. *Revue de Recherches en Economie et en Management Africain*, *2*(2).

Cheng, F., Lin, M., Yüksel, S., Dincer, H., & Kalkavan, H. (2020). A hybrid hesitant 2-tuple IVSF decision making approach to analyze PERT-based critical paths of new service development process for renewable energy investment projects. *IEEE Access.*

Chik, M. N. (2012). *Sukuk: Shariah guidelines for Islamic bonds.* Centre for Banking Studies, Central Bank of Sri Lanka.

Dixon, R. K., Li, J., & Wang, M. Q. (2016). Progress in hydrogen energy infrastructure development—addressing technical and institutional barriers. In *Compendium of Hydrogen Energy* (pp. 323–343). Woodhead Publishing. doi:10.1016/B978-1-78242-362-1.00013-4

Du, L., Dinçer, H., Ersin, İ., & Yüksel, S. (2020). IT2 fuzzy-based multidimensional evaluation of coal energy for sustainable economic development. *Energies*, *13*(10), 2453. doi:10.3390/en13102453

Duqi, A., & Al-Tamimi, H. (2019). *Factors affecting investors' decision regarding investment in Islamic Sukuk*. Qualitative Research in Financial Markets. doi:10.1108/QRFM-01-2018-0009

Eti, S., Kalkavan, H., Dinçer, H., & Yüksel, S. (2020). Predicting the role of Islamic banking on sustainable economic development: An analysis for Turkey with ARIMA model. In *Handbook of research on creating sustainable value in the global economy* (pp. 146–164). IGI Global. doi:10.4018/978-1-7998-1196-1.ch009

Güçlü, F. (2019). *The Rise of Environmental Consciousness in Islamic Finance: Green Sukuk*. Academic Press.

Haiyun, C., Zhixiong, H., Yüksel, S., & Dinçer, H. (2021). Analysis of the innovation strategies for green supply chain management in the energy industry using the QFD-based hybrid interval valued intuitionistic fuzzy decision approach. *Renewable & Sustainable Energy Reviews*, *143*, 110844. doi:10.1016/j.rser.2021.110844

Halim, Z. A., How, J., & Verhoeven, P. (2017). Agency costs and corporate sukuk issuance. *Pacific-Basin Finance Journal*, *42*, 83–95. doi:10.1016/j.pacfin.2016.05.014

Hossain, M. S., Uddin, M. H., & Kabir, S. H. (2020). Sukuk and bond puzzle: An analysis with characteristics matched portfolios. *Emerging Markets Finance & Trade*, 1–26. doi:10.1080/1540496X.2019.1706478

İsabet, E. Y., & Kazak, H. (2019). Sukuk As One The Islamic Financial Instrumentes: Investigation of The Murabahah Contracts In The Hybrid Sukuk In Terms Of Compliance With The Islamic Provision. *Journal of Life Economics*, *6*(1), 91.

Keshminder, J. S., Singh, G. K. B., Wahid, Z. A., & Abdullah, M. S. (2019). Green sukuk: Malaysia taking the lead. *Malaysian Journal of Consumer and Family Economics*, *22*, 1-22.

Klein, P. O., & Weill, L. (2016). Why do companies issue sukuk? *Review of Financial Economics*, *31*, 26–33. doi:10.1016/j.rfe.2016.05.003

Klein, P. O., Weill, L., & Godlewski, C. J. (2018). How sukuk shapes firm performance. *World Economy*, *41*(3), 699–722. doi:10.1111/twec.12509

Li, X., Zhu, S., Yüksel, S., Dinçer, H., & Ubay, G. G. (2020). Kano-based mapping of innovation strategies for renewable energy alternatives using hybrid interval type-2 fuzzy decision-making approach. *Energy*, *211*, 118679. doi:10.1016/j.energy.2020.118679

Li, Y. X., Wu, Z. X., Dinçer, H., Kalkavan, H., & Yüksel, S. (2021). Analyzing TRIZ-based strategic priorities of customer expectations for renewable energy investments with interval type-2 fuzzy modeling. *Energy Reports*, *7*, 95–108. doi:10.1016/j.egyr.2020.11.167

Moghul, U. F., & Safar-Aly, S. H. (2014). Green Sukuk: The Introduction of Islam's Environmental Ethics to Contemporary Islamic Finance. *Geo. Int'l Envtl. L. Rev., 27*, 1.

Morea, D., & Poggi, L. A. (2017). An innovative model for the sustainability of investments in the wind energy sector: The use of green sukuk in an Italian case study. *International Journal of Energy Economics and Policy, 7*(2).

Muhmad, S. N., & Muhmad, S. N. (2018). Potential Development of Sukuk in Competitive Market. *International Journal of Business and Management, 2*(2), 26–29.

Munir, S., Masruro, U., Fawaiq, A., & Merlinda, S. (2020). Green Sukuk: Indonesian Youth Investment Prospects For Environmental Sustainability. *Economics Business and Organization Research*, 140-147.

Qiu, D., Dinçer, H., Yüksel, S., & Ubay, G. G. (2020). Multi-faceted analysis of systematic risk-based wind energy investment decisions in E7 economies using modified hybrid modeling with IT2 fuzzy sets. *Energies, 13*(6), 1423. doi:10.3390/en13061423

Rahim, S. R. M., & Mohamad, Z. Z. (2018). Green Sukuk for Financing Renewable Energy Projects. *TUJISE, 5*(2), 129–144. doi:10.26414/m031

Ramadhan, I. A. (2020, March). Green Sukuk Issuance as an Investment Instrument for Sustainable Development. In *International Conference on Law, Governance and Islamic Society (ICOLGIS 2019)* (pp. 95-98). Atlantis Press. 10.2991/assehr.k.200306.189

Razak, S. S., Saiti, B., & Dinç, Y. (2019). The contracts, structures and pricing mechanisms of sukuk: A critical assessment. *Borsa Istanbul Review, 19*, S21-S33.

Shaikh, S., & Saeed, S. (2010). *Sukuk bond: The global Islamic financial instrument*. Academic Press.

Siswantoro, D. (2018, November). Performance of Indonesian green sukuk (islamic bond): A sovereign bond comparison analysis, climate change concerns? *IOP Conference Series. Earth and Environmental Science, 200*(1), 012056. doi:10.1088/1755-1315/200/1/012056

Smaoui, H., & Khawaja, M. (2017). The determinants of Sukuk market development. *Emerging Markets Finance & Trade, 53*(7), 1501–1518. doi:10.1080/1540496X.2016.1224175

Smaoui, H., & Nechi, S. (2017). Does sukuk market development spur economic growth? *Research in International Business and Finance, 41*, 136–147. doi:10.1016/j.ribaf.2017.04.018

Suherman, S. (2019). Identifikasi Potensi Pasar Green Sukuk Republik Indonesia. *Human Falah: Jurnal Ekonomi dan Bisnis Islam, 1*(6).

Suroso, D. S., Setiawan, B., Pradono, S. P., & Fitriyanto, M. S. (2020). *Cross-Country Study of Indonesia Green Sukuk and REDD*. Academic Press.

Ubay, G. G., & Karakuş, H. (2020). The Role of Technology-Based Renewable Energy Investments on Macroeconomic Stability. In *Strategic Outlook for Innovative Work Behaviours* (pp. 119–130). Springer. doi:10.1007/978-3-030-50131-0_6

Xie, Y., Zhou, Y., Peng, Y., Dinçer, H., Yüksel, S., & an Xiang, P. (2021). An Extended Pythagorean Fuzzy Approach to Group Decision-Making With Incomplete Preferences for Analyzing Balanced Scorecard-Based Renewable Energy Investments. *IEEE Access: Practical Innovations, Open Solutions*, *9*, 43020–43035. doi:10.1109/ACCESS.2021.3065294

Yüksel, S., Dinçer, H., Karakuş, H., & Ubay, G. G. (2020). The Negative Effects of Carbon Emission on FDI: A Comparative Analysis Between E7 and G7 Countries. In Handbook of Research on Sustainable Supply Chain Management for the Global Economy (pp. 20-35). IGI Global.

Zain, N. R. M., Hasan, A., Yusof, S. A. M., & Ali, E. R. A. E. (2021). Combating Climate Change in Malaysia: Green Sukuk and Its Potential. In Handbook of Research on Islamic Social Finance and Economic Recovery After a Global Health Crisis (pp. 264-279). IGI Global.

Zhao, Y., Xu, Y., Yüksel, S., Dinçer, H., & Ubay, G. G. (2021). Hybrid IT2 fuzzy modelling with alpha cuts for hydrogen energy investments. *International Journal of Hydrogen Energy*, *46*(13), 8835–8851. doi:10.1016/j.ijhydene.2020.12.211

Zhong, J., Hu, X., Yüksel, S., Dinçer, H., & Ubay, G. G. (2020). Analyzing the investments strategies for renewable energies based on multi-criteria decision model. *IEEE Access: Practical Innovations, Open Solutions*, *8*, 118818–118840. doi:10.1109/ACCESS.2020.3005064

Zhou, P., Luo, J., Cheng, F., Yüksel, S., & Dinçer, H. (2021). Analysis of risk priorities for renewable energy investment projects using a hybrid IT2 hesitant fuzzy decision-making approach with alpha cuts. *Energy*, *224*, 120184. doi:10.1016/j.energy.2021.120184

Zhu, L., Hu, L., Yüksel, S., Dinçer, H., Karakuş, H., & Ubay, G. G. (2020). Analysis of strategic directions in sustainable hydrogen investment decisions. *Sustainability*, *12*(11), 4581. doi:10.3390u12114581

Zolfaghari, P. (2017). *An introduction to Islamic securities (sukuk)*. Academic Press.

KEY TERMS AND DEFINITIONS

USD: American dollar.

Chapter 4
Financing Sustainable Development in an Emerging Economy:
The Private Pension System in Turkey

Fatih Kayhan
Kırklareli University, Turkey

ABSTRACT

The purpose of this chapter is to review the Turkish private pension system as a key determinant of sustainable development of the country. The private pension system is of great significance in order to promote sustainable development. This is attributable to the fact that high level of total savings which are secured through individual savings are conducive to investments and in turn finances sustainable development and growth. Turkey, as an emerging economy, is considered to be one of those economies with low level of total savings. In an effort to deal with this issue, governments paid attention to an increasing level of savings of households. Individual retirement system (private pension system) is devised so as to increase savings and to fund investments with national sources. Therefore, it is safely argued that the very rationale behind the introduction of private pension system is associated directly with the aim of increasing total savings in the economy and ensuring sustainable development.

INTRODUCTION

The purpose o this chapter is to review Turkish Private Pension System as a key determinant of sustainable development of the country. Private pension system is of great significance in order to promote sustainable development. This is attributable to the fact that high level of total savings which are secured through individual savings are conducive to investments and in turn finances sustainable development and growth.

Turkey, as an emerging economy, is considered to be one of those economies with low level of total savings. In an effort to deal with this issue, governments paid attention to increasing level of savings of

DOI: 10.4018/978-1-7998-8900-7.ch004

households. Individual retirement system (private pension system) is devised so as to increase savings and to fund investments with national sources. Therefore, it is safely argued that the very rationale behind the introduction of private pension system is associated directly with the aim of increasing total savings in the economy and ensuring sustainable development.

1. INDIVIDUAL RETIREMENT (PRIVATE PENSION) SYSTEM IN GENERAL

Firstly, it is necessary to know the level of savings ownership rate and savings instruments. Apart from the statistical information provided by Turkish Statistics Institution (*TUIK*), another study is needed for searching this issue. 'Turkey's Savings Trend Research', a quarterly-report released by ING Banking Group & Ipsos and Bilkent University, studies regularly level of savings and provides data in detail. According to the 4Q2020 report of this research, the ratio of savings ownership rate declined to 18% with 2.2% decrease when compared to 3Q2020 results when it was 20%. It is argued in the report that considering a long period of time, the latest rate is still remarkably high.

As clearly stated by Turkish Monitoring Authority official website, Individual Pension System was entered into force pursuant to Individual Pension Savings and Investment System Law No. 4632. The IPS is based on the principle of the collection and investment of savings and then making a lump-sum payment or regular payments to the individual. IPS does not offer health or other services provided by the Social Security Institution. It was devised as a supplement and not an alternative to the compulsory social security system.

Individual Retirement System can be classified into two kinds: Voluntary Participation (IPS) and Auto Enrollment System (AES). Introduction of these systems is necessary in order to reduce the burden of social security systems on the public budget in Turkey (public economy dimension), to fund large-scale projects by directing individual savings to investments (for strong and sustainable economy), and finally to protect participants' purchasing power in their retirement (insurance function).

Individual retirement (the system) is covered under insurance industry. Therefore, the system is considered as both savings system and insurance system. The very nature of the fact that insurance is not a means of creating wealth but a system designed to compensate the damage incurred and to ensure that the economic life of the insured value remain the same. In view of the fact that the system is considered a business line under insurance services, one must bear in mind that IPS is not introduced in order to create wealth in a very short time. When this very fact is not taken into account, some clients expect great returns in a very short period of time from the system, which is against the rationale behind the establishment of the system. It is a saving tool that will eliminate the risk of losing purchasing power in the retirement period (long run perspective) and especially at the end of the working age. Investors who ignore this fact tend to compare IPS with conventional, short-term financial (investment) instruments, and evaluate this product with wrong expectations and then leave the system. On the other hand, those who understand the product correctly with the right expectations and stay in the system (for long term) can use the advantage of the IPS against income losses in retirement (with returns above inflation).

This system was not created with the aim of providing a great fortune to participants in the short term. In the scientific method, wrong assumptions cause incorrect models, resulting in wrong policy implication. In Turkey, with the help of participants who understand IPS correctly will provide individual resources to the economy and maintain their purchasing power during their retirement days. It would be the best starting point for reform attempts and resolve some issues arising out of the application of the system.

Recent figures about Private Pensions in the World and Turkey are as follows: OECD Pension Funds in Figures Report (Report date is June 2020) estimates pension fund assets of Q1-2020 as USD 29.8 trillion for OECD countries.

'Pension Markets in Focus No.17, 2020', an OECD report shows that in OECD countries, total amount of assets increased steadily between 2009-2019.

As far as Turkey is concerned, recent pension figures are as follows (as explained by Pension Monitoring Authority of Turkey) as of April, 2021. IPS Summary Data shows that; number of participants (=clients) is 6.9 Million with total fund size of nearly TL 160 Billion. On the other hand, AES Summary Data shows that; number of participants (=clients) is 5.7 Million with total fund size of nearly TL 12.6 Billion. Total pension assets as a percent of GDP of the country has varied between 2.5% and 3% since 2017. Therefore, it can be safely argued that in spite of the fact that pension system has been growing steadily in Turkey, with respect to total Asset under Management (AuM) as a percentage of national income, there is still room for development.

In Turkey, IPS constitute main part of the private pension system. Kaya and Kahya (2017:206) underline that anyone who has the capacity to use their civil rights can be included in the system, regardless of whether they are affiliated with the social security institution or not.

In Turkey, pension system is constructed on 3 pillars: As Peksevim and Akgiray state (2019), the first pillar is social security system (*'SGK'*), which is mandatory social security as in the case of many countries. The second one is the individual pension system (*IPS*) as the third pillar and this is voluntary private pension system. The second pillar is both 'Auto Enrollment System (*AES* for short), and a number of special savings funds (*munzam sandık*).

In the world, regarding the impact of Covid 19, Mitchell (2020) argues that retirement savings plans around the world was confronted by significant difficulties even before Covid-19 that appeared at the beginning of 2020. With the start of the disease, new uncertainties face retirement plan savers, retirees, and plan sponsors. However, as far as Turkey is concerned, Covid 19 did not lead to decrease in total assets under management. In a similar way, there was no decline in the number of contributors as of May, 2021 when numbers of pre-and-after Covid-19 period are compared.

2. LITERATURE REVIEW

This section covers previous studies. In the literature, it is argued that IPS is conducive to capital markets with individual savings being directed to financial instruments, in turn capital markets enhances. There are financial instruments in the portfolio of pension mutual funds, which are the building blocks of the Private Pension System. These instruments are stocks, bonds, time deposit / participation accounts, mutual funds, gold, real estate investment funds and other investment instruments. With the contributions (premium) collected from the participants within the scope of IPS, government (public) domestic debt securities (Treasury Bills, Government Bonds, Eurobonds, etc.) and debt instruments issued by the private sector (private sector bonds, 'ÖST') and stocks are purchased and included in fund portfolios. In this way, demand for financial instruments (mostly capital market instruments) is instrumental in the development of capital markets. Therefore, we can say that private pension funds (financial instruments of the individual pension system) positively affect the development of the capital market in Turkey (and in the world).

Bayar (2017) studies the effects of private pension funds on the debt instruments market and the stock market in Turkey and argues that private pension funds have significant positive effects on the development of both markets in the long term. He also contends that private pensions can ultimately contribute to economic growth by promoting the sustainable development of capital markets.

Ertugrul et al. (2018) conduct a study evaluating the effectiveness of the state contribution incentive practice that entered into force in 2013 in Turkey, and according to the findings, the individual pension system, which is expected to complete the public-funded pension system, has a limited positive effect on saving rates.

Şahin et al. (2019) analyze the effects of private pension funds on the development of Turkish capital markets using monthly data for 2006-2017 and reach the following findings; in the long run, private pension funds support the development of the capital market (Nonetheless, this finding loses its validity in the short term).

As a result of their study using stocks and debt instruments market data in Turkey, Togan-Egrican and Kayhan (2020) reach the findings confirming that voluntary private pension system funds are positively associated -statistically significantly- with various capital market indicators.

Financial Stability Report (2020:17), which is released semi-annually in Turkey by Central Bank of Turkey, indicates that during Covid-19, savings increased considerably in the world economies due to restraints on national and international economic activity, which is caused by uncertainty as well as limits on free movement. Similarly, in Turkey, savings (demand deposit, term deposit, shares, cash and private pension fund contribution amouns) tended to incline. That is to say, the rate of savings over gross domestic product of the country increased remarkably due to cautiousness.

3. LEGAL FRAMEWORK

The Individual Pension Savings and Investment System Law No. 4632 was published in the Official Gazette No. 24366 on April 7, 2001 and entered into force 6 months after its publication. The main objective of this law is to create private (individual) retirement plans that are complementary to the public social security system. While this law is the basic legal document for this business line, other very important (secondary law) legislative arrangements regulating individual pension are as follows:

- Capital Markets Board, Guide on Pension Investment Funds
- Capital Markets Board Resolutions
- Capital Markets Board, Regulation on Principles Regarding the Establishment and Activities of Pension Investment Funds
- Capital Markets Board Communiqué on Principles Regarding Performance Presentation of Individual Portfolios and Collective Investment Institutions, Performance Based Remuneration and Collective Investment Companies Ratings and Ranking Activities (VII-128.5)
- With the content complementary to the Private Pension Savings and Investment System Law, A Regulation on the Establishment and Working Principles of Pension Companies has been issued by the Ministry of Treasury and Finance.

The Law amending the Private Pension Savings and Investment System Law and Certain Laws and Statutory Decrees in order to regulate the incentives for the private pension system was published in the Official Gazette No. 28338 on 29 June 2012. With this law, the State Contribution application was started. All relevant legislation can be found on the Pension Monitoring Center web page. The primary issue that should be known about the legal status of the private pension system and pension (mutual) funds is as follows: These funds have no legal personality, they are not companies, but they are 'group of goods' ('*mal topluluğu*') in legal terms.

In Turkey, institutions regulating and supervising the Private Pension System are as follows: (1) Ministry of Treasury and Finance / Insurance and Private Pension Regulation and Supervision Agency; (2) Capital Markets Board; (3) Pension Monitoring Center.

Basic information about the funds is included in the information forms on the Public Disclosure Platform ('KAP') so that the fund participants can reach them at any time, and if there are changes in the funds, operations departments of the pension company update them.

4. COMPANIES, ORGANIZATIONAL STRUCTURE, MARKETING CHANNELS

Although officially started with the approval of the first retirement plans on October 27, 2003, IPS (voluntary private pension system) in Turkey has grown rapidly from 2004 to 2021, which is the third step (Third Pillar). The Automatic Enrollment System (AES), which entered into force in 2017 as the second step element of retirement ("second pillar"), caused a new development in addition to (voluntary) IPS.

As stated in Pension Monitoring Authority official website (egm.org.tr), Pension Companies that can make individual pension contracts in Turkey - arranged in alphabetical order- are as follows;

1. Aegon Emeklilik ve Hayat A.Ş.,
2. Allianz Hayat ve Emeklilik A.Ş.,
3. Allianz Yaşam ve Emeklilik A.Ş
4. Anadolu Hayat Emeklilik A.Ş.
5. Avivasa Emeklilik ve Hayat A.Ş.
6. Axa Hayat ve Emeklilik A.Ş.
7. Bereket Emeklilik ve Hayat A.Ş. (*)
8. BNP Paribas Cardif Emeklilik A.Ş.
9. Cigna Finans Emeklilik A.Ş.
10. Fiba Emeklilik ve Hayat A.Ş.
11. Garanti Emeklilik ve Hayat A.Ş.
12. Groupama Hayat A.Ş.
13. Türkiye Hayat ve Emeklilik A.Ş. (State-Run Firm)
14. Katılım Emeklilik ve Hayat A.Ş.(*)
15. Metlife Emeklilik ve Hayat A.Ş.
16. NN Emeklilik ve Hayat A.Ş.

(*) These two companies only offer interest-free private pension products within the scope of Participation Insurance. Other companies can offer both interest-based and interest-free pension funds / plans.

As of 24 August 2020, state-run companies, Halk Hayat ve Emeklilik A.Ş., Vakıf Emeklilik ve Hayat A.Ş., Ziraat Hayat ve Emeklilik A.Ş., merged under the name of Türkiye Hayat ve Emeklilik A.Ş.

These 16 companies operating in the Life and Pension branch also offer personal accident insurance and life insurance products to customers in addition to private pension products. Companies offer individual products (private pension, life insurance, personal accident insurance) and corporate products, but the products offered as corporate are essentially individual products offered to the employees of corporate customers (automatic participation, group pension, group life and group personal accident insurance).

Although the organizational structure of pension companies varies from institution to institution, they generally consist of the following departments: 1. Sales & Marketing units consisting of agencies, bancassurance, corporate solutions and direct sales and customer services, 2. Actuarial, Financial Valuation departments, 3. Operation units consisting of private pension operations and life insurance operations units, 4. Financial services group (Accounting & Finance, Fund Management & Fund Operations, Financial Planning), 5. Human Resources, 6. Administrative Affairs & Purchasing, 7. (Premium) Collection ('Tahsilat'), 8. Information Technologies, 9. Legal Affairs, Internal Audit, Internal Control, Risk Management Group.

Organizational structure in companies is as follows: Head Office, Regional Directorate and agency units (captive agencies or independent-contracted agencies). However, companies that do not open a regional office can carry out their transactions and sales only with the head office and contracted agencies. With the rise of private banking and private segment customer sales channels in banking and other financial institutions in recent years, in the field of private pension, corporate companies are establishing sales and service channels specific to their private customers with contributions over a certain amount.

Bancassurance turned out to be the leading sales and marketing channel; via this channel, insurance companies can use the sales channel power of banks with widespread branch sales networks. Both sectors benefit from this cooperation because insurance companies can offer their products to bank customers through branches with less cost (manpower cost, sales, location cost), while banks receive commission from the sales of insurance products in return. Benefiting from the professional sales and sales support organizations of banks is an important opportunity for insurance companies to increase their sales.

The cooperation of financial institutions, banks and insurance companies, especially in the form of *financial supermarkets,* is also important in terms of the total capital gain of the relevant group, providing cost advantage.

Agencies, on the other hand, as a sales channel have qualified sales force, experienced sales teams and therefore are more successful in terms of loyalty and stability in insurance product sales, especially in financial crisis periods. Agencies manage their operations with personnel with more professional experience in the insurance profession and have the capacity to successfully carry out a longer-term sales relationship with customers.

In recent years, technological advances, intensive use of the internet in financial transactions, has increased the importance of digital marketing departments and functions as well as classical / conventional sales departments. The detailed and visual presentation and marketing of the products on the pension company website, through applications such as the E-Branch, provides great cost and sales advantages.

5. PRIVATE PENSION FUNDS AND STATE CONTRIBUTION

The Regulation on Principles Regarding the Establishment and Activities of Pension Mutual Funds published in the Official Gazette on March 13, 2013 ("Capital Market Board Regulation") and the Capital Market Board (CMB) Pension Mutual Funds Guide contain detailed regulations on pension mutual funds.

In practice, fund types are described as follows (according to CMB, Pension Mutual Funds Guideline regulations):

- Stock Funds – invest at least 80 percent of its portfolio into domestic or foreign stocks.
- Borrowing Instruments Funds – need to invest at least 80 percent of its portfolio into domestic or foreign public or private sector debt securities or reverse repos.
- Participation Funds (Non-Interest Funds) – need to invest all of its portfolio into lease certificates, precious metals, participation accounts, and other financial instruments that must be interest-free.
- Mixed Funds – need to invest at least 20 percent of the fund's portfolio into stocks, bonds, precious metals and lease certificates
- (*) Money Market Funds – need to invest in liquid money market & capital market instruments with maturity less than 184 days.
- Precious Metals Funds – need to invest at least 80 percent of the fund portfolio in gold & other precious metals and also money & capital market tools based on these.
- Index Funds – need to invest at least 80 percent of the fund in assets within the framework of indices approved by CMB.
- Fund Basket Funds – need to invest at least 80 percent of its portfolio in mutual funds, securities mutual funds (*'borsa yatırım fonu'*), foreign mutual funds.
- (*)State Contribution Funds – need to invest heavily in government debt instruments. This is a compulsory fund established for state contributions.
- Variable (Flexible) Funds – Those funds that are not classified under the above-mentioned categories (in terms of portfolio constraints) are considered variable funds.
- (*) Standard Funds – These are compulsory funds for companies to set up. The fund is managed in accordance with rules specified by the Ministry of Treasury and Finance. The fund heavily invests in government debt instruments.

In practice, among the above IPS funds, those written with an (*) sign (in terms of fund size or because they are required to be established by legislation) are the leading funds.

According to the legislation, IPS-Standard Fund, AES-Initial Fund and AES Standard Fund are mandatory funds to be established. In addition, it is one of the necessary funds to establish a Money Market (formerly known as 'Liquid Fund') Fund for operational transactions and also for directing them to investment in practice.

State Contribution Fund

With the introduction of the state contribution, pension companies started to set up Contribution Funds in 2013; Contribution Fund is the fund established within the scope of the Regulation on State Contribution in the Private Pension System.

State Contribution is the amount paid by the state to the retirement account of the participant at the rate of 25% of the contributions paid by the participant. The state contribution cannot exceed 25% of the gross minimum wage total for the relevant year. All participants who are citizens of the Republic of Turkey and who are within the scope of Article 28 of the Turkish Citizenship Law No. 5901 dated 29/5/2009, who pay a contribution to an individual or group individual pension contract, can benefit from the state contribution within the determined limit. Both IPS and AES participants are authorized to benefit from state contribution.

6. NON-INTEREST PENSION SYSTEM AND FUNDS (PARTICIPATION PENSIONS)

In Private Pension System, where individuals make savings through private pension companies, the important issue in terms of Islamic Law is the content of the funds.

In this system, there are pension mutual funds based on participation accounts, gold, stocks in accordance with the principles of participation banking and sukuks (lease certificates) that have been approved for compliance for individuals who are sensitive to interest. In addition, it is essential that investing in foreign currency, real estate and commodities by complying with certain rules / conditions will be within the scope of interest-free investment. In interest-free private pension, fund portfolio is structured with financial tools that have received the approval of the advisory board.

Financial instruments included in interest-free pension mutual funds are as follows:

*Participation accounts to be opened in participation banks,
*Stocks suitable for participation index,
*Gold and precious metals,
*Interest-free mutual funds,
*Sukuk, lease certificate and similar interest-free debt instruments issued in Turkey or abroad.

As explained in official website of Katilim Emeklilik ve Hayat A.Ş., a participation based pension company, private pension activities are conducted in line with 'Takaful' model: Takaful, which is called Participation Insurance, is mutual insurance. Takaful (in Turkish 'Tekâfül') means giving mutual guarantee, being a surety for each other. The main goal is to bring together the insured for the purpose of mutual aid and solidarity and to use the collected contributions by the people exposed to risk.

Kayhan (2020) analyze non-interest pension system and funds with SWOT analysis and argues that strength of non-interest pension system is the strong demand for this business from individuals especially in AES. Comparatively weaker side of the system is the lack of financial product for investment when compared to conventional rivals. There exists great potential for development of this market while the threat is seen as the risk of decline in the rate of growth of entry into this system due to economic conditions.

7. PORTFOLIO MANAGEMENT AND FUND PERFORMANCES

Pension companies defined as founding companies are entitled to receive a service income called fund management fee in return for fund operation service that they provide for participants. Fund management fee is shared between the founding company (= pension company) and the portfolio management company. The short name of this fee is 'FİGK' ('fund management fee'). Most of the FİGK (generally 80% and above) are booked as income (entry) to the founding company.

'FIGK', *'FTGK'* difference: Fund management fee deduction is the largest component of the *'fund total expense deduction'* (FTGK). The fund total expense deduction (FTGK) can be calculated when

other legal deductions related to the fund are added to the fund management fee (FIGK). The fund management fee is calculated as a certain proportion of the total fund value for each day and is collected from the fund monthly. Fund management fee rates are included in the fund prospectuses. The CMB applies a ceiling limit for the fund management fees so that the founders will not reflect a very high management fee to the fund.

Two criteria are taken into account when measuring the performance / return success of funds: *Benchmark* and *Peer Group*. When comparing fund performances, if peer group criterion is employed, it means that the success of the fund is measured by comparing the return of the fund with that of other fund(s) of the same type. On the other hand, the second criterion is the 'Benchmark'. Benchmark is the target performance measure.

According to the legislation, all funds except flexible funds have to set a benchmark. Also, flexible funds can set benchmarks. The content of the fund should be in line with the benchmark of the assets it invests. The benchmark is specified in the fund prospectus and basic information form (*'Tanıtım formu'*) approved by the CMB. The fund benchmark can be changed by a prospectus change. While measuring the success (average) of the fund, the benchmark and the peer group for any period, the following process is conducted: The ratio of the difference between the beginning and end of the period values (to the beginning values) is calculated and the success of the portfolio manager is evaluated by making a comparison among fund return, benchmark return and peer group return.

Fund performances are important in that when private pension funds have higher returns (over inflation), more participants shall enter the system. Kayhan et al. (2020) argue that fund portfolio returns and benchmark returns are in line significantly. The results are partly attributable to the regulations about pension fund management and portfolio structure. The paper also indicates that in the long run, the volatility of returns of the funds declines and returns tend to be in conformity with the main purpose of the private pension system.

Principles of portfolio management of private pension (mutual) funds are regulated under Article 20 of the CMB's Regulation on the Establishment and Activities of Pension Mutual Funds. Portfolio Managers are obliged to act within certain rules - except flexible (variable) funds - while managing their fund portfolios. The rules are determined under the regulations of the Capital Markets Board and the relevant regulations of the Ministry of Treasury and Finance for State Contribution funds. These rules can be changed when CMB deems necessary considering developments in financial markets, economy, pension system. When a certain part of a rule changes, companies are given reasonable time to adapt their fund portfolios to the new legislative restrictions.

TEFAS (Turkey Electronic Fund Trading Platform) website enables participant to review the fund returns comparatively.

8. FUND OPERATIONS

Fund operations can be conducted either by a special department in pension companies or by an outsourcing company, but it is the pension company that is liable for the smooth functioning of the fund operations.

Firstly, it should be pointed out that accounting system of the pension company and that of the funds are not the same, which gives rise to forming a separate accounting system.

Though the funds do not have a legal entity, they are separate groups of goods with a tax number. Likewise, the financial asset pool of the pension company is different from that of a pension fund.

Fund Operations (Service Departments) / Departments of pension companies carry out the daily operation of funds every day.

The process, is expressed in the technical term called *'fund integration process'*; in this process, the contribution amount paid by the participants is directed to investment on a daily basis: The contribution amount turns into a share according to the customer's fund purchase preference (order), while shares turns into the cash amount (in case of client order for the sale of shares) during the very same integration process.

The fund operations department, after this transaction, informs the Portfolio Management Company (PMC) about the total entry and exit amounts for each fund that is managed by PMC. With new money (total entry amount) financial products are purchased by the Portfolio Manager in the financial markets, and the notification of the transactions are sent to Takasbank (Central Settlement and Custody Bank). The new shares that each participant deserves and the assets belonging to him are notified electronically to the Takasbank on a file by fund operations department via files. At the end of day, mutual reconciliation between Fund Operation records (pension companies' own system records) and Takasbank is secured in order to refrain any failure of reconciliation.

The tax structure to which the company is subject and the tax structure to which the funds are subject are quite different. Pension funds are, in principle, not taxable. It is an incentive and advantage for the development of this system.

In addition, fund service / operation units in companies are responsible for issuing (without payment) the tax declarations (Corporate Tax, BA-BS forms) stipulated in the relevant laws for all active pension mutual funds.

9. RISK MANAGEMENT

The principles of risk management in Private Pension are explained in detail in the CMB Regulation and the Pension Funds (EYF) Guide published by the CMB. Risk Management has gained importance in the private pension system, as it is in all financial institutions, especially in banks, and its framework has been specified with comprehensive rules.

Value at Risk (VaR=RMD) explains the highest loss that the fund's total value can be exposed to, under normal market conditions and within a certain period of time. For example, if the VaR is calculated as 2% for a fund with a total fund value of TL 1 Million, it means; the total value of the fund may lose more than 20,000 TL with a probability of 1% within 1 day. Market risks included in the pension mutual fund portfolio, including risks from derivative instruments, should be taken into account in the VaR method. One of the "Relative VaR" or "Absolute VaR" method should be chosen, taking into account the fund investment strategy and risk profile.

On the other hand, the principles regarding the calculation of the Risk Value ('RD') of the fund, as explained in the Pension Funds Guide (in Turkish, '*EYF Rehber*'), are briefly as follows: While calculating the RD, the rules determined in the fund risk management system are taken into account and are checked periodically (usually weekly). Risk value ('RD') information of the fund is shown on the continuous information forms of the funds in the PDP (Public Disclosure Platform).

General method for calculating RD is as follows: RD is calculated by taking into account the volatility of the fund (volatility is calculated using weekly returns based on fund prices) Volatility is computed by taking into account the fund returns in the last 5 years in which the fund was operating, volatility is calculated using weekly returns, converted into an annual figure with a pre-determined formula. RD takes a value between 1 (lowest volatility = risk) and 7 (highest volatility = risk).

The main motivation (motive) in the RD regulations included in the Pension Fund Guide is to define the risks that pension funds will be exposed to in accordance with the legislation in order to protect the savings of the investors, and to periodically measure and professionally manage them within the method and system. Ultimate goal for risk management is that value of participants (and fund assets) is effectively protected against losses (risks).

10. AUTO ENROLLMENT SYSTEM (AES, 'OKS')

The Automatic Enrollment System, 'OKS' (AES), which was accepted in 2017 and put into practice in 2018, brought a significant change as the second step in private pension. It has enhanced the existing private pension system. The most important difference with voluntary IPS is that entrance to AES is mandatory for employees. AES is explained in EGM Web Portal with the following summary statements:

- Employees are included (directly, automatically) by their employers in the Automatic Enrollment System, where they can benefit from additional state contribution opportunities,
- Turkish citizens can participate,
- It is essential to stay in the system for a minimum of 10 years and it is mandatory for retirement. Those who stay in the system for 10 years and reach the age of 56 are entitled to retirement.
- There is 25% state contribution. Furthermore, TL 1000 additional state subsidy is transferred to the accounts of the employees entering the system with AES.

Kayhan and Islamoglu (2020) conduct a SWOT analysis about Auto Enrollment System in Turkey and argue that that strength of non-interest pension system is the increasing demand for this business from individuals in Turkey, while comparatively weaker side of the system is considered to be the lack of financial product for investment when compared to conventional rivals. They also maintain that there is great potential for development of this business line while the threat is seen as the risk of decline in the rate of growth of entry into this system due to economic conditions.

CONCLUSIONS AND POLICY RECOMMENDATIONS

For an economy, level of total savings is a key factor for ensuring sustainable development since high level of saving gives rise to increase in investment which is the leading force for economic development. The private pension system, which has the great potential to eliminate the savings-deficit complements the public social security system. Main points about the system are as follows:

* Private Pension System is not an alternative to the Public Social Security System. On the contrary, it has been established with a goal to complement the public social security system,
* The funds are managed by a portfolio management company, not by a pension company (founding company),

* Pension mutual fund is not a stock or bond, it can be seen as a financial instrument with a portfolio based on these or other financial instruments. Buying funds does not mean directly trading stocks and bonds, its operation is also substantially different.
* In fact, participants buy a pension plan, not a pension fund (The plans contain pension mutual funds, therefore, they indirectly buy pension funds.
* The funds are legally owned by the fund participants, not by the pension company. Hence, real owner of these funds are participants (via the shares '*katılma payları*').

The private pension system has the following goals in terms of the features described; (1) To create an accumulation that can finance large-scale investments close to 10 percent of the national income. (2) To protect purchasing power (of the clients) with an additional income in addition to social security during retirement, (3) To create resources for the non-financial sector (real sector), increasing production and employment, (4) To contribute to the development and deepening of capital markets and money markets by creating demand for public and private sector issues (money and capital market issues).

Finally, policy recommendations are as follows:

- The sale of the private pension system (voluntary IPS and AES) as an investment and savings tool should be made by only professional licensed persons, emphasizing that the system is a long-term insurance product rather than a short-term investment product, and withdrawals from the system during financial turbulence periods should be minimized without changing the voluntary basis. Asymmetric information will disappear when the product is correctly described and sold.
- Reports sent periodically by pension companies to customers (as required by official reporting obligations) should be very short, clear and concise, and should be prepared to contain at most 10 basic information. Technical terms and long & detailed explanations should be eliminated.
- The number of funds in the system should be reduced, the system should have a very simple structure. The complex system structure creates a negative image for the customers and prevents the development of the this business line. When examined from a behavioral point of view, it is a fact that customers always prefer simpler and more understandable products.
- Having the right to change the fund distribution frequently (in a year) may cause the investor (participant) to lose savings in short term via fund distribution changes. Changing the fund distribution frequently is not compatible with the rationale behind the system, there should be a maximum of 3 changes of fund distribution, because in the financial markets, big trend changes may not occur frequently in a year.
- Although Robotics-Consulting ('Robo-Advisor') systems have potential to create positive returns by way of the application of developing technology and artificial intelligence in finance, these new structures should be used only by professionals (portfolio managers).

REFERENCES

Bayar, Y. (2017). Individual Pension Funds and Capital Market Development in Turkey. *Review of Economic and Business Studies*, *9*(2), 95–109. doi:10.1515/rebs-2016-0036

Central Bank of Turkey. (2020). *Financial Stability Report*. Author.

Ertugrul, H. M., Gebesoglu, P. F., & Atasoy, B. S. (2018). Mind The Gap: Turkish Case Study of Policy Change in Private Pension Schemes. *Borsa Istanbul Review.*, *18*(2), 140–149. doi:10.1016/j.bir.2017.11.003

Kaya, F., & Kahya, M. (2017). *Sigorta ve Sigortacılık, Beta Yayıncılık, 1.* Baskı.

Kayhan, F. (2020). Türkiye'de Faizsiz Bireysel Emeklilik Sisteminin SWOT Analizi. *Finans Ekonomi ve Sosyal Araştırmalar Dergisi*, *5*(4), 745–754.

Kayhan, F., & Islamoglu, M. (2020). A Comprehensive Assessment of Auto-Enrollment System as a Second Pillar of Pension System: The Case of Turkey. *International Journal of Publication and Social Studies*, *5*(2), 167–177. doi:10.18488/journal.135.2020.52.167.177

Kayhan, F., Islamoglu, M., & Apan, M. (2020). Portfolio and Benchmark Returns of Private Pension Funds: A Comparative Analysis for Turkey. *Business & Management Studies: An International Journal*, *8*(5), 3891–3910.

Mitchell, O. S. (2020). *Building Better Retirement Systems in the Wake of the Global Pandemic*. National Bureau of Economic Research Working Paper 27261.

OECD. (2020b). *Pension Markets in Focus 2020*. Author.

Peksevim, S., & Akgiray, V. (2019). Reforming the Pension System in Turkey: Comparison of Mandatory and Auto-Enrollment Pension Systems in Selected OECD Countries. OECD.

OECD. (2020a). *Pension Funds in Figures 2020*. OECD.

Sahin, S., Ozdemir, Z. A., & Onal, Y. B. (2019). Türkiye'de Bireysel Emeklilik Sisteminin Sermaye Piyasasının Gelişimi Üzerine Etkisi. *Dokuz Eylul Universitesi Iktisadi ve Idari Bilimler Dergisi*, *33*(2), 517–541.

Togan-Egrican, A., & Kayhan, F. (2020). *Voluntary Private Pension Funds and Capital Market Development*. Working paper.

Turkey's Savings Trends Research. (n.d.). *4Q2020 Report*. ING & Ipsos & Bilkent University.

Chapter 5
Agriculture 4.0 for a Sustainable Food System:
A Holistic Model for the Transformation of Farms Towards a Sustainable Precision Agriculture

Manuel Antonio Fernández-Villacañas Marín
https://orcid.org/0000-0001-7397-6395
M&M Planning and Project Management, Spain & Technical University of Madrid, Spain

Ignacio Fernández-Villacañas Marcos
M&M Planning and Project Management, Spain & University of Alcala, Spain

ABSTRACT

This chapter aims to analyse the concept and implementation strategies of Agriculture 4.0 within the framework of the study of disruptive technologies and eco-innovation, which allows facing the needs derived from a sustainable food system. To do this, it strategically reflects on the design requirements of a holistic model for the transformation of agricultural holdings, aimed at the implementation of sustainable agrotechnology. The Third Green Revolution, its antecedents, orientations, and purposes, as well as the concept and functional aspects of sustainable food systems are analysed. Finally, a model of transformation of agricultural holdings towards the implementation of Sustainable Agriculture 4.0 is proposed, as well as a generic methodology applicable to specific projects located in specific areas, through formula for change and cost-benefit analysis.

INTRODUCTION

Achieving the Sustainable Development Goal by 2030 for a future humanity in which there is no hunger is a very ambitious goal that implies having more productive, efficient, transparent, sustainable, resilient, and inclusive food systems. In addition to covering the approximately 690 million people who still suffer

DOI: 10.4018/978-1-7998-8900-7.ch005

from hunger today, almost 9% of the world population (United Nations FAO, 2020), which will imply a very considerable increase in the demand for food that will be produced in the next 30 years. years, as the world population grows from the current 7.8 billion to the estimated 9.6 billion by 2050 (United Nations, 2021). It is estimated that their diet will require an increase in the world's food supply by at least 60%.

To face this situation (Hickey et al., 2019) it is necessary for the international community to do a decisive integral transformation of the current global agri-food system, considering current and future restrictions and challenges, such as the increasingly restricted availability of natural resources essentials such as fresh water and productive arable land. Likewise, the devastating effects of climate change (World Health Organization, 2020) and soil erosion will significantly modify the food production capacities of different regions and countries. In addition, the unbridled trend towards urbanization, with more and more people living far from agricultural areas, will make it necessary to find more efficient ways to produce and distribute food to large consumption centres.

In contemporary history, agriculture, as a fundamental support of the world food system, has undergone a series of revolutions that have increased its efficiency, performance, and profitability. The beginning of the First Green Revolution (Patel, 2012) occurred after the Second World War, beginning in the 1950s, and represented a major agricultural transformation that focused on advances derived from new farming methods, new practices and the implementation of new technologies, the partial mechanization of tasks, new irrigation systems by irrigation, as well as the use of pesticides, herbicides and chemical fertilizers, including the development of evolved varieties of cereals more resistant to pests and also to the harshest climates.

The figure of Norman Borlaug is recognized as the main promoter of the First Green Revolution (Swaminathan, 2009), a scientist who for several decades developed crosses of selected varieties of rice, corn, and wheat in developing countries, to achieve the most productive, seeking with their efforts to improve agricultural productivity in order to eradicate hunger and malnutrition in the most underdeveloped countries. This First Green Revolution in recent history achieved great success for decades and great recognition for achieving a sharp increase in production with high-yielding grains. However, the nutritional quality aspects were not considered with sufficient relevance, generating an expansion of the cultivation of cereal varieties that contained low quality proteins and excessive carbohydrate content, deficiencies in essential amino acids, and an unbalanced content of essential fatty acids. minerals, vitamins, and other essential elements for food quality (Sands et al., 2009). This nutritional impoverishment has generated the increasing incidence of certain chronic diseases, which affect not only the human diet but also the quality of meat products whose food base had been these new varieties of cereals.

At the beginning of the 1990s, new cultivation methods emerged based on the creation in the laboratory by means of certain techniques of genetically modified organisms, also called transgenics, which consist of the transfer between living organisms of genes responsible for certain desired characteristics, affecting their natural structure and, consequently, to its genome. Their application has increased exponentially since their appearance, but since their origin these transgenic crops have generated a huge discussion regarding their real and potential effects, both positive and negative, capable of significantly increasing the productivity of the land, but whose consequences on the Future health of individuals can be unpredictable (Montoya, 2007) (Muñoz, 1999). And likewise, the first outlines of so-called precision agriculture are beginning to be developed, as an innovation capable of reconfiguring farms of all sizes, allowing farmers to use only essential resources, and allocate them where they are needed. The concept of precision agriculture supposes a comprehensive management approach based on the collection and intensive processing of data, which allows orienting certain focused actions to improve the effectiveness, efficiency, and productivity of agricultural activities (Nowak, 2021). It is the Second Green Revolution, which, despite the methodological and technological differences with the First, has developed with the shared objective of ending hunger in the world. Although in both cases it has led to a significant increase in world agricultural production, the problem of hunger involves many more aspects related to the equitable global distribution of food and the proper functioning of international markets, in which the low purchasing power of a very significant part of the world's population makes it unfeasible to be able to access the global food market in a fair, free and direct way.

As we can see, there is a similarity in the approaches of those who promoted both Green Revolutions, the second being, in terms of its strategic approach, a continuation of the first. This similarity also extends to proven and potential consequences, both social and environmental at a global level, considering their own differences and nuances. However, the global trends that the agricultural sector will have to face in the coming years continue to pose a daunting challenge. It is necessary to continue increasing productivity and competitiveness, in an environment where sustainability and the fight against climate change and the scarcity of resources, especially water, are paramount. In this context, within the framework of scrupulous and very prudent bioethical respect, the agricultural sector must act strategically to adapt to changes in an increasingly dynamic and uncertain environment, and to the unrestricted use of available digital technologies, rapidly and continuous evolution. To achieve this, the main stakeholders must understand in detail how this new situation will affect them, as well as what challenges they will have to face by making use of new strategic and technological initiatives and applying new organizational models.

This chapter, within the framework of the study of disruptive technologies and eco-innovation, sets out to reflect on the concept and development strategies of Agriculture 4.0 that allows addressing the needs derived from a sustainable food system. To do this, a strategic reflection is proposed on the requirements of the design of a holistic model for the transformation of agricultural holdings, oriented towards the implementation of a new concept of sustainable precision agriculture. After the introductory section, the chapter is divided into three more sections. First, the Third Green Revolution, its antecedents and purposes are analysed. Subsequently, the concept of New Agriculture 4.0 and the development of methodologies and techniques that make its implementation possible are studied. Finally, as a result, the requirements of a model for the transformation of agricultural holdings towards the implementation of the new Sustainable Agriculture 4.0 are analysed, as well as a generic methodology is proposed to be able to carry out said transformation, in its application to specific projects.

RESEARCH METHODOLOGY

The research has been developed using a qualitative research methodology, both documentary and empirical, based on the opinions of a sample of professionals and public and private experts from the agricultural sector (productive and distributive) of Latin American countries with different levels of economic development: Mexico, Colombia, Ecuador, Peru, Guatemala, Nicaragua y El Salvador.

The research has been focused on the transformation of agricultural production and distribution systems, applying a descriptive and explanatory method on the research problem, generating new orientations, approaches, and trends to adequately orient the proposed solution and the new concepts that are being pledged. The literature reviewed has focused on the essential prior approaches related to the investigated problem. The review has been carried out with a sufficient level of detail, seeking its original aspects to guarantee objectivity and avoid interpretive biases. The agrarian reality on the investigated problem has been tried to perceive integrating it in a coherent, analytical, and enveloping way, guided by the qualitative research carried out.

METHODS: THE THIRD GREEN REVOLUTION TOWARDS A SUSTAINABLE FOOD SYSTEM

It is currently considered that we are immersed in the Third Green Revolution that is a consequence of the exponential development of digital technologies, that is, the digital revolution, and the demands of achieving the Sustainable Development Goals that have been established in the 2030 Agenda of the UN. It is about sustainable precision agriculture, also called Agriculture 4.0, or better still, Sustainable Agriculture 4.0, the result of which is a holistic approach that links farms, until now quite isolated, with the whole of the global agricultural food system. Advances in automation and robotics, in artificial intelligence and in sensor networks linked by the Internet of Things are exploited, applying this spatial approach at all levels of the agricultural process to better understand the biological systems at stake. It is a new agricultural revolution, the objective of which is to promote prosperity and at the same time maintain environmental sustainability, making it compatible with a significant growth in the global food supply.

On the other hand, systems consist of elements and interconnections and have a purpose or function (Meadows, 2009). The concept of the food system encompasses a wide variety of actors that carry out multiple interrelated activities of production, aggregation, transformation, distribution, consumption, and elimination of agri-food products, be it livestock, forestry or fishing. A food system develops in broader social, economic and environmental settings, in which they are integrated, and is made up of several subsystems (such as the agricultural subsystem, the productive inputs subsystem, the water resources subsystem, the waste, ...) and interacts with other essential systems (such as the food sanitation system, the commercial system, the logistics system, the energy system, ...) maintaining a dynamic balance between them, so that a transformation of one of them will necessarily affect the rest, and will require the achievement of a new equilibrium.

In 2015, the UN 2030 Agenda established the so-called Sustainable Development Goals, calling both the countries of the international community and all significant social actors to a more intense effort in the social, economic, and environmental spheres, above all, the fight against hunger and poverty, reducing human inequalities, as well as combating the deterioration of the environment. Although the achievements have been modest, since the promulgation of the 2030 Agenda, sustainability has become a critical aspect for civil society (Rosati & Diniz, 2019), but clearly insufficient to achieve global sustainability in such a degraded world (Caiado et al., 2018).

The qualification of a food system as sustainable implies guaranteeing its capacity for nutrition and food security for all its users, in addition to providing the social, economic and environmental bases to generate nutritional and food security so that future generations are not affected, being beneficial for society (social sustainability), profitable (economic sustainability), and generating a positive or neutral impact on the natural environment (environmental sustainability) (Nguyen, 2018). The Sustainable Development Goals call for major transformations in the agricultural sector and the food systems that depend on it, for which the global food system must be reformed to be more productive, more inclusive of poor and marginalized populations, resilient to the environment and able to offer health services. and nutritious diets for the entire population. Food security and nutrition issues are highly complex and dynamic and evolve over time. Assuming their sustainability implies that they are analysed with a multidisciplinary and inter-institutional character through an increasingly globalized orientation. A more holistic, integrated, and synergistic approach is needed, which considers the food system, considering all the elements, their relationships, and related effects. The analysis of the problems of nutrition and food security cannot be limited to one of the dimensions of the problem. The framework should be expanded, considering all the relevant causal variables and all the social, economic, and environmental impacts of the alternative solutions that can be generated to achieve transformative systemic changes. It is a new approach that must structure, host, and coordinate all the initiatives developed by all the parties involved, at the local, national, regional, and global levels, both public and private, in dimensions such as agriculture and livestock, education, health, marketing, logistics, environment, climate change, political situation, cooperative model, R&D, gender integration, available infrastructures, transportation, ... (Nguyen, 2018).

FAO's overarching goals in the development of sustainable food systems are to reduce poverty and ensure nutrition and food security for the entire population, in a way that does not compromise the ability of economic, social, and natural environments to generate nutrition and adequate food security for future generations (HLPE, 2014). These goals are at the core of the sustainable food system and are embedded in the broader performance related to the three dimensions of sustainability; economic, social, and environmental. The food system proposed by the FAO (United Nations FAO, 2021) is represented by a wheel that is represented in Figure 1 and describes the different elements of the food system, as well as the interactions between the different levels.

Figure 1. FAO Food system wheel – elements and interactions (United Nations FAO, 2021)

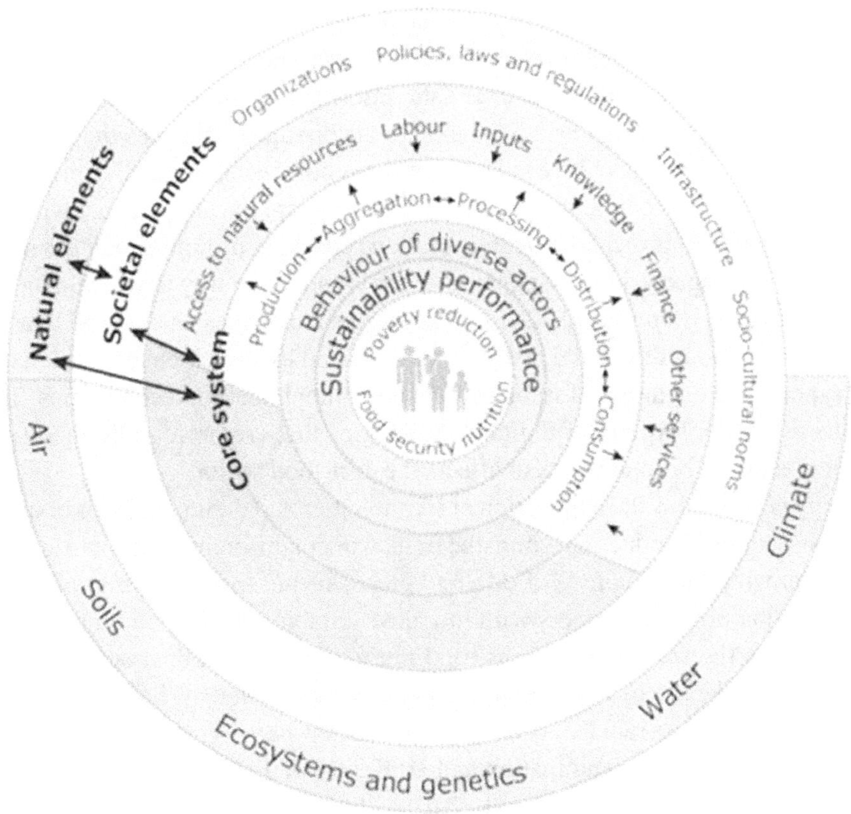

Moving from the inside out, this action is determined from the core of the system by the behaviour of all the actors who are interested in the food system, highlighting that of the members of society themselves. These behaviours take place within the structure of the food system, which includes the value chain layer through which food products flow (production, aggregation, processing, distribution, and consumption, including waste disposal) and the layer of services. that supports this flow (access to natural resources, work, supply of inputs, knowledge, financial aspects, and other services). All these activities are embedded in a social context, which includes related policies, laws and regulations, socio-cultural norms, available infrastructure, and organizations, as well as a natural environment, which includes water, soils, air, climate, as well as ecosystems and genetics.

Investigating, analysing, and understanding food systems through systems thinking is essential to identify the causes that generate their dysfunctions, as well as the aspects that make food systems vulnerable to climate change, thus allowing the formulation of more effective and efficient strategies to increase the Resilience and Productivity of global food systems (World Economic Forum, 2017).

The development of the sustainable food value chain advocated by FAO has a dynamic and holistic approach, based on methodologies that allow to measure, understand, and improve the sustainability of the value chains that make up global food systems, although there are complementary approaches that should be considered, such as the territorial approach (OECD & FAO & UNCDF, 2016) and the market systems approach (Humphrey, 2014).

RESULTS: THE NEW SUSTAINABLE AGRICULTURE 4.0

We define digital transformation as the set of systemic processes necessary for the application of digital technologies, fully understood, which implies the strategic, organizational, and cultural reinvention of the entity in which it is applied, and which uses data, information and intelligence to improve its performance and the necessary capacity to adapt quickly to the disruptive changes that are generated in the environment (Fernández-Villacañas, 2018). The result is a new technological scenario, in the face of which the environment changes continuously and increasingly rapidly, at a rate that will surely continue to accelerate exponentially. This technological revolution is causing profound changes in all sectors and in society, in economic prospects, in work planning models and in the way in which man-machine interactions should be oriented (Barreto et al., 2017). The engine that drives change and that is the cause of this continuous acceleration is the digital revolution, motivated by the global expansion of the Internet, and the strong exponential evolution of information and communication technologies. Thus, today we are fully in a process of deployment of digital technologies that are profoundly modifying society in general (Evtodieva et al., 2017) and that also affects the agri-food sector.

Agrotechnology is defined as the application of technologies and their principles of use to manage all aspects related to agricultural production within the framework of different spatial and temporal concepts, with the aim of improving the responses of different crops, saving time, and improving environmental quality. This application provides farmers with the most suitable methods and machinery to optimize production, focusing on the specific processes used in the sector. It is about supporting the farmer in their operations, making the use of resources more effective and efficient (Sáiz & Rovira, 2020).

Agricultural technologies are rapidly evolving towards a new paradigm, that of Agriculture 4.0 (Rose et al., 2021), in which digitization, automation and artificial intelligence play a very important role in improving agricultural production. (Bacco et al., 2019). This evolution represents the application of the engine of the fourth industrial revolution to the impulse of the third green revolution, and implies great opportunities, but also multiple challenges.

The development of Agriculture 4.0 within the framework of sustainability entails the promotion of an agriculture of environmental conservation in which the diversification and adaptation of the types of crops and the improvement of the soil, increasing its structure and organic matter, are fundamental promote rich microbial diversity, retain water and nutrients, better manage pests and diseases, and generally make agricultural soils more productive and resilient to climate change. Agricultural robotics is well suited to support these sustainable practices by enabling weed removal and precise handling of nutrients and pesticides, disease and weed removal, all using mechanical systems emphasized by local application of products. chemical products. Agricultural robots can also replace hard work, especially when labour availability is reduced, thus increasing economic sustainability (Santos-Valle & Kienzle, 2020).).

The conceptual result is the new Sustainable Agriculture 4.0. But beyond that, the agricultural technologies or agrotechnology that are being used around the world are much more varied and extensive. The following is a typological inventory of the main agrotechnology that are already used in the investigated countries of Latin America, which in any case are localized and specific.

Typology and Most Used Agrotechnological Tools

There are different types of technologies that can be applied in agriculture (Ugochukwu & Phillips, 2018) (Kamilaris et al., 2017), among which the following stand out:

- Agrochemical technologies, for the creation, development and use of nutrients, fertilizers, pesticides, and better phytosanitary procedures (Chhipa, 2017).
- Mechanical technologies, for the design and development of new, more effective, and efficient machinery, such as tractors, furrowers, sprinklers, collectors and all those advanced machineries that make the farmer's work more effective and efficient (Fountas et al., 2015).
- Biotechnologies, for the creation of seeds, fertilizers, nutrients, and pesticides, based on cellular modifications and biogenetics (Barrows et al., 2014).
- Information and communication technologies, creating new platforms and developing digital applications to improve the administration and monitoring of the processes of agricultural holdings (Sousa et al., 2016).
- Robotic technologies, for the development, together with mechanical technologies, of highly specialized, intelligent, and hyper-connected autonomous machinery, capable of connecting and integrating with satellites and drones to generate adequate information to carry out the sowing, fertilization and harvesting processes using Robots autonomous or semi-autonomous (King, 2017).

New solutions to farmers' daily problems must be able to increase productivity without causing emerging damage to the environment (Stuart et al., 2014), making use of instruments that allow, for example, to detect the most appropriate time to harvest. Without damaging the crop, have online information on the status of the crop using drones or sensors, on the humidity and temperature of the soil, etc. (Khanal et al., 2017). All this information online can allow farmers to improve their decisions about the use of seeds, fertilizers, water, or other resources, in terms of quantities and when they should be applied in each area. The result is a more effective prevention of pests, diseases, and the elimination of weeds, as well as an improvement in the efficiency of risks, which allow achieving a higher productivity of crops (Yao et al., 2017).

As the most used agrotechnological tools, we can highlight the following:

- Sensors: The Internet of Things (IoT) application allows the connection of sensors or other non-digital objects through the Internet, which allows the information generated to be managed and controlled remotely. The analysis of the information obtained allows verifying the humidity levels or pest prevention (Muangprathub et al., 2019) (Ouafiq et al., 2020).
- Applications: Farmers traditionally control their activities through manual field notebooks. The improvement lies in the development of applications for mobile phones or tablets, which allow optimizing activities such as the location of protected areas, meteorological information, storage and stock levels, documentation and personalized reports, crop planning, etc. (Mandi & Patnaik, 2019).
- Drones: These small remote unmanned or manned aerial vehicles allow farmers to monitor large areas quickly and without moving from one location, generating online aerial maps, and capturing information on the status of crops and soil. Drones are a very useful tool to assist farmers in many tasks on farms (Tsouros et al., 2019).
- Robots: During the last years, many types of robots, autonomous or semi-autonomous (working alone or under the control of farmers) capable of performing many of the agricultural tasks, even those that were more intensive in the hand of construction site. Among these activities we can highlight the automated harvesting, pruning, and planting systems, optimized diffusion of agrochemicals, monitoring in nurseries, vineyards and orchards, classification and packaging of crops, elimination of weeds, soil exploration, collection of fresh fruit, etc. (Manjunatha et al., 2016).
- Intelligent tractors and harvesters: An intelligent autonomous tractor has GPS to be able to operate in the field with total precision and autonomously, programming its routes (Poonia et al., 2020), which allows the farmer to dedicate his time to other tasks. In the case of combines, these programmed systems allow to establish planned unloading locations and locations, adjust vehicle speed, control location, and command the harvesting device to optimally synchronize the speed and direction of action of the harvesting machine combine harvester.
- 3D Maps: These devices allow you to identify and identify cracks in the ground, analyse weak or dry areas, recognize insect problems, and weed pest problems, capture, and evaluate images to differentiate healthy vegetation from those in poor condition. status, assess the health status of crops and soils, control the irrigation status, etc. (Potena et al., 2020).

DISCUSSION: THE TRANSFORMATION OF AGRICULTURAL FARMS TOWARDS SUSTAINABLE AGRICULTURE 4.0

According to FAO doctrine, the sustainability of food systems must be analysed holistically. The development of a sustainable food system (SFS), an agricultural production-distribution system, must simultaneously generate positive value in the economic, social, and environmental dimensions (Nguyen, 2018).

In relation to the economic impact, a food system is considered sustainable if the activities carried out by all the actors involved along the chain generate net economic benefits for all of them (wages for workers, taxes for governments, benefits for companies and system improvements). provision of food for consumers), in accordance with a fair and ethical distribution of the global economic surplus generated. Regarding social impact, a food system is considered sustainable when there are social benefits that exceed social costs (relative to nutrition, health, cultural traditions, working conditions, animal welfare, institutions, ...) and considering vulnerable groups. categorized by gender, age, race, ... as well as there are no opportunity costs related to other systems, adjacent or not to the food system, that rival it in the allocation of resources and generation of net social benefits. Regarding environmental impact, sustainability is determined by ensuring that the impacts of food system activities on the surrounding natural environment are neutral or positive, considering biodiversity, water, soil, animal and plant health, footprint carbon, water footprint, food loss and waste. and toxicity. Any proposed action to address a problem or weakness, or to take advantage of an opportunity, should be comprehensively evaluated in relation to all other dimensions to ensure there are no undesirable impacts, as well as to promote growth. inclusive socio-economic growth, green economic-ecological growth, and socio-ecological progress (Figure 2).

Figure 2. Sustainable Food Systems Development and Impacts (Nguyen, 2018)

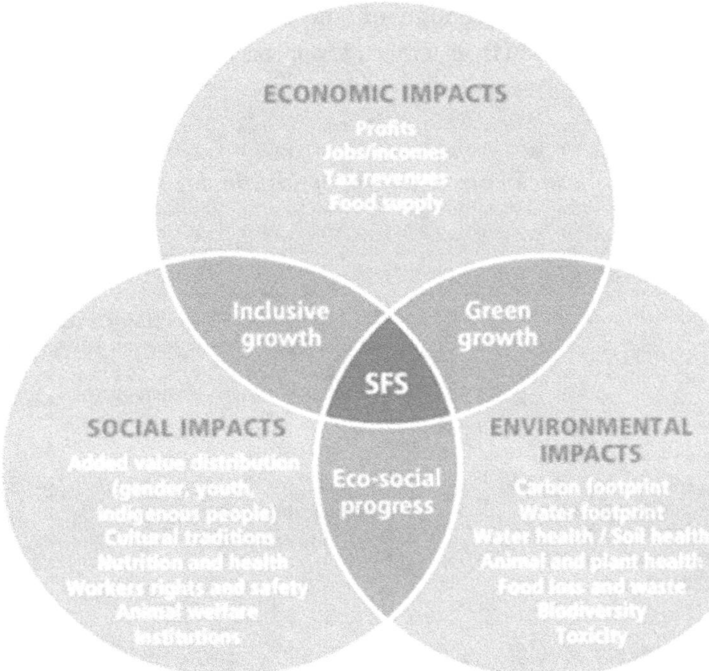

Main Strategic Aspects for the Implementation of Sustainable Agriculture 4.0: Working Hypothesis

While some authors consider that agriculture is experiencing important progress due to the application of the digital revolution in agriculture, others have highlighted the existence in many countries of a very low rate of adoption of agrotechnology (Nowak, 2021). Thus, in developing economies and their rural areas, due, among other factors, to the deficiency of their technological infrastructures, scarce digital culture and limited access to Internet services, there is a delay in adoption. Sustainable Agriculture 4.0.

Furthermore, there is often a certain isolation of the technological issue from its social matrix, not taking sufficiently into account that technology, which currently constitutes the main basis of economic development, must always be oriented towards social development. Therefore, agrotechnology must be understood as an effective means to improve the environment, and to integrate society and improve human relations (De Clercq et al., 2018).

As a result of a qualitative research carried out based on the opinions of a sample of professionals and public and private experts from the agricultural sector (productive and distributive) of Latin American countries (with different levels of economic development: Mexico, Colombia, Ecuador, Peru, Guatemala, El Salvador, and Nicaragua), six hypotheses have been put forward on the main areas of action necessary for the transformation towards Sustainable Agriculture 4.0. As main explanatory questions of the surrounding situation of the agricultural sector, potentially generalizable, stand out the digital transformation of the agricultural sector, innovation and training of young farmers, the development of new cooperative models, water and energy efficiency, the new role of global consumers and, finally, the promotion of research in agrotechnology and the transfer and application of the solutions generated.

These aspects, which are interconnected, are also considered as great challenges and opportunities for agriculture as a strategic sector of the economy of emerging countries, so it is necessary to evolve the strategic attitude of the actors involved from a reactive approach to a proactive one based on strategic and prospective anticipation (Figure 3).

Figure 3. Main strategic aspects for the implementation of Sustainable Agriculture 4.0
(Own elaboration)

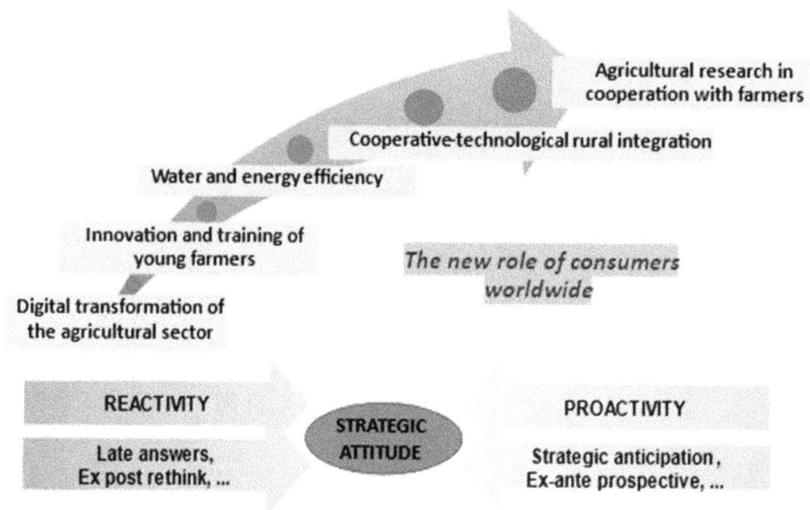

First, the digital transformation of the agricultural sector is a necessary condition to be able to move towards the application of specific Sustainable Agriculture 4.0 projects and enable a transcendent improvement in the productive planning of crops, the agri-food integration. supply chains, improving efficiency in the use of productive resources, such as inputs, energy, land, and water, and in the implementation of new, more sustainable strategies. It is a necessary, but not sufficient, condition to be able to prosper in the modernization of agriculture.

Second, the digital transformation of the agricultural sector will be of little use and that digital technologies were available to farmers, if there were no agricultural producers willing to implement transformative agriculture projects. Young and well-trained farmers are considered essentially those who can generate a more effective, efficient, and competitive agriculture. Thus, it is young people, both women and men, who are in the best position to promote the digitization of agricultural holdings and the commercialization of their products, being immersed from their birth in digital culture. Young people are the ones who will best be able to energize and transform agrarian systems through the application of new digital technologies and the implementation of new production and distribution strategies.

Third, the application of the transformation of Sustainable Agriculture 4.0 fully affects the implementation of a new way of using water for irrigation. Sustainable precision agriculture must be responsible for an irrigation management fully adapted to the specific needs of the different parcels of agricultural holdings. Achieving fully efficient water management is not only a problem of ecological sustainability, but also a problem of economic profitability given the increasing cost of the energy required to extract water from underground and marine sources, especially through the available reverse osmosis systems, and treat it for use. for irrigation. In this sense, the use of rural areas for the installation of renewable energy sources is another pending issue to be addressed in parallel in the agricultural world.

Fourth, the view is often raised that farmer tend to be highly individualistic and thus explain the difficulties in energizing strong associative structures in the agricultural sector. However, it is farmers who develop associative movements of all kinds in all countries, despite the territorial dispersion that their activity implies. It is necessary to make associative activity more effective and efficient in the face of the new challenges that Sustainable Agriculture 4.0 represents, for which the traditional agricultural association does not have enough strengths. The improvement of rural cooperative integration through more efficient socio-labour and technological models is another of the key aspects that is promoted.

Fifthly, it is a reality that agricultural research, as well as the efficient transfer and application of the solutions and technologies generated, have lagged in developing countries with respect to the changes that have occurred, in general, in the field. of technological research applied to other sectors. And the problem is not so much one of inadequacies in terms of the availability of research centres and scientific infrastructures, but of inefficient use, not focused on the real problems of the agricultural sector, and with a level of coordination that does not make optimal use. of its potential. And, above all, little international scientific cooperation in these matters. It seems that it is also necessary to promote inclusive and more pedagogical advice formulas towards farmers, to connect them effectively with the vanguard of the ecological, agrotechnological and energy transition.

Finally, it is worth highlighting the new role that consumers have in current food systems, as an increasingly critical factor to which crop production and planning strategies must be oriented. The consumer is no longer the taxable person who adapted to the offer that existed in the retail stores and made their purchase decisions based on the offer of the day and, above all, on prices. The consumer has become an active subject, aware of his food sovereignty, and that he exercises it by making decisions focused on quality, labelling traceability and prices, in relation to his buying habits. For this reason, consumers must guide producers today and be an essential element in farmers' decisions about what and how to produce. Although most consumers are not directly related to the agricultural productive sector, but through the intermediation of the commercial distribution sector, it is also true that more and more virtual market technological initiatives are emerging through direct relationship channels between the initial productive supply and the final demand.

The holistic vision, integration, coherence, collaboration, innovation and flexibility are key concepts to support the development of Sustainable Agriculture 4.0, whose adaptation will imply the implementation of new more holistic and ecological productive-distributive strategies, the design of a new organization and the management of change towards a new technological culture of agricultural cooperatives, as well as the establishment of new methodologies to be able to successfully address the complex processes of transformation of agricultural holdings and management of rural change (Figure 4).

Figure 4. Dimensions of the transformation towards Sustainable Agriculture 4.0 (Own elaboration)

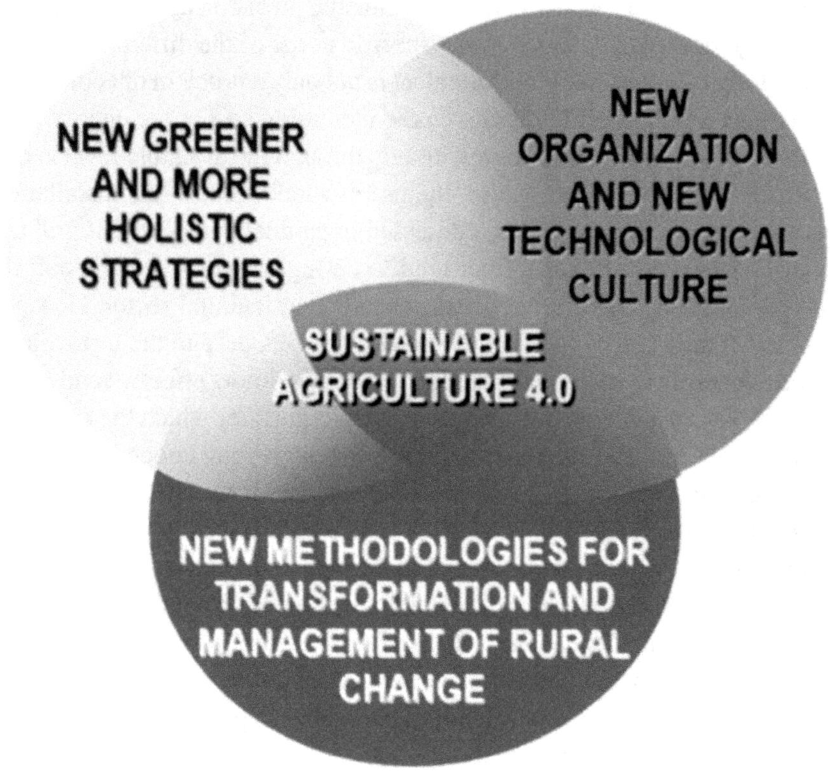

Indeed, the future of the agricultural sector in each region requires the establishment of new holistic public-private strategies, integrating international cooperation strategies, innovation strategies and the implementation of digital technologies for robotization and automation of agricultural holdings, optimization strategies productive of agriculture. farms, and strategies to optimize logistics and marketing processes for national and international markets. The implementation of these strategies requires the prior international transfer of already available and verified technological solutions, the reform of agricultural cooperatives, the improvement of the qualification of farmers and staff in the food processing industry, and the improvement of systems selection of markets and international market. The strategic impulse must accommodate the action and effort of the main agents and institutions towards the same objective, betting as has been reiterated on the social, economic, and environmental sustainability of the sector, as well as on the coordination of all initiatives. In terms of economic sustainability, initiatives should be geared towards increasing agricultural productivity, improving efficiency and being fully market and consumer oriented. In terms of social sustainability, projects should focus on dignifying and improving the lifestyles of farm workers, significantly increasing farm incomes, and developing more skilled, youthful, and diverse farm jobs. Finally, in terms of environmental sustainability, an improvement in the efficiency of the use of natural resources must be achieved, contributing to adaptation to climate change and its mitigation, as well as, ultimately, helping to preserve the natural environment.

The development and implementation of these new strategies requires changes in the organization as well as the implementation of a new agricultural technological culture, fundamentally through the design, planning, development and implementation of new agricultural logistics platforms 4.0 (Fernández-Villacañas, 2019) (Fernández-Villacañas, 2020), for international marketing, as well as above all, through the development of training programs in sustainable agrotechnology and the prioritized creation of "school farms" for the new training of young farmers. The strategic objective is the imbrication in these agricultural learning laboratories of the knowledge and skills of Sustainable Agriculture 4.0 that allow to enhance and catalyse the traditional knowledge of the agricultural sector. It is also necessary to develop and apply a new socio-technological model of digital cooperativism, through the promotion of collaborative systems in agricultural production and commercialization, which induce a positive social change towards the rural transformation of the entire region where it is applied.

Finally, it is necessary to identify and apply new methodologies to address the transformation processes of rural change management and agricultural holdings towards Agriculture 4.0. As has already been analysed, the possibilities and opportunities derived from the application and development of Agriculture 4.0 within the framework of sustainability to achieve an efficient and sustainable global food system are evident. However, the approach must evolve through a perfectly planned transformation process towards an emerging strategic, holistic, systemic, and collaborative concept that allows the implementation of specific transformation projects in specific areas and specific rural locations. Two traditional transformation methodologies that have been applied successfully for decades are considered very suitable. In the first place, as a methodology of organizational and social transformation, it is considered that Formula for Change initially proposed by Gleicher establishes a framework with which to solve complex organizational problems, and that was later modified and adapted by Dannemiller to integrate the organizational aspects with technological innovation processes, facilitating and understanding the connection of organizations with the results sought. Second, Cost-Benefit Analysis methodology is considered, which allows comparing the traditional approaches to agricultural holdings and their commercial distribution and logistics systems, with the new emerging approaches to sustainable, technological, systemic, and holistic agriculture. This immense transformation requires the dynamic analysis in detail of the economic, social, and environmental, both monetary and non-monetary, both present and future costs and benefits, before making decisions about its implementation and about the use of large resources limited in terms of opportunity, susceptible alternative uses. Let us see both.

Methodology of Application of Formula for Change

Formula for Change (Beckhard, 1975) (Dannemiller & Jacobs, 1992) (Cady et al., 2014) is a model used to systematically implement programs, both in the public and private spheres, to overcome organizational inertia and improve efficiency and productivity. It is a specific approach to the management of linked human resources, which recognizes that change in an organizational environment depends on the corporate culture, the personality of the individuals and the motivations of the people involved.

Based on the original model for the change proposed by Gleicher, Dannemiller proposed a version that has been the most used in practice:

D x V x F x CL > R

In the equation, D represents dissatisfaction with current conditions, V the vision of how things could be, F indicates that the action plan is acceptable and establishes the first steps to take, and CL represents the creative leadership necessary to achieve that the transformation process works. All these multiplying factors must exceed R, which represents the resistance, the feasibility of the change and its durability.

The objective is to evaluate the real possibility that transformation projects towards Sustainable Agriculture 4.0 can prosper with certain guarantees, so that the evaluation made of the first three variables shows the ability to overcome resistance to change.

- First, the level of dissatisfaction with the current situation must be high enough so that those involved recognize the need for change, have time to reflect and accept it.
- Second, in relation to the vision of how things could be, all the human resources of the different organizations, even if the change comes from senior management, must share the objective to be achieved, ensuring their individual contribution to the transformation program. and its role in the future once the process of change has occurred.
- Third, in relation to the action plan, for the transformation to get off to a good start, the first small steps necessary to achieve the stated objective must be known and accepted by the personnel involved, without generating anxiety or discouragement.

In relation to these three variables, since they are multiplying in the formula, any non-compliance with any of them will mean the failure of the transformation program. Since human nature is very risk averse, most people prefer to avoid losses rather than seek gains or improvements. Resistance to change is therefore a natural reaction that justifies ensuring that projects are carefully prepared and adequately supported.

Cost-Benefit Analysis Methodology

Cost-Benefit Analysis methodology (Nas, 2016) (Boardman et al., 2018) is a set of techniques that allow estimating the net comprehensive benefit of any large initiative or project in terms of the impact generated on the different stakeholders (Figure 5). It is a tool through which it is not intended to replace political decisions, but rather seeks to support, improve, and provide them with objective content, eliminating the arbitrariness and high risk faced by large decisions based on subjectivity or intuition.

Every major initiative or project represents a change in the socioeconomic and environmental system of the different organizations and entities it affects, with respect to what would happen if the initiative or project were not carried out. To evaluate ex ante, the transformation initiative must be carried out or not, as well as under what conditions, all the changes that will occur represented by a multitude of variables in time relative to two situations must be studied and monitored, before and after the transformation and realization of the corresponding investments. Therefore, estimates on the implementation of a specific Sustainable Agriculture 4.0 project in a certain region and location should be analysed, in relation to the social, economic, and environmental impacts, as well as the opportunity costs of the alternative

Figure 5. Cost-Benefit Analysis methodology (Own elaboration)

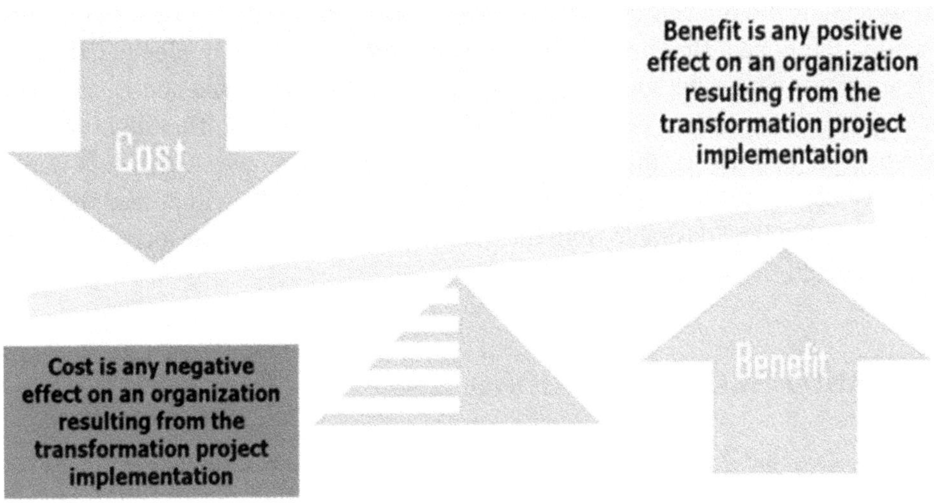

investments analysed. If the situation in aggregate terms of organizational and social benefits and costs is better with the new project than without it, then it must be accepted. But if this aggregate situation worsens, it should be rejected until studying and knowing the changes that would be necessary to introduce in the project to improve the net benefits of the initiative. On the other hand, if in some variables the new situation is beneficial, but for other variables it is detrimental, the decision would then depend on the weight given to relative gains and losses, which implies the incorporation of multi-criteria decision techniques (Atkinson & Stiglitz, 2015).

The generic operation of Cost-Benefit Analysis methodology is based on the estimation and sum of the equivalent monetary value of the proposed project, of the benefits and added costs of the different alternatives (monetary and non-monetary), economic, social and environmental. To reach a conclusion about its suitability, all aspects of the initiative, both positive and negative, must be expressed in terms of a common unity. Therefore, the economic, social, and environmental costs and benefits must transform their units of measurement and, together with the traditional analysis, also be estimated as opportunity costs, thus determining what remains to be gained from having rejected an option, and studying whether the network The balance of economic, social, and environmental costs and benefits exceeds those of the initial situation. This would bring improvements that are not only quantified in economic terms but also in comprehensive sustainability, and if all the information affects possible alternative courses of action other than the specific proposed Sustainable Agriculture 4.0 project.

As we can see, the use and valuation of opportunity costs is fundamental, since market prices, although in general a valid reference, are nevertheless frequently distorted by the market itself in competitive situations other than perfect competition, such as oligopolies or monopolies. And even in many cases there are not even those markets from which to extract the prices that are necessary for the evaluation. In such cases, it is necessary to use the so-called shadow prices that better reflect the non-economic marginal cost. The estimation of shadow prices is a sensitive issue since it affects the results of the project evaluation in a very relevant way. But the main distortions derive from rigidities in the price and quantity adjustment mechanisms, from the use of market power in the exchange of certain agricultural products, as well as from the not always consistent application of taxation (De Rus, 2008).

Finally, determining the net economic, social, and environmental benefit of the implementation of Sustainable Agriculture 4.0 projects and investments requires comparing the flow of benefits and costs that are generated throughout their life cycle with respect to an already implemented solution that is takes as a reference to establish the study. To update this flow of financial and non-financial benefits and costs, a social discount rate should be used that reflects the minimum profitability threshold required for the innovative project, that is, the opportunity cost of the funds invested. If this implementation presents a positive net benefit, its comprehensive sustainability benefits will exceed the corresponding costs and, therefore, it will be desirable from the point of view of public and social interest in the field of Sustainable Agriculture 4.0.

SOLUTIONS AND RECOMMENDATIONS

The concept of Sustainable Agriculture 4.0 has been exposed, whose implementation allows meeting the needs derived from a sustainable food system. It has strategically reflected on the most important requirements for the transformation of agricultural holdings, and for this a holistic model has been proposed, specifying the new strategies to be proposed, the changes to be introduced in the organization and a new agricultural technological culture, and Finally, two methodologies have been proposed to address the transformation processes of rural change management and agricultural holdings towards Sustainable Agriculture 4.0: Formula for Change and Cost-Benefit Analysis. This solution proposal provides a generic procedure for the planning, development and implementation of specific initiatives or projects for the transformation of Sustainable Agriculture 4.0 in specific regions and rural areas.

FUTURE RESEARCH DIRECTIONS

The study carried out on Sustainable Agriculture 4.0, aimed at providing a sustainable food system, shows interesting lines derived from research in relation to the implementation as concrete projects in specific regional areas and in specific rural localities. Within these multidisciplinary lines of research, it would be necessary to develop adapted solutions for the application of the proposed model on the different agrotechnology, the existing environmental, social, and economic restrictions, and the socio-economic, economic-ecological and socio-ecological guidelines, that should serve as a guide. in the planning, definition, and implementation of transformation processes. It is considered necessary to deepen the applied studies of Sustainable Agriculture 4.0, which could be related to the reforms, changes, and policies that nations will foreseeably propose to alleviate the economic and social crisis that is being induced by the Covid-19 pandemic.

On the other hand, it is necessary to advance in the definition of a model of a vocational training centre for the training of young farmers, which is developed under the "farm-school" approach adapted to the specific economic, social, and environmental circumstances of each. regional area and rural location. These centres should also serve as a laboratory for the application of agrotechnology that facilitate technology transfer processes.

CONCLUSION

A conceptual review has been carried out of the aspects that determine Sustainable Agriculture 4.0 that can respond to the needs of a sustainable food system. As the most significant conclusions obtained from the research, we can highlight the following:

First, FAO projections show that the world is not on track to end hunger by 2030 and meet established nutritional goals. In addition, it is foreseeable that both the nutritional status and the food security of the most vulnerable social groups will be further undermined because of the health and socioeconomic repercussions of the Covid-19 pandemic. Everything indicates that there will be major global food problems by 2050, if there are no disruptive changes that significantly improve food production and distribution.

Second, the approach proposed until a few years ago of sustainable food systems has great limitations to improve nutritional status and food security as necessary, since, although the objectives pursued are established with holistic and systemic criteria, the objectives of the projects that develop them and focus their application almost exclusively on agrotechnical aspects.

Third, the development of Sustainable Agriculture 4.0 represents an orientation that can generate sustainable food systems in emerging countries, providing new opportunities that can attract young farmers and entrepreneurs to the sector, alleviating some of the causes of urban concentration that it is generating the abandonment of the rural habitat, thus contributing to a very significant advance in the economic dimension of sustainability.

Fourth, digitization is undoubtedly the main catalyst for the disruptive improvement of all productive and distributive agricultural activities. Resource management can be significantly optimized and become smart, forward-thinking, individualized, hyper-connected, and forward-thinking.

Finally, Sustainable Agriculture 4.0 will create, if applied correctly, highly productive, prospective, and adaptable systems to changes, such as those derived from climate change, generating greater food security, productivity, profitability, and sustainability. Its potential benefits are clear, but major transformations will be needed in agricultural systems, rural economies, farming communities, and natural resource management. This is challenging and will require a systematic and holistic approach to make the most of the new approach.

REFERENCES

Atkinson, A.B., & Stiglitz, J.E. (2015). *Lectures on Public Economics Updated edition.* Economics Books, Princeton University Press.

Bacco, M., Barsocchi, P., Ferro, E., Gotta, A., & Ruggeri, M. (2019). The digitisation of agriculture: a survey of research activities on smart farming. *Array*, 3-4.

Barreto, L., Amaral, A., & Pereria, T. (2017). Industry 4.0 implications in logistics: an overview. *Proceedings of the Manufacturing Engineering Society International Conference MESIC 2017.*)10.1016/j.promfg.2017.09.045

Barrows, G., Sexton, S., & Zilberman, D. (2014). Agricultural Biotechnology: The Promise and Prospects of Genetically Modified Crops. *The Journal of Economic Perspectives*, *28*(1), 99–120. doi:10.1257/jep.28.1.99

Beckhard, R. (1975). Strategies for Large System Change. *Sloan Management Review*, *16*(2).

Boardman, A. E., Greenberg, D. H., Vining, A. R., & Weimer, D. L. (2018). *Cost-Benefit Analysis. Concepts and Practice* (5th ed.). Cambridge University Press. doi:10.1017/9781108235594

Cady, S. H., Jacobs, R. J., Koller, R., & Spalding, J. (2014). The Change Formula: Myth, Legend, or Lore? *OD Practitioner*, *46*(3).

Caiado, R. G., Filho, W. L., Quelhas, O. L. G., de Mattos, D. L., & Avila, L. V. (2018). A literature-based review on potentials and constraints in the implementation of the sustainable development goals. *Journal of Cleaner Production*, *198*, 10.

Chhipa, H. (2017). Nanofertilizers and nanopesticides for agriculture. *Environmental Chemistry Letters*, *15*(1), 15–22. doi:10.100710311-016-0600-4

Dannemiller, K. D., & Jacobs, R. W. (1992, December). Changing the Way Organizations Change: A Revolution of Common Sense. *The Journal of Applied Behavioral Science*, *28*(4), 480–498. doi:10.1177/0021886392284003

De Clercq, M., Vats, A., & Biel, A. (2018). *Agriculture 4.0: The future of farming technology*. World Government Summit in Collaboration with Oliver Wyman.

De Rus, G. (2008). *Análisis Coste-beneficio. Evaluación económica de políticas y proyectos de inversión* (3rd ed.). Ariel Economía.

Evtodieva, T. E., Chernova, D. V., Voitkevich, N. I., Khramtsova, E. R., & Gorgodze, T. E. (2017). Transformation of logistics organization forms under the conditions of modern economy. In E. G. Popkova (Ed.), *Contributions to Economics*. Springer.

Fernández-Villacañas, M.A. (2018). *Las plataformas logísticas 4.0 y la mejora del comercio global: Creando ventaja competitiva logística y desarrollo sostenible*. VI Simposio Internacional Online de Logística y Competitividad, High Logistics Simposios, Medellín, Colombia.

Fernández-Villacañas, M. A. (2019). Desarrollo e Implementación de Plataformas Logísticas de Carga 4.0. *Proceedings of Primer Congreso Internacional de Tecnología e Innovación en Logística 4.0.*

Fernández-Villacañas, M. A. (2020). The New Concept of Logistics Platforms 4.0: Creating Competitiveness Within the Paradigm of Global Sustainable Logistics. In U. Akkucuk (Ed.), *Handbook of Research on Sustainable Supply Chain Management for the Global Economy* (pp. 36–62). IGI Global. doi:10.4018/978-1-7998-4601-7.ch003

Fountas, S., Sorensen, C. G., Tsiropoulos, Z., Cavalaris, C., Liakos, V., & Gemtos, T. (2015). Farm machinery management information system. *Computers and Electronics in Agriculture*, *110*(1), 131–138. doi:10.1016/j.compag.2014.11.011

Hickey, L. T., Hafeez, A. N., Robinson, H., Jackson, S. A., Leal-Bertioli, S. C. M., & Tester, M. (2019). Breeding crops to feed 10 billion. *Nature Biotechnology*, 37. PMID:31209375

HLPE High Level Panel of Experts on Food Security and Nutrition. (2014). Food losses and waste in the context of sustainable food systems (No. 8). CFS Committee on World Food Security HLPE, Food and Agriculture Organization of the United Nations (FAO).

Humphrey, J. (2014). *Market systems approaches – A literature review*. The BEAM Exchange. Available at: www.beamexchange.org

Kamilaris, A., Kartakoullis, A., & Prenafeta-Boldù, F. X. (2017). A Review on the Practice of Big Data Analysis in Agriculture. *Computers and Electronics in Agriculture*, *143*(1), 23–37. doi:10.1016/j.compag.2017.09.037

Khanal, S., Fulton, J., & Shearer, S. (2017). An overview of current and potential applications of thermal remote sensing in precision agriculture. *Computers and Electronics in Agriculture*, *139*, 22–32. doi:10.1016/j.compag.2017.05.001

King, A. (2017). Technology: The Future of Agriculture. *Nature*, *544*(7651), S21–S23. doi:10.1038/544S21a PMID:28445450

Mandi, K., & Patnaik, N. M. (2019). Mobile apps in agriculture and allied sector: An extended arm for farmers. *Agriculture Update*, *14*(4), 334–342. doi:10.15740/HAS/AU/14.4/334-342

Manjunatha, S.B., Biradar, D.P. & Aladakatti, R. (2016). Nanotechnology and its applications in agriculture: A review. *Journal of Farm Sciences, 29*(1).

Meadows, D. (2009). *Thinking in systems: A primer*. Earthscan.

Montoya, D. (2007). Nuevas necesidades en ingeniería para el desarrollo de la biotecnología. *Revista Colombiana de Biotecnologia*, *9*(2).

Muangprathub, J., Boonnam, N., Kajornkasirat, S., Lekbangponga, N., Wanichsombat, A., & Nillaor, P. (2019). IOT and agriculture data analysis for smart farm. *Computers and Electronics in Agriculture*, *156*, 467–474. doi:10.1016/j.compag.2018.12.011

Muñoz, E. (1999). *Biotecnología y desarrollo en distintos contextos culturales. Influencias e impactos*. Consejo Superior de Investigaciones Científicas.

Nas, T. F. (2016). *Cost-Benefit Analysis. Theory and Application* (2nd ed.). Lexington Books.

Nguyen, H. (2018) *Sustainable food systems. Concept and framework*. Agricultural Development Economics Division (ESA), Food and Agriculture Organization of the United Nations FAO. Last access 22/05/2021 web http://www.fao.org/3/ca2079en/CA2079EN.pdf

Nowak, B. (2021). Precision Agriculture: Where do We Stand? A Review of the Adoption of Precision Agriculture Technologies on Field Crops Farms in Developed Countries. In *Agricultural Research*. Springer. doi:10.100740003-021-00539-x

Organisation for Economic Co-operation and Development (OECD), UN Food and Agriculture Organization (FAO) & UN Capital Development Fund (UNCDF). (2016). Adopting a Territorial Approach to Food Security and Nutrition Policy. OECD Publishing.

Ouafiq, E., Elrharras, A., Mehdary, A., Chehri, A., Saadane, R., & Wahbi, M. (2020). IoT in Smart Farming Analytics, Big Data Based Architecture. In Human Centred Intelligent Systems. Springer.

Patel, R. (2012). The Long Green Revolution. *The Journal of Peasant Studies*, *40*(1), 1–63. Advance online publication. doi:10.1080/03066150.2012.719224

Poonia, A. S., Banerjee, C., Banerjee, A., & Sharma, S. K. (2020). Comparative Analysis of Internet of Things Monitoring Devices in Smart Agriculture Using Deep-Learning Techniques. In Soft Computing: Theories and Applications. Springer. doi:10.1007/978-981-15-0751-9_129

Potena, C., Khanna, R., Nieto, J., Siegwart, R., Nardi, D. & Pretto, A. (2020). AgriColMap: Aerial-Ground Collaborative 3D Mapping for Precision Farming. *IEEE Robotics and Automation Letters, 4*(2), 1085-1092.

Rosati, F., & Diniz, L. G. (2019). Addressing the Sustainable Development Goals in sustainability reports: The relationship with institutional factors. *Journal of Cleaner Production*, *215*, 1. doi:10.1016/j.jclepro.2018.12.107

Rose, D. C., Wheeler, R., Winter, M., Lobley, M., & Chivers, C. A. (2021). Agriculture 4.0: Making it work for people, production, and the planet. *Land Use Policy*, *100*(January), 104933. doi:10.1016/j.landusepol.2020.104933

Sáiz, V., & Rovira, F. (2020). From Smart Farming towards Agriculture 5.0: A Review on Crop Data Management. *Agronomy (Basel)*, *10*(2), 1–21. doi:10.3390/agronomy10020207

Sands, D. C., Morris, C. E., Dratz, E. A., & Pilgeram, A. L. (2009). Elevating optimal human nutrition to a central goal of plant breeding and production of plant-based foods. *Plant Science*, *177*(5), 377–389. Advance online publication. doi:10.1016/j.plantsci.2009.07.011 PMID:20467463

Santos-Valle, S., & Kienzle, J. (2020). *Agriculture 4.0 – Agricultural robotics and automated equipment for sustainable crop production.* FAO.

Sousa, F., Gian, N. & Home, R. (2016). Information Technologies as a Tool for Agricultural Extension and Farmer-to-Farmer Exchange: Mobile-Phone Video Use in Mali and Burkina Faso. *International Journal of Education and Development using Information and Communication Technology, 12*(3), 19-36.

Stuart, D., Schewe, R., & McDermott, M. (2014). Reducing nitrogen fertilizer application as a climate change mitigation strategy: Understanding farmer decision-making and potential barriers to change in the US. *Land Use Policy*, *36*(1), 210–218. doi:10.1016/j.landusepol.2013.08.011

Swaminathan, M. (2009). Norman E. Borlaug (1914–2009). *Nature, 461*(894). doi:10.1038/461894a

Tsouros, D. C., Bibi, S., & Sarigiannidis, P. G. (2019). A Review on UAV-Based Applications for Precision Agriculture. *Information (Basel)*, *10*(11), 349. doi:10.3390/info10110349

Ugochukwu, A. I., & Phillips, P. W. B. (2018). Technology Adoption by Agricultural Producers: A Review of the Literature. In N. Kalaitzandonakes, E. Carayannis, E. Grigoroudis, & S. Rozakis (Eds.), *From Agriscience to Agribusiness. Innovation, Technology, and Knowledge Management.* Springer. doi:10.1007/978-3-319-67958-7_17

United Nations. (2021). *Peace, dignity, and equality on a healthy planet: Population*. Last access 22/05/2021 web https://www.un.org/en/global-issues/population

United Nations FAO. (2020). *The State of Food and Agriculture*. FAO.

United Nations FAO. (2021) Developing sustainable food systems and value chains for climate-smart agriculture (B10 2 Food systems and value chains: definitions and characteristics) (Update). Food and Agriculture Organization of the United Nations (FAO).

World Economic Forum. (2017). *Shaping the future of global food systems: A scenarios analysis*. WEF.

World Health Organization. (2020). *Climate change and human health. Risks and responses*. UN.

Yao, X., Zhu, D., Yun, W., Peng, F., & Li, A. (2017). A WebGIS-based decision support system for locust prevention and control in China. *Computers and Electronics in Agriculture*, *140*, 148–158. doi:10.1016/j.compag.2017.06.001

Chapter 6
Food Loss and Waste:
A Sustainable Supply Chain Perspective

S. Su Baysal
Dalhousie University, Canada

M. Ali Ülkü
https://orcid.org/0000-0002-8495-3364
Dalhousie University, Canada

ABSTRACT

Sustainable production and consumption of food are vital for sustainable development. About one-third of all food produced for humans are either lost or wasted causing increased food insecurity and immense economic and social costs. In a world where famine has been an alarming issue, any action to reduce food loss and waste (FLW) is crucial. This chapter reviews, from a sustainable supply chain perspective, the extant literature on food supply chains and discusses FLW issues, especially within the context of sustainable consumption of fruits and vegetables. A framework for sustainable food supply chains (SFSCs) from both production and consumption ends are discussed. In doing so, such current disruptive intelligent technologies as blockchain and the internet of things are emphasized as potential enablers for SFSCs. Mainly driven by consumers' awareness of the pressing issues in the world and consumption behaviour, mitigating FLW in SFSCs would not only result in efficient land and water use but also positively impact climate change and livelihoods towards sustainable development.

INTRODUCTION

Thanks to the growing global population, climate change, deteriorating natural resources, urbanization at an unsustainable rate and widening inequities in food accessibility, hunger in the world is on the rise. Although public and corporate awareness of Food Loss and Waste (FLW) has increased, much needs to be done. According to the Food and Agriculture Organization of the United Nations (FAO), about one-third of all food produced globally is either lost or wasted and almost 690 million people are on the verge of starvation. To that end, Goal # 12 (Sustainable Production and Consumption) of the United

DOI: 10.4018/978-1-7998-8900-7.ch006

Nations Sustainable Development Goals (SDGs, hereafter), among others, aims to develop policies by which producers grow sustainably and efficiently and consumers shift to nutritious and safe diets with lower environmental footprints (FAO, 2021a).

From farming and production to packaging and distribution to retail and catering to the valorization of lost or wasted food, the food industry is an extremely complex Supply Chain (SC) networks of local and global small business entities ranging from small farm producers to giant retail stores. Add to this, the food industry is rife with disinformation such as "green-washing" that impede consumers from better transition to truly sustainable ways of food consumption and lifestyles.

There is a delineation between Food Loss (FL) and Food Waste (FW). As defined in FAO (2021b), FL is "the decrease in the quantity or quality of food resulting from decisions and actions by food suppliers in the chain, excluding retailers, food service providers and consumers. Empirically, it refers to any food that is discarded, incinerated or otherwise disposed of along the food supply chain from harvest/slaughter/catch up to, but excluding the retail level, and does not re-enter in any other productive utilization, such as feed or seed." And FW "refers to the decrease in the quantity or quality of food resulting from decisions and actions by retailers, food service providers and consumers." That is, FW happens when *edible* food is discarded prematurely or unnecessarily; and FL happens when food is damaged or destroyed before it reaches the consumer.

As reported by the World Food Program USA (WFPUSA, 2021), global hunger is not due to a lack of food but rather inefficiencies in the food system; roughly 1.3 billion tons of food worth about $1 trillion is lost or wasted each year. Interestingly, FW is found to be more prevalent in high-income countries, whereas FL happens more in low-income countries. It turns out that an average US citizen wastes about 110 kgs of food each year, while farmers in Africa lose about 40% of all the food they grow. These striking facts let alone point to the need to distinguish FL and FW and how appropriate mitigation strategies and actions should be taken for each, from the sustainability angle.

Having gained more interest since the late 20th century, the concepts of "sustainability" and "sustainable development" continue to evolve by including equity and justice in their scopes. Because of regulatory and consumer pressures and climate change, more and more companies seem to adopt these "buzz words" in their missions statements. On the other hand, an increasing amount of research effort has been directed to the study of sustainable supply chain management (e.g., Khan et al., 2020). One of the sustainability goals is to achieve a minimum level of well-being that depends on people's ability to adequately deal with inefficiencies that can lead them to poverty (Jackson, 2014). Ensuring that this goal is attainable for all humanity and mitigating food insecurity should be demonstrated primarily with food justice. In this period when production is gradually decreasing, the inverse rate increase in consumption does not provide equal opportunities for all individuals over the globe. Our materialistic world goes all the way into our fridges and disappears at the bottom of the garbage. Parfitt et al. (2010) reported that "As much as half of all food grown is lost or wasted before and after it reaches the consumer." Therefore, FLW must be eliminated by all means to alleviate hunger in the world.

As human beings consume more than they produce, the effective proof of this consumption is transformed into the most unconscious conduct. For example, 230 million tons of net production in Sub-Saharan Africa and 222 million tons of FLW in industrialized countries are almost equivalent to each other (Ishangulyyev et al., 2019). On the other hand, United Nations (2021b) reports that the agricultural supply needs to be increased by 70% to meet demand in 2050.

Any sustainability initiative should align with United Nations' Sustainable Development Goals (UN-SDGs). More than 700 million people, 10% of the world population, still cannot meet their basic needs by suffering poverty today. According to a United Nations (2021) report, 6% of this will remain under these conditions until 2030, and the COVID-19 pandemic has caused 70 million more people to be pulled to the extreme poverty limit. It is the order in the world that keeps us alive, but this order has already been broken that pushes one side to death while the other side lives unconsciously. The goal of Zero Hunger (UNSDG#2) could not be achieved by 2030 with the recent trends and lifestyle of consumers. Unfortunately, 840 million people still would fight hunger by 2030. Already, 135 million people do not have the necessary nutritional supplements; unfortunately, this number has doubled with the COVID-19 outbreak (United Nations, 2021a).

Pioneered by Elkington (1998), the Triple Bottom Line (TBL) approach to reinforce operational sustainability has become one of the main objectives of Sustainable Supply Chain Management (SSCM). TBL is analyzed in three dimensions; environmentally, economically and socially. TBL and "3Ps" known as the 'planet, profit, people' feature can often intersect in a common set (Ahi & Searcy, 2015). TBL is appropriate for producing the essential imaginative force and inventiveness with discovering new types of incorporating benefit, social and environmental business (Gold et al, 2013). Ülkü and Engau (2021) emphasized the importance of including a fourth pillar, culture, in the traditional TBL approach. This expansive approach termed the Quadruple Bottom Line (QBL) brings about the fourth P, the "purpose." In this way, decision-makers and firms expand their visions 'to be useful' while also creating a control mechanism in their profit-oriented resolutions. More than ever, the values that are appreciative in the eyes of the consumer rather than the annual turnover of the companies determine the future qualification and status of that company with a single social media post or news. The conceptualization of a Sustain-

Figure 1. Quadruple bottom lines and factors impacting FSSCs

able Food Supply Chain (SFSC) must embrace each and all of the QBL pillars while considering such constraints as those emerging from SDGs, government regulations, food security, quality and justice, and smart technologies enabling efficient farming and logistics and nudging consumers to sustainable consumption (see Figure 1).

Two terms that are pertinent to this study of the FLW from a Sustainable Supply Chain (SSC) perspective are Sustainable Supply Chain Management (SSCM), and Sustainable Supply Chain Analytics (SSCA).

Seuring & Müller (2008, p.1700) define Sustainable Supply Chain Management (SSCM) as "... *material, information, capital flows, and cooperation among companies along the supply chain which take goals from the dimensions of sustainable development... into account which are derived from customer and stakeholder requirements. In sustainable supply chains, environmental and social criteria need to be fulfilled by the members to remain within the supply chain, while it is expected that competitiveness would be maintained through meeting customer needs and related economic criteria."* The latter term, SSCA, is defined by Ülkü and Engau (2021, p.7) as "...*the set of activities and analytic tools that are big-data-driven in order to provide both optimal solutions and innovative foresight to complex supply chain decision-making problems with the goals of achieving cultural, economic, environmental, and social sustainability pillars.*"

The studies appear to concur that fresh fruits and vegetables (products of the soil) contribute most to the avoidable food waste and are generally critical in influencing the all-out turnover on each role piece of the distribution chain (Kaipia et al., 2013). The waste of fresh fruits and vegetables is experienced at every stage of the supply chain. First, 20% of waste occurs during production. A further 3% loss is added during transportation and another 1% with packaging. In addition to the 12% waste in distribution and market shelves; the percentage hits its maximum on the customers' hands by 28% (Chen et al, 2020). In short, 64% of fresh food seem to have never existed.

Food and Agriculture Organization of the United Nations (FAO) indicates the numbers into two categories: food loss and food waste index. While the food loss index is on a production level, the food waste index measures retail and consumption levels. According to the FAO's worldwide calculations, the food loss from farms to transportations is at %13.8 and about more than $ 400 billion per year. The index of food losses is higher when it comes to fresh fruits and vegetables (FAO, 2021a). The loss is not just the fruit and vegetables themselves. For example, an average of 13 litres of water is spent growing a tomato (FAO, 2018) also its seed, fertilizer, transportation process means both labour and finance. In the U.S, for the restaurant industry alone, between 4% and 10% of food waste occurs before it reaches the customer's plate (Sakaguchi et al, 2018).

Figure 2 displays aggregate data collected by FAO between 1997 and 2017 as boxplot statistics on FLW happening during various supply chain stages for fruits and vegetables. From pre-harvesting to consumption, the median % of FLW hovers up to 20%. The most variability with largest range is observed during pre-harvesting and the least in packaging. Thus, the focus in this chapter is particularly on sustainability issues regarding fruits and vegetables (fresh produce) from a supply chain perspective.

Retail stores, for example, aim to improve the customer experience in the fastest and simplest way as well as reducing food waste; for online shopping, home delivery, and transparency created with the concept of sustainability, their visions must constantly renew themselves for not staying behind the times. With the food waste strategies and awareness created just by retail stores in the US, up to $20 billion can be benefited from economic, social and environmental factors (Teller et al, 2018). One of them is to promote fruits and vegetables called "ugly" and affirm them by normalizing customers' habits.

Figure 2. FLW world data (1997-2017) boxplot aggregate statistics on all supply chain stages for fresh produce (Source: http://www.fao.org/platform-food-loss-waste/flw-data/en/ 1997-2007)

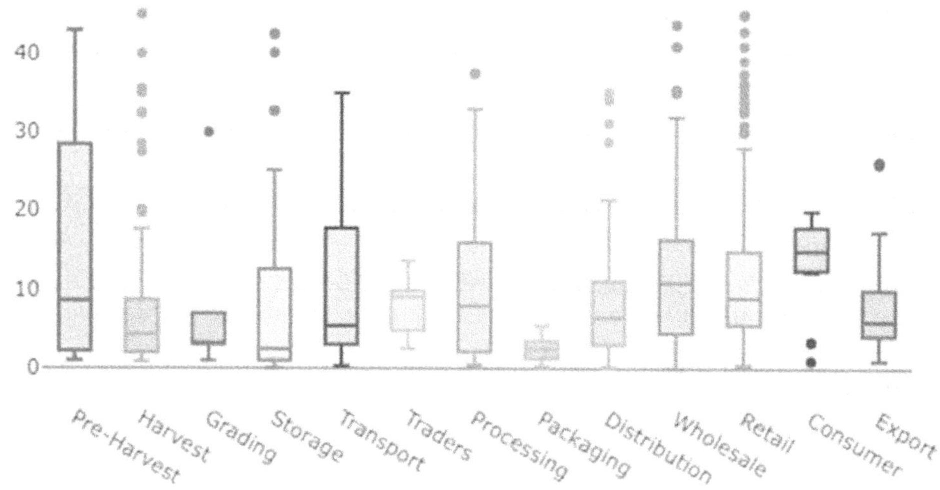

BACKGROUND ON FOOD SUPPLY CHAINS AND A FRAMEWORK

By the end of the last century, it provided more accessible and widespread sales networks with the logistics opportunities offered in the international food trade. This brought about changing consumer demands and consumption patterns. Consumers' shopping volumes increase with self-identification, personal values, economic freedom, social and cultural perceptions, as well as psychological changes and commodification instinct. As the most important factors for conscious consumers; in addition to freshness, nutritional values, taste, smell and texture, safety in foods, many value judgments such as traceability, sustainability, ecological impacts, and fair trade also reveal their importance. Traceability aims to eliminate the question marks of the consumer to a large extent by the obligation to present the product's historical information, its origin and reliability. Collaborative and cooperative responsibility has thus become beneficial to both sides (Badia et al, 2015).

One of the most important issues in SFSC is food security which is to support all people's right to a healthy life with physical, social and economic access to safe and nutritious foods always. What is meant to draw attention with the words "all people and always" is the basic need for nutrient distribution with equality and continuity, and that this need is the right of every segment. This reveals the importance of the concept of sustainability in food production, which has been highlighted over and over, so that future generations will not be deprived of their most basic rights (Mbow et al., 2020).

The food supply chain has shown up in our lives in the simplest way since people entered the age of production. Since it will continue as long as people exist, it constitutes an inevitable part of our daily lives. Behind the scenes, an adventure from the production process of raw materials to their processing and packaging, preserving them under certain conditions and distributing them all over the world to our table is discussed.

FLW may occur in all of the stages from supply (pre-harvest to harvest) to production (processing, packaging) to distribution (warehousing) to retailing and household consumption. There are many possible causes FLW can happen throughout the Food Supply Chain (FSC), these could be due to external

Figure 3. A pictorial of a Sustainable Food Supply Chain

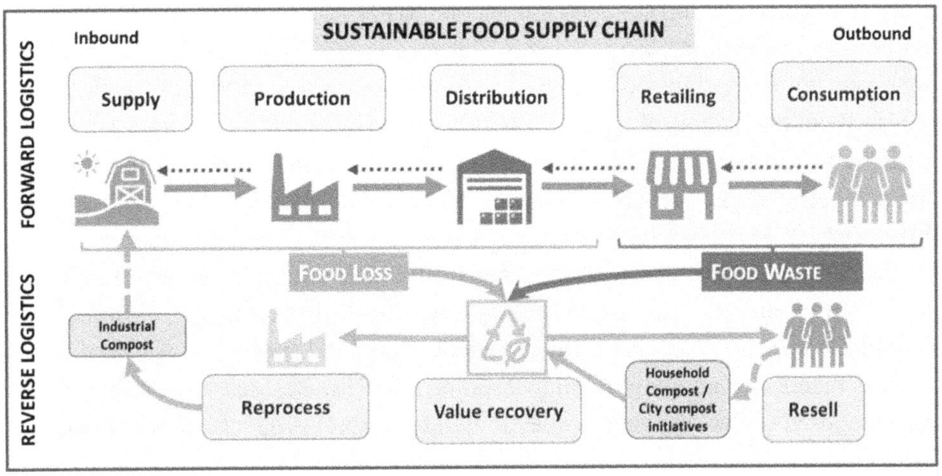

issues such as the impacts of climate change (drought, floods, fires etc.), collaboration issues (see, Annosi et al., 2021) between FSC members (e.g., poorly designed contracts), operational issues such as overproduction, poor packaging, machine malfunctioning, inadequate transport vehicles, and incorrect pricing and mismanaged inventories of fresh produce, behavioural issues related to consumers' unsustainable consumption patterns, and cultural issues such regarding it a table manner to leave some food in the plate or preferring "all-you-can-eat" options (c.f., Luo et al.,2021).

A Sustainable Food Supply Chain (SFSC) strives to improve performance on all four pillars of QBL, namely economic, environmental, cultural, and societal impacts. Figure 3 displays how SFSC, from a closed-loop perspective, integrates forward logistics (from raw material to consumption) and reverse logistics (from loss and waste to valorizing and putting it back in the supply chain, or composting and disposing to landfills as the last resort). Therefore, Figure 3 provides a framework for identifying stages at which FLW occurs, how those SFSC members need to be in cooperation, how the loops can be closed within the food system, and what type of logistics models (forward and reverse, inbound and outbound) may work for an integrated SFSC.

One can conjecture that a unit of FW is more costly than a unit of FL. Hence, while it is of utmost importance to ameliorate inefficiencies on the supply side of an SFSC, it is at least as important to have a culture of sustainable consumption on the consumer side. The consumer's perception of the value of the product matters in purchasing decisions, which collectively form the demand for the product, and thereby, the need for a business. While selling the product to the consumer, its packaging and label are sold as well. In addition to attracting the customer to the product, the purpose of packaging and labelling for food supply is more than that.

The right packaging methods are a formative element on the quality, freshness, safety, obtainability of a product during transportation and shelf life. With tracking and data systems, the classical process has been replaced by "Smart Packaging (SP)" and its impact on reducing food loss and waste due to incorrect temperature in the holding conditions cannot be disregarded (Chen et al, 2020). Preserving the life and quality of fresh products is an important criterion that reveals the moisture and atmospheric composition in the packaging process. It has been observed that varying the gas concentrations as well as the correlative humidity in the package and the air-wet package extend the fresh product quality and shelf life during transportation and storage (Jalali et al, 2017).

Conscious consumers examine product labels during the purchase process. Not all consumers are vigilant about product information such as the expiry date. According to Wilson et al (2019); confusion and unawareness about date labels of ‘sell by,’ ‘best by,’ ‘use by,’ and ‘best before’ accounts for 20% of consumer waste, which amounts to $29 billion per year, let alone, in the US. In other words, consumers should distinguish between these labels for quality or safety with the difference at “best if used by” and “use by” (Wilson, 2019).

Contrary to what is believed, “best before labelling” which opens the door to many discussions is not about the consumption reliability of the product, but labelling that is put on limitation to indicate its ultimate quality. Based on this conceptual confusion alone, in Europe, the waste of 8.8 million tons of food product is mentioned annually (Collart & Interis, 2018). Besides the regulations of the labels, consumers are supposed to ask themselves questions. The correct product and label inquiry occur with the conscious consumer.

On the other hand, consumers do not necessarily look at not only the expiry date. Some other product labelling such as “organic certified” or “ sustainably sourced” may also affect their purchase decisions. Many consumers take the power to choose the naturally grown fresh fruits and vegetables that enter their homes, and they refuse to incorporate chemical pesticides. For consumers who make green purchases, eco-friendly packaging and labels are the most important factor. While many producers give the right to these labels and dedicate their company’s mission and vision to organic cultivation; some companies deceive their customers just for more profit under organic fruits and vegetable fraud.

The industry is aware that buyers prefer environmentally-friendly farming techniques, yet they do not necessarily tolerate the regular results of the ecologically cordial creation: the flawed appearance of items. Past studies have shown that most buyers say they purchase natural items to stay away from pesticides. Nonetheless, because buyers will pay more for ideal natural apples than for perfect formulaic apples, the rate markdown because of cosmetic damage has higher dollar esteem in the rebate of natural apples than for the conventional apples (Yue et al., 2009).

According to the U.S Department of Agriculture data, the largest loss under the name of FW is experienced in the retail and consumer segment; recall Figure 2. The department argues that supermarkets have the authority to prevent waste, to influence which production is placed and to affect the consumer experiences in a positive or negative change of opinion during their shopping (Bolos et al., 2019). The major obstacle in many studies on this subject to date is that although the consumer is inclined to reduce waste, it is the consumer’s habits, lifestyle, beliefs, and prejudices that suppress this information (Närvänen et al., 2018).

INTELLIGENT TECHNOLOGIES FOR FOOD SUPPLY CHAIN SUSTAINABILITY

One of the mandates of SSCM is efficient and effective collaboration between SFSC members. To make the food supply chain sustainable, the interests of various parties in the supply chain need to be coordinated, especially in the consumption and production sectors (Govindan, 2017). Environmental and business sustainability go hand in hand. Such operational tactics as synchronizing inbound and outbound deliveries and using environmentally-friendly materials advance the performance of companies (Mele

& Russo-Spena, 2015). Advents in communication and production technologies have immense potential in reducing FLW; recall Figure 1.

With increased globalization, regulations, technological progress, and the availability of diverse cultural and behavioural data from mobile devices, studying SFSCs becomes more prevalent. Among the existing possibilities, the ones that are actively used are Blockchain, Industry 4.0, Internet of Things, and Big Data Analytics (c.f., Rahimi et al., 2020). For instance, Blockchain technology, a distributed database of records or shared public and private ledgers among agents, offers non-localization, security, auditability, and smart execution (Saberi et al., 2019) that are much needed for SFSCs. Tracking social and environmental potential dimensions may also inform sustainability policies on traceability, health & safety procedures, and waste reduction (Joubert, & Jokonya, 2021). Especially the food supply chain processes are questioned by consumers and governments for their transparency.

Under China's Paris Agreement, blockchain-based activities allow carbon and greenhouse gas emissions reduction assets besides the assurance of human rights and work ethics under 'smart contracts' (Saberi et al., 2019). Accessible information per green product could be complex, but not impossible to verify. An example of this is realized with Endorsement of the Forestry Certification, from manufacturing to final points under 740 million acres of forest blockchain technology, providing the user with the guarantee information data of the construction of the table they own from that tree (Rosencrance, 2017). The same principle can aim to benefit the fresh food supply chain. With the most basic usage, it means a large percentage of the positive impact that continues from preventing fraud of organic produce to reducing food loss during transportation.

Industry 4.0 (I4.0) refers to the digitization and intelligence of business operations & processes. As a result, I4.0 and sustainability are associated with the power to enhance the lifestyle for future generations. Cloud networking, CPS, big data analytics, and other internet-based technology are driving I4.0 (Luthra et al, 2020). The 4th Industrial revolution called cyber-physical systems is evolving step by step to the 5th Industrial revolution which is mass customization & autonomous manufacturing (Nahavandi, 2019). Increasing trading volumes, shorter product life- cycles, greater product complexity, and globally faster production are all factors to consider as the remedy of this rising technology system (Morrar et al., 2017).

Another emerging paradigm, the Internet of Things (IoT), provides the communication and sensing of physical and virtual objects in indoor and outdoor environments with a network connection between each other. This information technology, which is introduced with the layers' basis, has been used since 1999 (Abdel et al., 2018). The most important gains for the food supply chain; real-time visibility, highest quality, delivering in shorter times and preparing in optimal condition factors that have a large share in the safety part of fresh food produce (Zhao et al., 2015). Those layers shape sensing, communication, and application. The quality of life of crops and livestock are maintained by the sensing layer's records. Temperature, humidity, carbon dioxide in fields, greenhouses and during transportation can be observed by the wireless sensors. The communication layer is intended to distribute information to various stakeholders at the time. It may also be the case that the system can manage goods with such identifier programs: Serialized Global Trade Item Number or Global Individual Asset Identifier. The application layer assures supported applications to the farmers, retailers, government with the radio frequency identification (RFID) tags. Apart from the supply chain partners, consumers will also have access to the most basic databases such as product warranty, usage date, product history and transactions (Zhao et al., 2015; Abdel et al., 2018). Smart agriculture has emerged mostly with IoT applications; by tripling its market size, a \$15.3 billion global value can be reached by 2025. IoT devices can be classified according to their focus area (De Clercq et al., 2018). For example, the agricultural IoT devices, Arable and Semios, maintain water and crop management with the input from soil (Arable, 2021; Semios, 2021); Smart Elements interrelate with a farmer based on microclimate level, solar radiation or leaf wetness (Incyt, 2021), and Growlink regulates light, humidity and soil terms (Growlink, 2021).

On the other hand, Big Data Analytics (BDA) has the core discipline of 5Vs: volume, variety velocity, veracity, and value. The "volume" stands for revealing the greater size of its data capacity and alleviating any data confusion or deficiencies in other devices. Generating its data from diverse sources & backgrounds refer to variety. Velocity's response is based on realizing data generation and delivery speed in real-time, while the concept of veracity draws attention to the importance of quality and reliability level. Targeting the minimum risk ratio by dragging its data to a concrete decision stage with saving time and money is positioned under the name of value (Nguyen et al, 2018). BDA, such as other software systems, is at the forefront by collecting data from multiple sources (i.e., from industry data collected by sensors or equipment to external social media) and providing reliable data to plan, execute or create 'green opportunities'. BDA, which still has many hypotheses on it, can emerge with its capacity to lead to innovations in the green product and fresh produce categories through the predictive analysis system (Bag et al., 2020).

Advance sensor and drone systems may further enhance farming sustainability. A good application of this is by UniSA, whereby convenient and cost-effective soil moisture can be observed via RGB (red, green and blue colour mode) camera technology (University of South Australia, 2021).

Insights and Recommendations

If change is to begin, that change must start from the smallest building block of society. According to a study conducted on children between the ages of 5 and 11, either way, the acceptance or rejection of suboptimal fruits and vegetables is a reflection of the consumer behaviour of family members. In short, the acceptance of atypical produce can find an explanation with the concept of 'familiarity' coming from an early age education (Makhal et al, 2020).

Moreover, some studies show that cultural differences and social judgments affect our shopping habits. In this part, material power and status ego come into play. In some cultures, how much food you put in front of your guest becomes a symbol of your economic power, and in fact, these 'ego wars' effectuate a significant percentage of food waste (Aschemann et al, 2015). In addition, overstocked products and maintaining these products in wrong conditions (e.i. extremely low or high refrigerator temperature, wrong handling of fresh produce or storing directly in contact with other products) are a major part of household waste (Aschemann et al, 2015). Some of the easiest behavioural measures put forward by The United States Environmental Protection Agency, such as preserving vegetables and fruits in different parts, keeping bananas, apples, and tomatoes out of contact with each other, and significantly affecting the freezing process to all leftover meals (Sakaguchi et al, 2018) are the effective details that every consumer can make a difference by applying.

According to Waste and Resources Action Programme (WRAP)'s technical research in the field (2015), 20% to 25% of wastes are caused by incorrect packaging operations, packaging of large quantities of products that are difficult to consume on time, thus going beyond best-before dates. Other post-purchase behavioural problems include not using the product's packaging for its intended purpose, ignoring label suggestions for storage conditions.

Regarding food security and justice, some redistribution plans and community-focused actions have been successful in mitigating FLW within supply chains. Plans that facilitate the redistribution of food near shelf-life limits and food supply to consumers, such as "community refrigerators'', play a crucial role in reducing waste, especially where communities confront restricted availability and affordability of food (Martindale & Schiebel, 2017).

Logistics is not just about reaching the customer's hands. This is a two-way street, and distribution always comes back with reuse, recycling, value recovery or rethink and refuse (recall Figure 3). One can think of this as the water cycle in the atmosphere, each drop of water that falls on the ground collects in the clouds again, and this system creates the balance. This mindset is also required for a sustainable food supply. It is now time to evolve into being a "minsumer"; to minimize consumption to what meets one's minimalistic needs while minimizing the impact of consumption on the environment and others' lives.

A critical amount of the waste happening through the whole "farm-to-fork" chain might be straightforwardly or by implication brought about by the reluctance to purchase and devour outwardly imperfect products. These outwardly problematic food varieties stray from utilization principles, either because of odd shape or shading, in light of different defects, or as a result of being near the best-before date, without any effect on taste quality (Rohm et al., 2017).

Producers dispose of almost 30% of fruits and vegetables as a result of "aesthetic imperfections" (Rohm et al, 2017). For instance, about 30% of carrots harvested does not even reach a place on retailers' shelves because of beauty standards (FAO, 2018). Almost every vegetable and fruit have a set diameter, size or colour scale. Fresh produce, which has matured in the field and set off, must exactly match these measurements if it wants to reach the customer's table and if that cucumber is a few inches shorter than the scale then it has no chance. However, the only reason why they cannot be grown by similar standards is drug-free and hormone-free farming. Nothing produced naturally can compare to uniform factory production. A portion of fresh food cannot be classified under poor quality just because a vegetable or fruit is supposedly described as 'ugly'. The pale colour or the shape of an apple does not prevent it from being an apple, nor does its taste different from other perfect-looking apples. If that apple is not to be exhibited in the museum, compliance with beauty standards should not be more significant than its nutritional value.

The shape and look of the product impact consumers' purchase behaviour. For example, Intermarché (2021) a supermarket chain in France, conducted a consumer experiment in 2014 on "inglorious fruits and vegetables," in which some products were described as ugly and about to be thrown away, for sale as %30 cheaper with a simple campaign. Although appreciated, even an almost 50% price drop has not changed the minds of "all" customers, in this "inglorious fruits and vegetables" experiment on consumer perceptions. Buyers required a higher markdown to buy items that were seen as ugly or uncertain, with discounts extending from 24% for the cucumber with its appearance divergence to 67% for an apple with a brown coloured spot (Rohm et al, 2017). The taboo of 'quality is always more expensive' causes some, if not every customer, to change their shopping patterns. Because consumer psychology can automatically approach a product with a question mark when less money is spent on than regular. In this section, the question marks in the minds of consumers and how that distrust can be solved in different ways than dropped prices. The level of shoppers choosing apples with abandon was equivalent to or lower than 15%. Changes in quality assumptions might be expected to adequately diminish waste practices by customers (Jaeger et al, 2018). According to Dusoruth & Peterson (2020), the crucial insight is to educate the community that ugly produce is safe and accurate to eat. Past experience is one of the most influential factors in consumer behaviour. Impulses and reminders have a significant impact on consumers' decision mechanisms. Studies argue that there is a correct link between nudging and decision-making targets examine this in three groups. 1) detailed information appealing to visual memory, 2) targeting emotions and feelings, communicating with striking slogans, and 3) also seen as being driven by design through external factors: the ability to experience and test the product at the nearest distance (Bolos et al, 2019).

By keeping the experience of the consumer at the forefront, fruits and vegetables that are not defined as attractive in fresh produce can be presented to the customer in three categories; processed fresh food, direct sales and other usages.

For the processed fresh food category, liquid conversion and piece-packaging processes can be used. Processing is not just about chemical change to a product; many fruits and vegetables are open to fresh consumption in both liquid and solid forms. The main purpose here is to provide the product to the customer in a new form by changing its shape or form without disturbing its chemistry (repurposing). For example, organic juices, fruit compotes, and detox juices have been in a massive revamp lately. Perhaps because of its imperfect shape, colour, or merely rotting fruit or vegetable in its current state, it may be offered for sale and receive demand at a low percentage. Because the customer no longer has the chance to pick up that 'ugly' fruit and examine it, then the product in that form must not appeal visually. Therefore, consumer psychology will start to make choices without taking the established aesthetic codes as the basis. The same logic applies to fresh fruits and vegetables that are shredded and repackaged. When an amorphous potato is cut into cubes and served, the negative effects of being "shapeless" in the eyes of the consumer disappear. At the same time, cut and served fruits and vegetables are in more demand due to their convenience or 'open and finish!' principle: Smaller portions do not need restoring by consumers and maybe more economical.

In the second category, direct sales, tasting perception and influence are the most important factors. In addition to product promotion stands prevalent in grocery stores, tasting stands could also be established for imperfect fresh produce. First, the so-called "perfect" apple should be presented to the consumer, and then the "crooked, not as bright and red as the other, maybe a little rotten" apple should be presented to taste. The remaining benchmark is at the customer's discretion. Similar to quite susbstitutable products (e.g., Pepsi and Coca Cola), most people probably would not be able to tell the difference in the taste of those products unless they see their brand names. The same is true in this scenario; if the customer experiences that imperfect produce themselves and sees that it is no different from the other, then only a delicious apple will remain from the bias in mind. Another option is guiding the customers with external factors during their purchase by referring the journey of the product from its origin to the shelves of the supermarket and exemplify in which various ways they can consume it with the best experience. For example, a decaying tomato can be turned into tomato paste, or an un-demanding apple or grape can be used to make natural vinegar at home. Evidently, direct experience with the product (touching, tasting, smelling etc.) creates long-lasting memory with consumers and impact their re-purchase decisions – for instance, products bought online are more likely to be returned compared to those bought in a brick and mortar store (Dailey & Ülkü, 2018). While consumers are looking for their most basic needs (staple products) which by design are located in the further corners of the supermarket, they will be exposed to see other products on the way and end up buying more than what they planned for. The creation of the mentioned new sections lays the foundation for this mentality. A section description that says 'apple for vinegar making' on it will ensure that the consumer is positively inclined to the idea of making vinegar at home with that 'imperfect' apple.

Last but not least category is the other usage; preventing produce from being waste and evaluate it different, innovative and sustainable forms. One of these could be insemination. The kernel is the name of the miracle it possesses for the return of the fruit it belongs to, and its life is hidden in it. Plants first protect their seed then form the fruit itself. The only thing that is necessary for a fruit and vegetable to maintain its continuity is its seed. The seed of each fruit and vegetable that is seen as waste is buried under the soil for about 10 cm, and the re-seeding and producing process is started.

Rootstocks, grafting methods and the purpose of grafting are the key to transforming the 'imperfect' fruit and vegetable into its best version. The grafting process in vegetables enables the cultivation of the varieties naturally and regains their life by repairing the damages in the species. At the same time, this method eliminates the disease that occurs from the soil or adverse temperature conditions and provides more efficient water and nutrient intake, while preventing the use of drugs in the species and setting them into the natural recovery process.

Another effective example for the "other usage" section would be turning avocado waste into 'green' plastic. From the grocery bags to the container of the food we order, to the straw of the coffee we drink, plastic has settled in our lives. On the other hand, the use of plastic is tried to be minimized due to its damages to nature, and for this, it is discussed how to obtain 'plastic' with materials that are 'natural' and whose destruction in nature is the most harmless. According to a Mexican company, BioFase, avocado seeds are one of these natural materials. The company says that it is not very ethical to produce bioplastic from potatoes or corn while there is such a high hunger rate in the world, and that is why they chose manufacturing this turnout from avocado seeds, which are called waste after the fruits' consumption and guacamole production (AIM2Flourish, 2021). Since BioFase turns the refuse part of avocado into their main material, they keep their production at the same cost scale compared to normal plastic, and they create the necessary circulation by recycling the seeds that go to waste.

FUTURE RESEARCH DIRECTIONS

As the climate change presses and causes more than ever FL, and as the unsustainable forms of lives in high-income countries produce FW that could be otherwise used to supplant hunger in the under-developed parts of the globe, studying issues related to security, sustainability, and justice in the food systems are utmost important. Therefore, analytical and theoretical studies related to food supply chains will keep current on the agenda of scholars, decision- and policy-makers. Inherent in its complex nature, such studies necessitate an interdisciplinary approach (e.g., Kaufman and Ülkü, 2018). From an SFSC perspective there exist many research venues.

With SDG target 12.3, the United Nations (2021b) states, "By 2030, halve per capita global food waste at the retail and consumer levels and reduce food losses along production and supply chains, including post-harvest losses." Food waste per capita (from processing through to consumption) and food loss that occur from production up to, but excluding retailing level, will be pivotal in measuring supply chain performance. Research on integrating FLW metrics in data-driven decision models for SFSCs is all but one venue that warrants immediate attention.

Another future research venue is related to embedding SFSC research in the water-energy-food nexus and circular economy (e.g., Ciccullo et al., 2021; Ülkü et al., forthcoming). Last but not least, more research needs to be done as to how consumers can be better influenced to purchase and consume food sustainably and how technology can be used to that end (e.g., Currie et al., 2020).

CONCLUSION

This chapter, with a focus on the fresh produce supply chain, reviewed food loss and waste from the lens of QBL sustainability. Such emerging technologies as Blockchain, I4.0, IoT and BDA were emphasized as tools to increase visibility, efficiency, collaboration quality and speed of SFSCs. For instance, across Canada, almost 60% of valuable nutrient goes to the green trash cans (Divert NS, 2021). These green 'recycle' bins have been used in Nova Scotia for over 20 years, but this has been throwing away more than it seems: labour, hours, water, money, and most importantly, the future. A lot of issues have been tried to be addressed in the chapter, there is a part that does not seem to be an obstacle to realize the fact that the perception of beauty is imposed on fruits and vegetables. Much more effective steps than the price drop strategies are elucidated in the processed fresh food, direct sales and other usage categories that can be applicable on products called ugly or shapeless. Potential effects of social media perceptions, perfectionism, socio-economic ego, and culture were also mentioned in this matter. If the change is to be started, it should start within the smallest yet most powerful "buyer" unit in a supply chain: the consumer. The critical action then is to educate new generations about the impacts of FLW and encourage them into more sustainable ways of food consumption via this simple fact: Mitigating FLW would not only result in efficient land and water use but also positively impacts climate change and livelihoods towards sustainable development.

ACKNOWLEDGMENT

This research was supported by the CRSSCA- Centre for Research in Sustainable Supply Chain Analytics, at Dalhousie University, Canada. Grant number: CRSSCA #68024-21001-SCR.

REFERENCES

Abdel-Basset, M., Manogaran, G., & Mohamed, M. (2018). Internet of Things (IoT) and its impact on supply chain: A framework for building smart, secure and efficient systems. *Future Generation Computer Systems*, *86*, 614–628. doi:10.1016/j.future.2018.04.051

Ahi, P., & Searcy, C. (2015). Assessing sustainability in the supply chain: A triple bottom line approach. *Applied Mathematical Modelling*, *39*(10-11), 2882–2896. doi:10.1016/j.apm.2014.10.055

AIM2Flourish. (2021). *BIOFASE: The solution from Avocado Seed*. Retrieved from https://aim2flourish.com/innovations/biofase-the-solution-from-avocado-seed

Annosi, M. C., Brunetta, F., Bimbo, F., & Kostoula, M. (2021). Digitalization within food supply chains to prevent food waste. Drivers, barriers and collaboration practices. *Industrial Marketing Management*, *83*, 208–220. doi:10.1016/j.indmarman.2021.01.005

Arable. (2021). *How we turn data into insights*. Retrieved from https://arable.com/

Aschemann-Witzel, J., De Hooge, I., Amani, P., Bech-Larsen, T., & Oostindjer, M. (2015). Consumer-related food waste: Causes and potential for action. *Sustainability*, *7*(6), 6457–6477. doi:10.3390u7066457

Badia-Melis, R., Mishra, P., & Ruiz-García, L. (2015). Food traceability: New trends and recent advances. A review. *Food Control*, *57*, 393–401. doi:10.1016/j.foodcont.2015.05.005

Bag, S., Wood, L. C., Xu, L., Dhamija, P., & Kayikci, Y. (2020). Big data analytics as an operational excellence approach to enhance sustainable supply chain performance. *Resources, Conservation and Recycling*, *153*, 104559. doi:10.1016/j.resconrec.2019.104559

Bolos, L. A., Lagerkvist, C. J., & Nayga, R. M. (2019). Consumer choice and food waste. *Choices*, *34*(1), 1–7.

Cicculo, F., Caglianoa, R., Bartezzaghia, G., & Perego, A. (2021). Implementing the circular economy paradigm in the agri-food supply chain: The role of food waste prevention technologies. *Resources, Conservation and Recycling*, *164*, 105114. doi:10.1016/j.resconrec.2020.105114

Collart, A. J., & Interis, M. G. (2018). Consumer imperfect information in the market for expired and nearly expired foods and implications for reducing food waste. *Sustainability*, *10*(11), 3835. doi:10.3390u10113835

Currie, B. A., French, A. D., & Ülkü, M. A. (2020). Sustainability, Big Data, and Consumer Behavior: A Supply Chain Framework. In I. Rahimi, A. H. Gandomi, S. J. Fong, & M. A. Ülkü (Eds.), *Big Data Analytics in Supply Chain Management* (pp. 109–132). CRC Press. doi:10.1201/9780367816384-7

Dailey, L. C., & Ülkü, M. A. (2018). Retailers beware: On denied product returns and consumer behavior. *Journal of Business Research*, *86*, 202–209. doi:10.1016/j.jbusres.2018.01.064

De Clercq, M., Vats, A., & Biel, A. (2018). Agriculture 4.0: The Future of Farming Technology. *World Government Summit.* Retrieved from https://www.oliverwyman.com/our-expertise/insights/2018/feb/agriculture-4-0--the-future-of-farming-technology.html

Divert, N. S. (2021). *Fighting Food Waste*. Retrieved from https://divertns.ca/education/fighting%20 food-waste

Dusoruth, V., & Peterson, H. H. (2020). Food waste tendencies: Behavioral response to cosmetic deterioration of food. *PLoS One*, *15*(5), e0233287. doi:10.1371/journal.pone.0233287 PMID:32469982

Elkington, J. (1998). Partnerships from cannibals with forks: The triple bottom line of 21st-century business. *Environmental Quality Management*, *8*(1), 37–51. doi:10.1002/tqem.3310080106

FAO. (2018, March 15). *Beauty (and taste!) are on the inside*. Retrieved from http://www.fao.org/fao-stories/article/en/c/1100391/

FAO. (2021a). *Sustainable Development Goals: 12.3.1 global food losses*. Retrieved from http://www.fao.org/sustainable-development-goals/indicators/12.3.1/en/

FAO. (2021b). *Foof Loss and Food Waste*. Retrieved from http://www.fao.org/food-loss-and-food-waste/flw-data)

Gold, S., Hahn, R., & Seuring, S. (2013). Sustainable supply chain management in "Base of the Pyramid" food projects—A path to triple bottom line approaches for multinationals? *International Business Review*, *22*(5), 784–799. doi:10.1016/j.ibusrev.2012.12.006

Govindan, K. (2017). Sustainable consumption and production in the food supply chain: A conceptual framework. *International Journal of Production Economics*, *195*, 419–431. doi:10.1016/j.ijpe.2017.03.003

Growlink. (2021). *The brain of your smart farm.* Retrieved from https://growlink.com/sensors/

Incyt. (2021). *Weather and microclimate mornitoring.* Retrieved from https://incyt.io/pages/weather-monitoring

Intermarché. (2021). *Inglorious Fruits and Vegetables.* Retrieved from: https://www.youtube.com/watch?v=p2nSECWq_PE&ab_channel=Marcel

Ishangulyyev, R., Kim, S., & Lee, S. H. (2019). Understanding food loss and waste - Why are we losing and wasting food? *Foods*, *8*(8), 297. doi:10.3390/foods8080297 PMID:31362396

Jackson, T. (2014). Sustainable consumption. In *Handbook of Sustainable Development* (p. 279). Edward Elgar Publishing. doi:10.4337/9781782544708.00029

Jaeger, S. R., Machín, L., Aschemann-Witzel, J., Antúnez, L., Harker, F. R., & Ares, G. (2018). Buy, eat or discard? A case study with apples to explore fruit quality perception and food waste. *Food Quality and Preference*, *69*, 10–20. doi:10.1016/j.foodqual.2018.05.004

Jalali, A., Seiiedlou, S., Linke, M., & Mahajan, P. (2017). A comprehensive simulation program for modified atmosphere and humidity packaging (MAHP) of fresh fruits and vegetables. *Journal of Food Engineering*, *206*, 88–97. doi:10.1016/j.jfoodeng.2017.03.007

Joubert, R.-L., & Jokonya, O. (2021). A systematic literature review of factors affecting the adoption of technologies in food waste management. *Procedia Computer Science*, *181*, 1034–1040. doi:10.1016/j.procs.2021.01.298

Kaipia, R., Dukovska-Popovska, I., & Loikkanen, L. (2013). Creating sustainable fresh food supply chains through waste reduction. *International Journal of Physical Distribution & Logistics Management*, *43*(3), 262–276. doi:10.1108/IJPDLM-11-2011-0200

Kamble, S. S., Gunasekaran, A., & Gawankar, S. A. (2020). Achieving sustainable performance in a data-driven agriculture supply chain: A review for research and applications. *International Journal of Production Economics*, *219*, 179–194. doi:10.1016/j.ijpe.2019.05.022

Khan, S. A. R., Yu, Z., Golpîra, H., Sharif, A., & Mardani, A. (2020). A state-of-the-art review and meta-analysis on sustainable supply chain management: Future research directions. *Journal of Cleaner Production*, 123357.

Luo, N., Olsen, T. L., & Liu, Y. (2021). A Conceptual Framework to Analyze Food Loss and Waste within Food Supply Chains: An Operations Management Perspective. *Sustainability*, *13*(2), 927. doi:10.3390u13020927

Luthra, S., Kumar, A., Zavadskas, E. K., Mangla, S. K., & Garza-Reyes, J. A. (2020). Industry 4.0 as an enabler of sustainability diffusion in supply chain: An analysis of influential strength of drivers in an emerging economy. *International Journal of Production Research*, *58*(5), 1505–1521. doi:10.1080/00207543.2019.1660828

Makhal, A., Robertson, K., Thyne, M., & Mirosa, M. (2020). Normalising the "ugly" to reduce food waste: Exploring the socialisations that form appearance preferences for fresh fruits and vegetables. *Journal of Consumer Behaviour*, 1–15.

Martindale, W., & Schiebel, W. (2017). The impact of food preservation on food waste. *British Food Journal, 119*(12), 2510–2518. doi:10.1108/BFJ-02-2017-0114 PMID:29720739

Mbow, C., Rosenzweig, C. E., Barioni, L. G., Benton, T. G., Herrero, M., . . . Diouf, A. A. (2020). Food security. In *Climate Change and Land: an IPCC special report on climate change, desertification, land degradation, sustainable land management, food security, and greenhouse gas fluxes in terrestrial ecosystems*. Geneva: Intergovernmental Panel on Climate Change (IPCC). Retrieved from https://www.ipcc.ch/site/assets/uploads/sites/4/2019/11/08_Chapter-5.pdf

Mele, C., & Russo-Spena, T. (2015). Eco-innovation practices. *Journal of Organizational Change Management, 28*(1), 4–25. doi:10.1108/JOCM-08-2013-0146

Morrar, R., Arman, H., & Mousa, S. (2017). The fourth industrial revolution (Industry 4.0): A social innovation perspective. *Technology Innovation Management Review, 7*(11), 12–20. doi:10.22215/timreview/1117

Nahavandi, S. (2019). Industry 5.0—A human-centric solution. *Sustainability, 11*(16), 4371. doi:10.3390u11164371

Närvänen, E., Mesiranta, N., Sutinen, U. M., & Mattila, M. (2018). Creativity, aesthetics and ethics of food waste in social media campaigns. *Journal of Cleaner Production, 195*, 102–110. doi:10.1016/j.jclepro.2018.05.202

Nguyen, T., Li, Z., Spiegler, V., Ieromonachou, P., & Lin, Y. (2018). Big data analytics in supply chain management: A state-of-the-art literature review. *Computers & Operations Research, 98*, 254–264. doi:10.1016/j.cor.2017.07.004

Parfitt, J., Barthel, M., & Macnaughton, S. (2010). Food waste within food supply chains: Quantification and potential for change to 2050. *Philosophical Transactions of the Royal Society of London. Series B, Biological Sciences, 365*(1554), 3065–3081. doi:10.1098/rstb.2010.0126 PMID:20713403

Rahimi, I., Gandomi, A. H., Fong, S. J., & Ülkü, M. A. (Eds.). (2020). *Big Data Analytics in Supply Chain Management: Theory and Applications*. CRC Press. doi:10.1201/9780367816384

Rohm, H., Oostindjer, M., Aschemann-Witzel, J., Symmank, C., & Almli, V., De Hooge, I. E., ... Karantininis, K. (2017). Consumers in a sustainable food supply chain (COSUS): Understanding consumer behavior to encourage food waste reduction. *Foods, 6*(12), 104. doi:10.3390/foods6120104 PMID:29186883

Rosencrance, L. (2017). *Blockchain technology will help the world go green*. Bitcoin Magazine.

Saberi, S., Kouhizadeh, M., Sarkis, J., & Shen, L. (2019). Blockchain technology and its relationships to sustainable supply chain management. *International Journal of Production Research, 57*(7), 2117–2135. doi:10.1080/00207543.2018.1533261

Sakaguchi, L., Pak, N., & Potts, M. D. (2018). Tackling the issue of food waste in restaurants: Options for measurement method, reduction and behavioral change. *Journal of Cleaner Production, 180*, 430–436. doi:10.1016/j.jclepro.2017.12.136

Semios. (2021). *Water management.* Retrieved from https://semios.com/solutions/water-management/

Seuring, S., & Müller, M. (2008). From a literature review to a conceptual framework for sustainable supply chain management. *Journal of Cleaner Production, 16*(15), 1699–1710. doi:10.1016/j.jclepro.2008.04.020

Teller, C., Holweg, C., Reiner, G., & Kotzab, H. (2018). Retail store operations and food waste. *Journal of Cleaner Production, 185*, 981–997. doi:10.1016/j.jclepro.2018.02.280

Ülkü, M. A., & Engau, A. (2021). Sustainable Supply Chain Analytics. In W. Leal Filho, A. M. Azul, L. Brandli, A. Lange Salvia, & T. Wall (Eds.), *Industry, Innovation and Infrastructure. Encyclopedia of the UN Sustainable Development Goals*. Springer. doi:10.1007/978-3-319-71059-4_117-1

Ülkü, M. A., Skinner, D. M., & Yıldırım, G. (forthcoming). Towards Sustainability: A Review of Analytical Models for Circular Supply Chains. In Circular Economy Supply Chains - From Chains to Systems. Emerald.

United Nations. (2021a). *Take Action for the Sustainable Development Goals*. Retrieved from https://www.un.org/sustainabledevelopment/poverty/hunger/

United Nations. (2021b). *United Nations Goal 12: Ensure Sustainable Consumption and Production Patterns*. Retrieved from https://www.un.org/sustainabledevelopment/sustainable-consumption-production

University of South Australia. (2021, March 15). Standard digital camera and AI to monitor soil moisture for affordable smart irrigation. *ScienceDaily*. Retrieved from www.sciencedaily.com/releases/2021/03/210315110210.htm

WFPUSA. (2021). World Food Program USA-*Food Waste vs. Food Loss: Know the Difference and Help #StopTheWasteToday*. Retrieved from https://www.wfpusa.org/articles/food-loss-vs-food-waste-primer/

Wilson, N. L., Miao, R., & Weis, C. S. (2019). When in Doubt, Throw It Out! The Complicated Decision to Consume (or Waste) Food by Date Labels. *Choices, 34*(1), 1–7.

WRAP. (2021). *Consumer Attitudes to Food Waste and Food Packaging*. Retrieved from https://www.wrap.org.uk/sites/files/wrap/Report%20%20Consumer%20attitudes%20to%20food%20waste%20and%20packaging_0.pdf

Yue, C., Alfnes, F., & Jensen, H.H. (2009). Discounting Spotted Apples: Investigating Consumers' Willingness to Accept Cosmetic Damage in an Organic Product. *Journal of Agricultural and Applied Economics, 41*, 29-46.

Zhao, X., Fan, H., Zhu, H., Fu, Z., & Fu, H. (2015, April). The design of the internet of things solution for food supply chain. In *2015 International Conference on Education, Management, Information and Medicine*. Atlantis Press. 10.2991/emim-15.2015.61

ADDITIONAL READING

Astill, J., Dara, R. A., Campbell, M., Farber, J. M., Fraser, E. D., Sharif, S., & Yada, R. Y. (2019). Transparency in food supply chains: A review of enabling technology solutions. *Trends in Food Science & Technology*, *91*, 240–247. doi:10.1016/j.tifs.2019.07.024

Bouchery, Y., Corbett, C. J., Fransoo, J. C., & Tan, T. (Eds.). (2016). *Sustainable supply chains: A research-based textbook on operations and strategy* (Vol. 4). Springer.

Coderoni, S., & Perito, M. A. (2020). Sustainable consumption in the circular economy. An analysis of consumers' purchase intentions for waste-to-value food. *Journal of Cleaner Production*, *252*, 119870. doi:10.1016/j.jclepro.2019.119870

Grabs, J., & Carodenuto, S. L. (2021). Traders as sustainability governance actors in global food supply chains: A research agenda. *Business Strategy and the Environment*, *30*(2), 1314–1332. doi:10.1002/bse.2686

Hák, T., Janoušková, S., & Moldan, B. (2016). Sustainable Development Goals: A need for relevant indicators. *Ecological Indicators*, *60*, 565–573. doi:10.1016/j.ecolind.2015.08.003

Jaeger-Erben, M., Jensen, C., Hofmann, F., & Zwiers, J. (2021). There is no sustainable circular economy without a circular society. *Resources, Conservation and Recycling*, *168*, 105476. doi:10.1016/j.resconrec.2021.105476

Saberi, S., Kouhizadeh, M., Sarkis, J., & Shen, L. (2019). Blockchain technology and its relationships to sustainable supply chain management. *International Journal of Production Research*, *57*(7), 2117–2135. doi:10.1080/00207543.2018.1533261

Smith, B. G. (2008). Developing sustainable food supply chains. *Philosophical Transactions of the Royal Society of London. Series B, Biological Sciences*, *363*(1492), 849–861. doi:10.1098/rstb.2007.2187 PMID:17766237

Vinuesa, R., Azizpour, H., Leite, I., Balaam, M., Dignum, V., Domisch, S., ... Nerini, F. F. (2020). The role of artificial intelligence in achieving the Sustainable Development Goals. *Nature Communications*, *11*(1), 1–10. doi:10.103841467-019-14108-y PMID:31932590

Wang, C., Ghadimi, P., Lim, M. K., & Tseng, M. L. (2019). A literature review of sustainable consumption and production: A comparative analysis in developed and developing economies. *Journal of Cleaner Production*, *206*, 741–754. doi:10.1016/j.jclepro.2018.09.172

KEY TERMS AND DEFINITIONS

Agriculture: Farming; the science, art, and business of cultivating soil and producing crops, and raising livestock.

Big Data Analytics: All the data processing and analysis techniques to examine large amounts of data in order to provide insights by uncovering hidden data structures (correlations, patterns) and to inform business decisions and public policies.

Consumer Perception: Consumers' awareness of, impressions and opinions about a product, brand, and business.

Intelligent Technology: Application of scientific knowledge performing decision-making functions that would formerly have required human intervention, e.g., artificial intelligence.

Quadruple Bottom Line: The four pillars of sustainability in measuring business performance: culture (purpose), economic (profit), environment (planet), society (people).

Serialized Global Trade Item Number: Fourteen-digit number to identify/categorize a tradable item such as apples.

Supply Chain: The set of all parties and activities involved in fulfilling a customer request.

Sustainable Consumption and Production: Doing more and better with less.

Sustainable Development: Meeting the needs of today without sacrificing those of the future.

Chapter 7
Technological Application to Managing a Municipal Urban Garden

Fernando Almeida
https://orcid.org/0000-0002-6758-4843
Polytechnic Institute of Gaya, Portugal

Nuno Miranda
Polytechnic Institute of Gaya, Portugal

Bruno Vieira
Polytechnic Institute of Gaya, Portugal

ABSTRACT

The implementation of urban gardens, which are increasingly appearing in cities, aims to respond actively to the growing demand for urban spaces for the installation of urban gardens, creating conditions for the practice of sustainable agriculture in an urban context. Through these initiatives, it is intended to ensure that the needs of the population are met and to maximize the benefits arising from the practice of urban agriculture, both for the environment and for people's quality of life. Technology is a facilitating element in the process of acquiring and maintaining these urban gardens. This chapter presents an app that can be used by farmers to manage the production of consumer goods in this space, providing information about the status of crops, products to be grown, and types of required maintenance. This app simplifies the production process and also increases the sustainability of agriculture activities considering the economic, social, and logistical dimensions.

INTRODUCTION

The progressive global migration of the rural population to large urban centers has contributed to the emergence of environmental, social, and economic problems (Østby, 2016). With this migratory flow to

DOI: 10.4018/978-1-7998-8900-7.ch007

large cities, populations seek to improve their quality of life without considering the possible irreversible damage to the environment. A key concern in the planning and management of cities is to ensure sustainable development. Sachs and Ki-moon (2015) advocate that sustainable development seeks to respond to the ecological limits of the planet, which are not infinite, and it is also necessary to ensure the existence of natural resources for future generations.

Sustainable cities are those that align their living, production, and consumption patterns based on a combination of economic and socio-environmental aspects. As Cohen (2017) argues, instead of promoting disorderly growth and consumption, they adopt public policies and actions that positively impact sustainability. This can be realized in different areas of intervention in cities such as mobility, education, energy, and environment (Saad et al., 2017; Tafidis et al., 2017; Trindade et al., 2017). It becomes unequivocal that sustainability is a purpose for all humanities since consumption habits are driving natural resources to depletion, besides destroying flora and fauna species.

The rapid growth of cities has led to the massification and industrialization of production processes. This situation has led to the loss of food quality, which leads people to increasingly value the food from organic and healthy agriculture (Kearney, 2010). As a result of this process, urban agriculture presents progressively as an important alternative for feeding the population that allows ensuring sustainable development in urban space (Chumbler et al., 2016). Several differences can be found between urban agriculture and agriculture in rural areas, of which location stands out as an evident differentiating element between these two concepts, and a few improvements, where the objective of food production may be considered the most common element. Despite the differences between these types of agriculture, they can be seen as complementary. Rural agriculture provides food in large quantities, while urban agriculture exploits a few niche markets that are often overlooked in rural agriculture (Hamlin et al., 2016).

Agriculture has been enhancing the city with its multi-functionalities, which goes beyond food production and benefits other elements of the urban environment, such as services, green areas, buildings, leisure spaces, among others. Several authors have debated the benefits resulting from this activity. Heather (2012) mentions that at the environmental level, urban agriculture promotes the development of green spaces, the recovery of degraded areas, and the reduction of pollution. At the economic level, urban gardens contribute to the reduction of probity, because they enable the harvesting of food, of good quality and for personal consumption (Krikser et al., 2019).

Agriculture has gone through several phases of evolution since the introduction of mechanization, genetics and, more recently, information technology. Fourth revolution technologies (i.e., agriculture 4.0) have led to emerging technologies such as sensors, remote connection, statistics, and artificial intelligence being adopted in agriculture to improve product performance and to increase the visibility and efficiency of the entire value chain (Braun et al., 2018; Rose and Chilvers, 2018). Smartphones and tablets promise to revolutionize agriculture due to their wide range of functionalities and features. The adoption of these technologies turns possible to attract ordinary citizens to agricultural activity and there has been a progressive demand for the use of this equipment by new farmers and agricultural technicians through the use of applications that facilitate daily work on their farms. In this sense, this chapter presents the development process of a technological application for mobile devices that enables the management of a municipal urban garden. The developed application uses exclusively open-source software and provides users with advice on products to be produced, when they should be produced, production times and harvesting processes, which are fundamental for users who are not familiar with the production processes in agriculture. Throughout this work, the technological options and the added value that this project offers to the community will be discussed, especially for people living in cities, but who wish to experience urban agriculture. Furthermore, the challenges posed to the processes of economic, social. and environmental sustainability by the adoption of this new technological solution are addressed.

BACKGROUND

The Information Technology

Over the past decades, the provision of telecommunications services around the world has undergone significant changes. There have been changes in the institutional structure that served as the basis for the provision of telecommunications services, especially after the privatization of these sectors that were previously in the possession of the state. However, these transformations are not solely about the institutional field. As Hui (2012) and Stone (2015) point out, this change in the telecommunications sector has gone together with technology. In the first phase, digital technology, and later the application of fiber optics, promoted a revolution in the capacity and speed of information transmission through the system using data, voice and image.

The term information technology came to be adopted in place of informatics. Simultaneously, concepts related to data processing and information systems emerged as telecommunications became the basis of computing, where database managers became available and provided data communication services, through application software among other services. According to De Jong (2016), computing is understood as a means to produce, transmit, store, adhere and use various information, which can be used in various contexts.

Information technology has become a term increasingly used in the use of equipment, applications, services and basic technology falling into the categories of computers, telecommunications and multimedia data. For Neuman (2010), information technology has evolved a lot with the rapid development of technology, and with this development comes solutions made available by computers. The trend is that information technology is important in society, where the process of digitization of various contents has become a norm of life and survival. Shackelford (2020) adds that we are currently living in the information age and the potential of new technologies is undeniable.

The opportunities that exist in the information technology field can be seen through the innovations made by the business sector. Information technology is an important element that can have positive impacts on the business world and enables an improvement in the strategic and functional context, enabling the company to improve its planning process and interact with its stakeholders in a more productive way (Ommen et al., 2016; Samuel, 2014). The use of information technology has been on the rise in all sectors, presenting optimism and enabling scientific advances in the fields of knowledge. Antonucci et al. (2017) state that information technology makes it easier to incorporate new habits in people and build new social relationships.

The Use of Information Technology in Family Farming

Information is essential for the development of urban and rural enterprises. Prosperity, progress, and development of a country will depend on its ability to acquire, produce, access and use the relevant information for its development (Momaya, 2019). Similarly, access to information is also crucial to enable people to know their social rights and benefits and sources of support to overcome social exclusion (Maldonado et al., 2006). Based on these considerations, the need to make available efficient and accessible applications for the management of small rural establishments emerges. Therefore, the adoption of technology in different situations provides a scenario full of opportunities for building less asymmetric environments, because technology can contribute to the generation of new knowledge.

Goulet (2020) highlights the importance of technologies for family farm management, identifying them as a fundamental part of the knowledge and techniques of a rural enterprise or as a way to achieve successful entrepreneurship. The use of new technologies for family farmers enables conditions for them to explore new opportunities and practices that require a more sophisticated level of production management. In addition to the ease of searching, accessing, storing, and disseminating information, these technologies will increasingly serve to exchange information among farmers. With the help of these tools, Kalita (2017) points out that family farmers can meet their needs namely through access to food distribution channels.

The use of information and communication technologies allows greater control of the processes that occur in rural management, opening new ways to obtain greater and better production. At this level, Urquhart et al. (2008) consider that information and communication technologies are key instruments in the fight against poverty because if used correctly they provide capacity for the populations of developing countries and for disadvantaged societies to overcome barriers to development and thus know how to deal with social problems.

Several authors look at the benefits that information and communication technologies can provide in rural areas. Yekinni et al. (2019) refer to the potential for transforming often scattered information into knowledge that can be exploited by farmers. Mishra (2008) looks at the potential of technologies as a vehicle for information that can contribute to social change. Finally, Qing-tao et al. (2013) and Rosemary et al. (2014) highlight the role of technology in reducing losses in agriculture, particularly in irrigation systems and pest control.

However, several challenges associated with family farming arise. Infrastructure is a major problem in rural areas since the lack of connectivity and Internet access does not allow these communities to access content (Ruiz-Martínez & Esparcia, 2020). In most countries, broadband is mainly accessed in large urban centers and does not reach rural areas. Therefore, it becomes relevant that rural areas are not seen as a distant and backward place, but as places where diversity manifests itself and in which technology can be used to improve the living conditions of its population.

Another challenge that restricts the adoption of information and communication technologies in rural management is the low educational qualification of rural producers (Stelmach, 2011). Education is related not only to the ability to obtain and process information but also to the use of management techniques. Mastering certain practices require more expertise than others. According to Mwangi & Kariuki (2015), the educational level and experience of producers are personal characteristics that determine decisions about the adoption of these practices. These factors directly affect the adoption of new technologies. Increasingly, the viability and effectiveness of new technologies require a management process that is not based on pure mastery of traditional farming and breeding knowledge and practices. For this reason, Ra et al. (2019) highlight that human capital is a relevant factor in explaining technology adoption by farmers.

The realization of management practices aligned with the adoption of technologies can provide the farmer with important data and information for the managerial decision-making process and greater efficiency in the productive and financial process (Adolph et al., 2020). However, the development of management techniques that take into account the particularities of family farming has not been assumed as a priority. Most applications are not intuitive to use and is too complex in the face of the specific needs of family farming. They typically request greater technical skills to be used.

The Diffusion of Innovation

The strategic innovation process goes much further than the development of new technologies, products, and services. It also involves the creation of new business models, new ways to meet consumer needs, new organizational processes, and new ways to compete and cooperate in the business environment (Meissner, 2019; Steiber & Alänge, 2015). The importance of innovation is perceived as essential by organizations to survive in an increasingly competitive and globalized landscape.

Innovation diffusion can be seen as the process by which innovations spread, through market channels from the first introduction to different consumers, countries, markets, and firms. Peres et al. (2010) look at diffusion as the process where new ideas are invented, spread, and adopted, producing social consequences and changes. This conception meets the principle that without diffusion, an innovation has no economic impact.

The adoption of an innovation is part of a complex decision-making process. Throughout this process, the organization goes through several phases where it acquires initial knowledge about the innovation, goes on to implement a new business idea, until it confirms its decision to implement. From a problem or need, innovation is recommended as a possible solution. Therefore, investments in innovation are justified when there is a need to obtain more satisfactory results (Hall et al., 2016).

In the innovation decision-making process, it is necessary to deal with the uncertainty inherent in the decision to adopt a new alternative instead of using previously existing and consolidated ideas. Brophey et al. (2013) note that it is the aspects of perceived novelty in an innovation and the associated uncertainty that differentiates innovation decision-making from other types of decision-making. The decision about an innovation is not an instantaneous or impulsive action. Rather, as Fu et al. (2020) point out, it is a process that occurs over time and consists of a series of interconnected actions.

Innovation diffusion occurs within a social system. The social structure of the system affects the diffusion of innovation in various ways as recognized in Troshani & Doolin (2007). The role of opinion leaders and change agents, innovation decisions, and the consequences of innovation emerge in the diffusion process. All these points involve the relationship between the social system and the diffusion process that occurs within it.

Several categories of adopters of an innovation emerge. This definition takes into consideration the time it takes an individual to adopt the innovation and consequently several categories of individuals emerge such as innovators, early adopters, early majority, late majority, and latecomers (Goffin & Mitchell, 2017). Innovators are individuals with an adventurous nature who wish to try new things and accept the risk involved in new experiences; early adopters are opinion-forming elements; the early majority are mostly followers; the late majority are typically suspicious and skeptical towards innovations and do not wish to expose themselves to any risk; while in the latecomers are found individuals who place little value on innovations and place greater importance on other factors such as price.

MAIN FOCUS OF THE CHAPTER

This chapter's main objective is to present and describe the process of implementing a technological solution in the urban agriculture field. One of the ways that urban agriculture has become popular in large cities is through municipal gardens. These cultivation spaces aim to provide municipalities with a community space that allows for a strong ecological, social, and economic connection between residents and a sustainable agricultural activity. Urban gardens enable the practice of different agricultural activities, allowing the cultivation of healthy food according to seasonality, adding quality to urban daily life, and savings to the household economy (Diekmann et al., 2020). Horticultural users can produce vegetables for self-consumption or recreation, install basic support structures and perimeter fences on their plot, and enjoy the compost resulting from the composting process.

The technological solution was to build an app for Android devices that allows the management of a family vegetable garden. To identify the critical success factors, a literature review was conducted about the determining factors in the adoption of an agriculture app. A key element highlighted by Rose and Chilvers (2018) is the visibility of harvests and increased efficiency of the entire value chain. Costopoulou et al. (2016) complement the notion of visibility by focusing on the relevance of business and financial data. Access to information on agricultural services is another aspect considered fundamental, which contributes to the massification in the use of these applications (World Bank, 2012). Finally, some authors also highlight the importance of considering aspects related to the application use process such as customization (Lantzos et al., 2013), and the impact of the application such as increased marketing efficiency (Abishek et al., 2016; Belakeri et al., 2017) and reduced transport costs (Belakeri et al., 2017). Finally, we also looked at the app's potential to improve innovation processes whether at the level of incremental, radical, or disruptive innovation as recognized by Tian et al. (2020).

After the identification of the critical success factors, research questions were defined that will serve as a reference to evaluate the impact of the app in the management of an urban garden. The following research questions were defined:

- RQ1. Does the app provide visibility into crops and promote increased efficiency throughout the value chain?
- RQ2. Does the app provide information on business and financial data?
- RQ3. Does the app provide information on agricultural extension services?
- RQ4. Can the app be customized by users?
- RQ5. Does the app turn agricultural marketing more efficient?
- RQ6. Does the app contribute to the reduction of transport costs?
- RQ7. What is the app's potential in promoting innovation management processes?

SOLUTIONS AND RECOMMENDATIONS

Functional Requirements

Functional requirements include the needs, features or functionalities expected in the process of developing a software solution. Consequently, a functional requirement expresses an action that must be performed through the system. Sommerville (2015) points out that a functional requirement allows expressing the behavior of the system, that is, the process or transformation that software or hardware components perform on the inputs to generate the outputs. Within the scope of this application, the following functional requirements were defined:

- FR1: Authentication and user registration in the app;
- FR2: Search for urban gardens for planting. The user can consult the area of each plantation and the associated rental price. The rental of the land can be done every six months or annually;
- FR3: Planting vision, through which it is possible to analyze the condition of crops and access information from sensors on temperature and humidity. Furthermore, the app gives suggestions about the suitable time to grow specific crops and recommends new crops that can be cultivated in each period. These suggestions are released considering the season;
- FR4: Consultation of statistics that allows analyzing the production history (e.g., average growth time, average water spent, etc.);
- FR5: Access to the production guide, which allows less experienced users to know the products of each season, the type of seed, the care to be taken in monitoring production, etc.

Figure 1 presents the UML use cases. The app allows the user to search for farmland to rent and proceed with planting, as well as keep better control over the state of his plantation and have access to statistics from the sensors present on the land. It also provides access to planting guides provided by an agronomist.

Figure 1. UML use cases
Source: authors

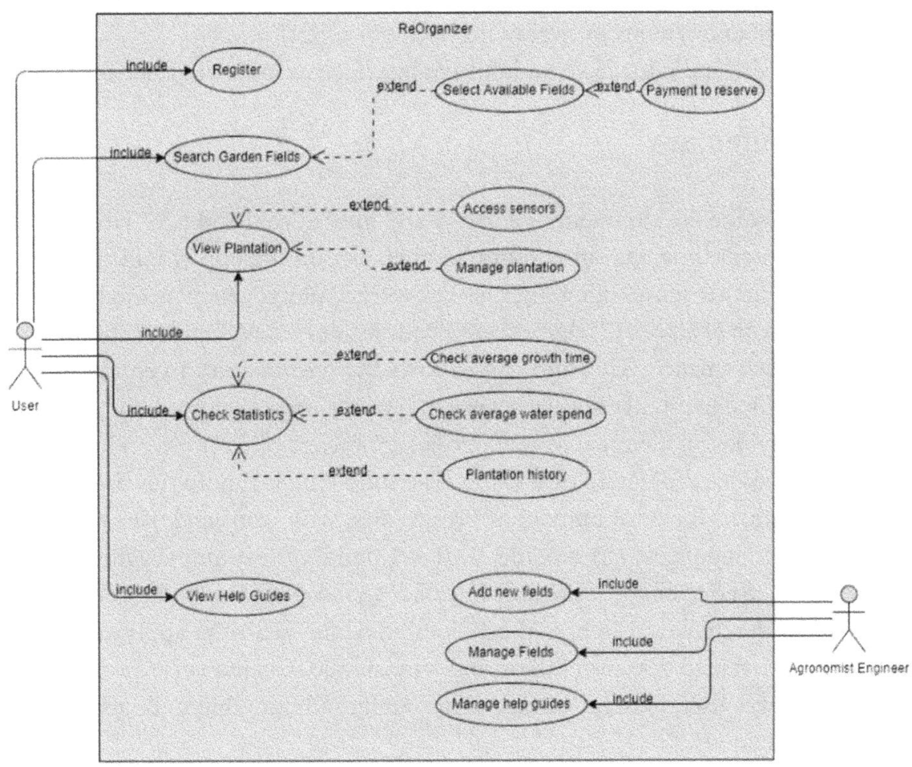

The system also allows interaction with the administrator of the application, who in this case must be an agronomist engineer appointed by the City Council. This role in the app has the mission of writing the production guide and can add and manage urban gardens.

Non-Functional Requirements

In addition to functional requirements, it is also relevant to consider non-functional requirements. These requirements, although not directly related to the functionality of a system, support the operational environment in which the functional requirements are executed. Sommerville (2015) emphasizes that the non-functional requirements have a relevant role during the development of a system, acting as criteria for the selection and composition of software architecture, among the various design alternatives. In this project the following non-functional requirements were incorporated:

- NFR1: Backup – data should be backed up on a daily basis so that there is no loss of information by users;
- NFR2: Performance – the efficiency of the application should be a relevant factor. In this sense, the execution of queries and necessary calculations should use the least amount of resources;
- NFR3: Interoperability – the app should allow connection and communication with the database and with the temperature and humidity sensors. Furthermore, interoperability has been considered at the level of integration with external applications that enable the app to provide post-harvest facilities as well (e.g., transportation services after harvest, oil-extraction for mint cultivation, sugar production after sugarcane harvest, etc.);
- NFR4: Privacy – access to the app should be protected by user authentication and all functionalities offered by the app requires that this process be carried out. In this regard, it should not be possible to carry out anonymous browsing in the app;
- NFR5: Usability – the app must have an intuitive interface to make it easier for the user to use it.

Technological Architecture

Initially, it was considered the development of a native app for Android and iOS. However, this solution would increase development time and, in addition, any change in the app would imply replication by both operating systems. In this sense, other alternatives were explored, such as the hybrid development model. According to Huynh et al. (2017), hybrid applications allow development for multiple platforms without major loss of performance. This approach allows the programmer to create a core of the program in a native way, which explicitly allows access to devices on the smartphone, and another multi-platform part, which provides an identical user experience for users of Android and iOS devices. Figure 2 presents the UML deployment diagram of the app. The following technologies have been adopted: (i) Ionic - framework that allows the development of Progressive Web Apps and uses HTML5, CSS3, and JavaScript programming languages; (ii) NodeJS - allows building fast and scalable network applications and was essentially used to connect the user interface to the database. Node.js uses a non-blocking event-driven I/O model that makes it light and efficient, ideal for real-time applications with intensive data exchange across distributed devices (Tilkov and Vinoski, 2010); and (iii) MariaDB – open-source relational database that was used for data storage. This database is suitable to be used in the context of non-intensive data applications.

Figure 2. UML deployment diagram
Source: authors

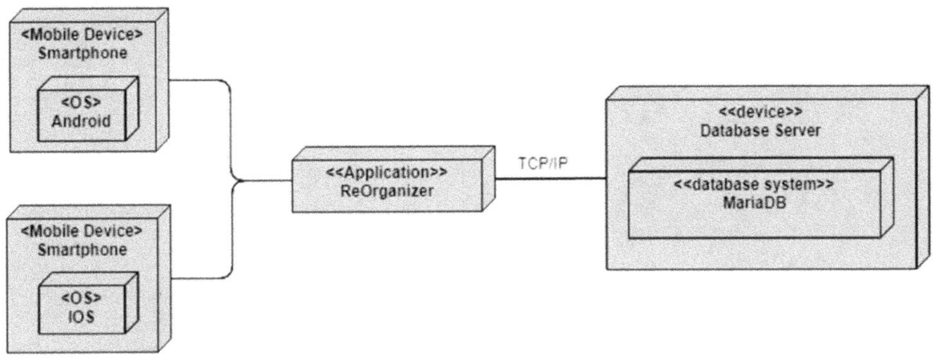

Prototype

A basic and essential operation offered by the system is the user's registration and log-in (Figure 3). When registering, the user must enter his e-mail address and password. An email is sent to the user to confirm the validity of his account. According to Glaser (2014), this procedure is essential in applications aimed at a wide audience to avoid fake email addresses and potential spam messages.

Figure 3. User's registration and log-in
Source: authors

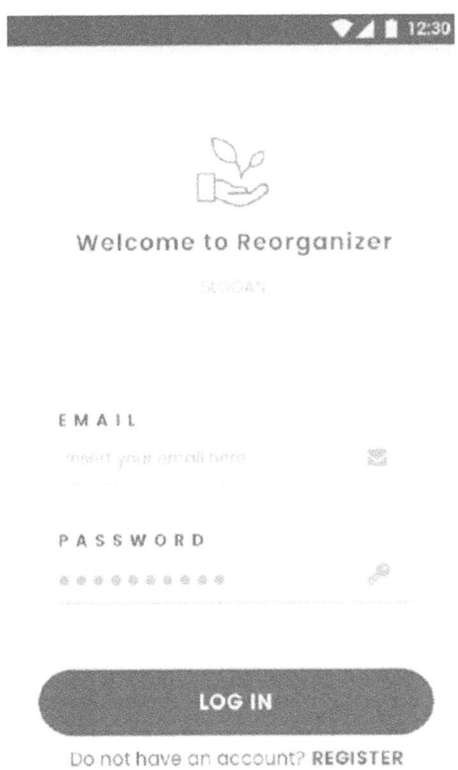

Figure 4. Menu layout
Source: authors

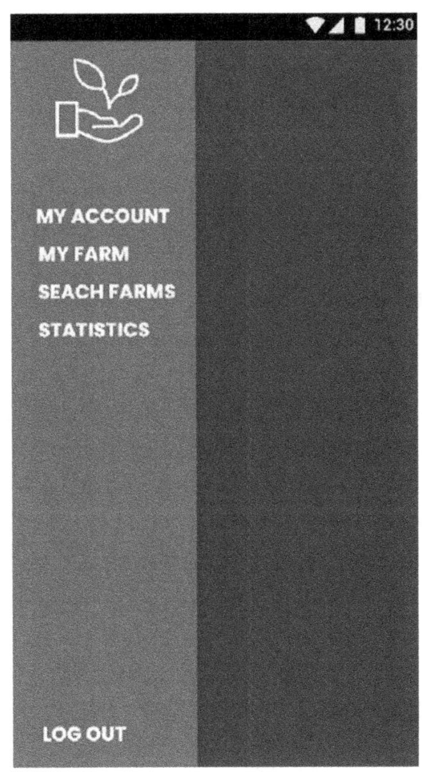

Figure 5. Acquisition of a new platform
Source: authors

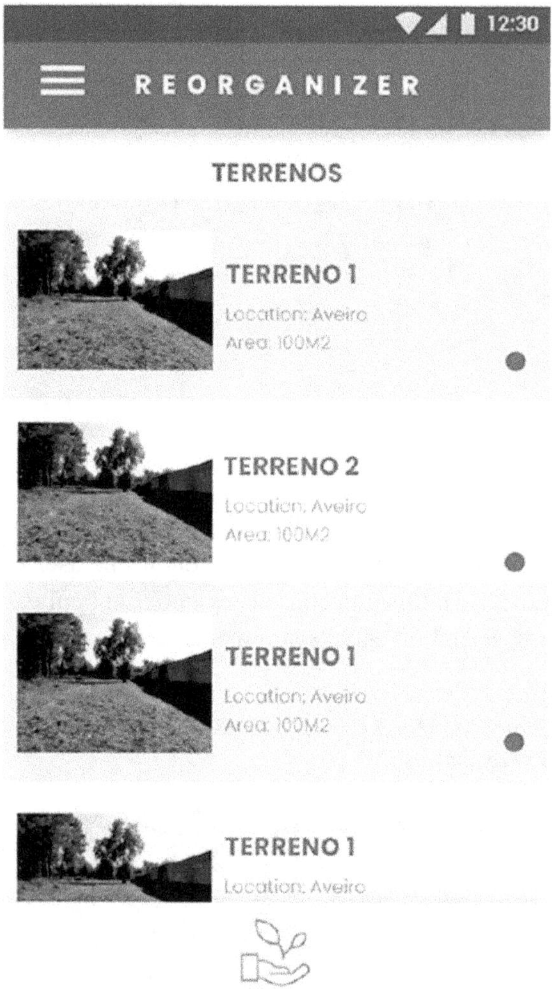

Figure 6. Harvest process
Source: authors

After that, the user can access the main panel of the app (Figure 4) which consists of a navigable side menu that offers four main operations: (i) my account; (ii) my farm; (iii) search farms; and (iv) statistics. At the bottom of the window is the log-out button.

One of the key operations is the acquisition of a new plantation. The plantations are managed by the administrator of the app. Each plantation has information about its area, products that can be planted and costs associated with the land as depicted in Figure 5. After the acquisition of a plantation, it becomes possible to include products in that plantation until the whole area is planted. The user can have a plantation consisting of one or several products. This decision is conditioned by the area and typology of the land.

Another fundamental operation offered by the plantation is the gathering of a harvest as it is presented in Figure 6. In the first phase, it is necessary to select the plantation and the harvest. Subsequently, it is checked whether the harvests are ready to be collected. After that, an update of the plantation status is made.

DISCUSSION

The analysis and discussion of the results provided by this study were based on a focus group. This is a qualitative research technique, derived from group interviews, which collects information through interactions between individuals. The focus group differs from the individual interview in that it is based on the interaction between people to obtain the necessary data for the research. It is a method that is based simultaneously on communication and interaction (Krueger and Casey, 2014). The main objective is to gather detailed information on a specific topic and in which a specific group of participants is pre-selected. Queirós et al. (2017) state this method seeks to collect information that can provide an understanding of perceptions and attitudes about a product or service.

In this study, focus groups were used to explore and measure users' reactions to the app. This approach enables immediate ideas to be gathered to improve the app. Additionally, it can helps to identify and validate customer's requirements. In the focus group, user reaction to the requirements (i.e. functional and non-functional) implemented in the app was examined. Two group interviews were considered in this process: (i) the first interview lasted 45 minutes and aimed to explain the context of the app and its operation; and (ii) the second interview lasted 90 minutes and allowed to collect evidence and opinions

Table 1. Themes identified by the thematic analysis

Requirement	Themes
FR1	Panel 1: easy registration but difficulties in logging-in Panel 2: easy registration and intuitive login Panel 3: complex registration but intuitive login
FR2	Panel 1: no proximity indication Panel 2: search for rental period Panel 3: complex navigation
FR3	Panel 1: alert dispatch Panel 2: low resolution of cams, more sensors, and alert dispatch Panel 3: deeper focus and alert dispatch
FR4	Panel 1: high relevance Panel 2: personalized statistics Panel 3: high relevance
FR5	Panel 1: usefulness of functionality Panel 2: usefulness of functionality, and complement with videos Panel 3: not relevant
NFR1	Panel 1: easy to use Panel 2: easy to use Panel 3: difficulties in accessing the restore
NFR2	Not applied (N/A)
NFR3	Panel 1: compatibility issues Panel 2: compatibility issues Panel 3: compatibility issues
NFR4	Panel 1: available Panel 2: available Panel 3: available
NFR5	Panel 1: common standards to other apps Panel 2: common standards to other apps Panel 3: larger buttons

Source: authors

on the relevance of the app. Between the two focus group sessions, there was a 30-day time frame in which a trial account was associated with each user.

The focus group was formed by a panel of 6 users. This panel was made up of a diverse group of individuals considering their age group and level of qualifications. Three groups were formed: (i) Panel 1: 2 individuals with a taste for urban agriculture but no specific knowledge in this area; (ii) Panel 2: 2 individuals under 30 years of age with a specific background in agronomy; and (iii) Panel 3: 2 individuals over 60 years of age with no specific educational background in agriculture but more than 30 years of professional experience in the field. This diverse and complementary panel allows you to evaluate the impact of the various functional and non-functional requirements offered by the app.

Most of the considerations gathered through the focus group were common among the various panels, but there are some exceptions that it is relevant to explore. In the registration and login process, panel 1 indicated some difficulties in the login because they are used to using their Gmail and Facebook accounts; while panel 3 considered that excessive information was requested in the registration of the app. To mitigate this situation, and even keeping the same registration model, only a few fields were selected as mandatory fields.

The search for urban gardens was the greatest difficulty felt by users. The indication of available places on the map is relevant, but not sufficiently enough to ensure that users find the best urban garden for them. Navigation, as indicated in panel 3, may become too complex. This has been corrected by two changes: (i) indication of the user's proximity to each urban garden considering its GPS location; and (ii) customized search considering the rental period. This navigation on a map considering the proximity of the user is advocated by Ishikawa et al. (2008) as a good practice in mobile navigation. Additionally, alerts were included in the app to capture user focus when an important event occurs (e.g., approaching the harvest period, adverse weather conditions). Access to the production guide was a feature that did not deserve full agreement. Nevertheless, the authors decided to keep it relevant for less experienced users as indicated by Panel 1.

In the exploration of the non-functional requirements, there was full agreement on the behavior between panels 1 and 2. Only in Panel 3, some changes should be noted. This group experienced some difficulties in the process of restoring the app which was resolved with additional support and also by the inclusion of larger buttons for a touch interface. Finally, compatibility problems were encountered due to the difficulty of using XML data in some sensors. A suggestion to solve this problem is given by Almeida et al. (2011) in which the adoption of open standards is suggested. Open standards should be understood as absolutely essential in an increasingly globalized and interconnected world. For interoperability to happen, it is necessary that everyone agrees on how this interoperability will occur. This means that the more standardized the interoperability mechanisms are, the less effort will be required to create interfaces for interoperability.

RQ1. Does the app provide visibility into crops and promote increased efficiency throughout the value chain?

A key factor of precision agriculture as recognized in Miles (2019) and Njoroge et al. (2018) is access to information. Technology plays a crucial role in obtaining and analyzing data from multiple sources. For this purpose, sensors that can obtain real-time data on an Internet of Things network should be considered (Karim et al., 2017). Several sensors can be used, such as height, temperature, humidity, and volumetric sensors.

The app enables real-time monitoring of the status of each harvest. The "my farm" functionality enables the user to see which products have been planted on his land, the weather report, and the current state of his plantation. Therefore, the user knows through the sensors the development state of the planted products and can make timely decisions about anticipating the harvesting process. Currently, the system only includes interaction with temperature and humidity sensors. In future versions, it is planned to include access to cameras that provides a real-time view of the farm and complement it with the introduction of other sensors (e.g., height, volumetrics) that allow having more precise information about the evolution of some products).

RQ2. Does the app provide information on business and financial data?

Another feature offered by the app is access to statistics. Mishra and Homa (2019) highlight the role that statistical analysis of agricultural data can play in crop planning and productivity. In this sense, access to this information becomes vital for users. The app considers several statistics, such as: average temperatures and precipitation, production areas used, average growth time, average water spent, and profitability of the land. The analysis of these statistics is important for the user to understand the behavior of his production. It is also possible to access the history of the plantations.

RQ3. Does the app provide information on agricultural extension services?

Information technologies have provided significant advances in conventional agricultural practices. Rose and Bruce (2018) point out that the applications help to support decision-making based on the data collected. Decision support systems in agriculture have mainly supported precision farming, which will increase productivity levels and reduce negative environmental impacts. Therefore, agricultural decision support systems aim to analyze the impact of productivity considering various elements such as human resources, financial, energy, climate, water, among others (Rupnik et al., 2019).

Given the high degree of technology and volume of data required to operate a support decision system, these systems tend to be used primarily by experienced and advanced users. However, in this study, the app targets a different audience, consisting mainly of users with little or no experience in agriculture and who wish to embark on the adventure of creating an urban garden for the first time. In this situation, a decision support system should fundamentally guide users to identify the best products considering soil and climate characteristics.

RQ4. Can the app be customized by users?

Lantzos et al. (2013) consider that an app to support the management of a farm should be customized according to farmer crop and process needs. Therefore, the processes defined in the app should consider the specificities of each farm. The app allows choosing the land for the plantations. The layout of each plot can be quite different depending on the total area, soil conditions, and products that can be grown there. According to the selection of the land, the quantity and type of products that can be grown there are different. This is an essential aspect of customizing the app.

Furthermore, the app provides other types of customizations. One of them is the recommendation/service system that is available and that is pertinent in giving useful indications about the best products to be planted, the irrigation process, etc. However, not all users, especially those with more experience in agriculture will need this help, so this feature can be disabled by the user. Additionally, access to cameras and sensors is another feature that can be customized by the user.

RQ5. Does the app turn agricultural marketing more efficient?

Web and mobile technologies have brought new opportunities to traditional market sectors such as agriculture. In fact, despite the growing adoption of emerging technologies in agriculture, in some cases, their impact is restricted to agricultural production. This view is confirmed by Abishek et al. (2016) when highlighting that many agricultural producers do not use marketing to create value for their customers and, consequently, for their business. However, Mirzaei et al. (2016) stress that it is essential that a farmer is not exclusively focused on production. Therefore, it is equally important that there is an understanding of the end customer and segments the market according to the company's value proposition.

Information technology can play an important role in improving market knowledge and increasing marketing efficiency. Although the app does not include product sales functionalities, there are elements provided by the app that allows the farmer to have real-time knowledge of the production process and his crops. This information is of vital importance for the market segmentation process, allowing more efficient management between supply and demand. In the future, the app will be integrated with an e-commerce tool (e.g., Opencart, Magento) that will allow the farm's products to be placed directly online for sale.

RQ6. Does the app contribute to the reduction of transport costs?

The use of smartphones in agriculture contributes to having more accurate and timely market information (Belakeri et al., 2017). This information is also important to reduce the frequency and costs of transport (Belakeri et al., 2017). By adopting production sensors, production surveillance cameras, real-time information on productivity, the state of each plantation, harvest dates, etc., it is possible to minimize the number of trips to the urban garden. This situation has contributed both to the reduction of transport costs and sustainability of the environment.

RQ7. What is the app's potential in promoting innovation management processes?

The app assumes a disruptive potential of creating a new market in the urban agriculture field different from what is offered by traditional production models without the support of technology. According to Nagy et al. (2016), a disruptive solution is usually simpler and cheaper than what already exists in the market. Another characteristic of this app is that it aims to focus on a target audience with little knowledge of agriculture, but who want to take their first steps in this field. This growing concern about organic farming is a factor that can enhance the adoption of this app.

Another key factor highlighted by Kumaraswamy et al. (2018) about disruptive innovation is that the product created needs to offer obvious improvements to the user. This can be achieved in different ways. For example, having a product that is simpler to use, requiring less time and less work from the user to perform a certain task; or alternatively, a product that is more complex and does not generate a great deal of time savings, but proves to be more comfortable or use far fewer resources. This app opts for the first strategy by assuming itself as a product with a very low user learning curve. Additionally, it contributes to a more sustainable agriculture since the use of technology is a relevant element to reduce waste and make more informed and complete production choices.

FUTURE RESEARCH DIRECTIONS

The development of information and communication technologies has brought rapid changes in society. The various stakeholders in this area need to be very careful not to become outdated and disconnected. The impact of new technologies is not only in cities but also in rural areas. In fact, cities can also offer small rural spaces that are important for the quality of life of their populations. One area of research seeks to look at this phenomenon of urban and municipal gardens and explore their impacts on the quality of life of citizens (Audate et al., 2018).

Another research field is looking at the role of technology in creating the concept of Agriculture 4.0 (Arvanitis & Symeonaki, 2020; Klerkx et al., 2019). Slowly this is an area that has taken up space in farming and animal agriculture. With this, the expectation is that technicians and producers will be better prepared for a sustainable production and a demanding market. The application of Agriculture

4.0 requires investment in people, processes, and technologies. People, so they know how to deal with the new demands and challenges that this revolution requires processes since Agriculture 4.0 changes the way we deal with agribusiness data and information; and technologies, since it requires the use of hardware and software that enable its application, such as sensors, mobile devices, among others.

Agriculture 4.0 is based on digital content through the processing and analysis of the vast amounts of data being produced in all areas that contribute to agricultural development, such as biotechnology, climate change, and agricultural sciences. Knowledge that will be applied in all links of the production chain, from pre-production, through production, to the post-production phase. In agriculture, big data is often seen as a combination of technology and analytics that can collect and compile new data and process it in a more useful and timely manner to aid decision-making (Li & Niu, 2020; Spanaki et al., 2021). Data mining is the computational process of discovering patterns in large data sets, involving methods at the junction of artificial intelligence, machine learning statistics, and database systems. The real-time information provided by big data analytics can help optimize performance analysis to show how farmers are using their data and what adaptations are needed to account for emerging weather events, or disease/pest outbreaks. This potential offered by big data can be leveraged by both large-scale farmers and small-scale urban gardeners.

CONCLUSION

Urban gardens arise in the context of government action at the local level (e.g., municipalities) that cities should develop strategies that enhance urban sustainability in different dimensions. Therefore, although urban agriculture is not the complete and definitive solution to urban problems, it can become an essential part of strategic measures to improve the quality of life in cities. These gardens are mainly for the disadvantaged population, although it is not exclusive to this social group. The production carried out is mainly subsistence, since the plots to be cultivated are small and the space is shared by several citizens.

Technology can play a significant role in the management of these spaces. Many of the owners of these urban agriculture spaces have no prior knowledge of agriculture and often have other professional occupations as well. Therefore, participation in urban agriculture is a part-time activity that arises for personal and health reasons, as they want to produce and consume healthy food from organic farming.

This chapter presents a technological solution for the management of an urban vegetable garden based on the construction of an app. This app enables the search for urban gardens for planting in the geographic area of each user. The app enables the user to follow the production process and access information about the growing process of the plantations based on temperature and humidity sensors. The app gives access to historical information that allows the user to analyze its performance. Equally relevant is the presentation of suggestions about the species to be planted, the most suitable time of year to plant them, and the care to be taken in their maintenance. Finally, the user has access to a guide on organic agriculture, which is particularly relevant for users less familiar with agriculture.

REFERENCES

Abishek, A., Bharathwaj, M., & Bhagyalaksgmi, L. (2016). Agriculture marketing using web and mobile based technologies. *Proceedings of the 2016 IEEE Technological Innovations in ICT for Agriculture and Rural Development (TIAR)*, 41-44. 10.1109/TIAR.2016.7801211

Adolph, B., Allen, M., Beyuo, E., Banuoku, D., Barrett, S., Bourgou, T., Bwanausi, N., Dakyaga, F., Derbile, E. K., Gubbels, P., Hié, B., Kachamba, C., Naazie, G. K., Niber, E. B., Nyirengo, I., Tampulu, S. F., & Zongo, A. F. (2020). Supporting smallholders' decision making: Managing trade-offs and synergies for sustainable agricultural intensification. *International Journal of Agricultural Sustainability*, 1–18. doi:10.1080/14735903.2020.1786947

Almeida, F., Oliveira, J., & Cruz, J. (2011). Open Standards and Open Source: Enabling Interoperability. *International Journal of Software Engineering and Its Applications*, *2*(1), 1–11.

Antonucci, T. C., Ajrouch, K. J., & Manalei, J. A. (2017). Social Relations and Technology: Continuity, Context, and Change. *Innovation in Aging*, *1*(3), 1–9. doi:10.1093/geroni/igx029 PMID:29795794

Arvanitis, K. G., & Symeonaki, E. G. (2020). Agriculture 4.0: The Role of Innovative Smart Technologies Towards Sustainable Farm Management. *The Open Agriculture Journal*, *14*(1), 130–135. doi:10.2174/1874331502014010130

Audate, P., Fernandez, M., Cloutier, G., & Lebel, A. (2018). Impacts of Urban Agriculture on the Determinants of Health: Scoping Review Protocol. *JMIR Research Protocols*, *7*(3), e89. doi:10.2196/resprot.9427 PMID:29588270

Belakeri, P., Prasad, C., Bajantri, S., Mahantesh, M., Maruthi, S., & Rudresh, G. (2017). Trends of Mobile Applications in Farming. *International Journal of Current Microbiology and Applied Sciences*, *6*(7), 2499–2512. doi:10.20546/ijcmas.2017.607.295

Braun, A. T., Colangelo, E., & Steckel, T. (2018). Farming in the Era of Industrie 4.0. *Procedia CIRP*, *72*, 979–984. doi:10.1016/j.procir.2018.03.176

Brophey, G., Baregheh, A., & Hemsworth, D. (2013). Innovation Process, Decision-Making, Perceived Risks And Metrics: A Dynamics Test. *International Journal of Innovation Management*, *17*(3), 1–22. doi:10.1142/S1363919613400148

Chumbler, M., Negro, S., & Bechler, L. (2016). *Urban Agriculture: Policy, Law, Strategy, and Implementation*. American Bar Association.

Cohen, S. (2017). *The Sustainable City*. Columbia University Press. doi:10.7312/cohe18204

Costopoulou, C., Ntaliani, M., & Karetsos, S. (2016). Studying Mobile Apps for Agriculture. *IOSR Journal of Mobile Computing and Application*, *3*(6), 44–49.

De Jong, K. (2016). *Evolutionary Computation: A Unified Approach*. Bradford Books.

Diekmann, L., Gray, L., & Baker, G. (2020). Growing 'good food': Urban gardens, culturally acceptable produce and food security. *Renewable Agriculture and Food Systems*, *35*(2), 169–181. doi:10.1017/S1742170518000388

Fu, J. S., Shumate, M., & Contractor, N. (2020). Organizational and Individual Innovation Decisions in an Interorganizational System: Social Influence and Decision-Making Authority. *Journal of Communication*, *70*(4), 497–521. doi:10.1093/joc/jqaa018

Goffin, K., & Mitchell, R. (2017). *Innovation Management: Effective strategy and implementation*. Red Globe Press. doi:10.1057/978-1-137-37344-1

Goulet, F. (2020). Family Farming and The Emergence of an Alternative Sociotechnical Imaginary in Argentina. *Science, Technology & Society*, *25*(1), 86–105. doi:10.1177/0971721819889920

Hall, B. H., Moncada-Paternó-Castello, P., Montresor, S., & Vezzani, A. (2016). Financing constraints, R&D investments and innovative performances: New empirical evidence at the firm level for Europe. *Economics of Innovation and New Technology*, *25*(3), 183–196. doi:10.1080/10438599.2015.1076194

Hamlin, R., Knight, J., & Cuthbert, R. (2016). Niche marketing and farm diversification processes: Insights from New Zealand and Canada. *Renewable Agriculture and Food Systems*, *31*(1), 86–98. doi:10.1017/S1742170514000489

Heather, K. L. (2012). *The Environmental Benefits of Urban Agriculture on Unused, Impermeable and Semi-Permeable Spaces in Major Cities With a Focus on Philadelphia*. Retrieved from https://repository.upenn.edu/mes_capstones/46

Hui, M. (2012). The Research on Development Trend of Telecommunications Industry. *IERI Procedia*, *3*, 232–236. doi:10.1016/j.ieri.2012.09.038

Huynh, M., Ghimire, P., & Truong, D. (2017). Hybrid app approach: Could it mark the end of native app domination? *Issues in Informing Science and Information Technology Education*, *14*, 49–65. doi:10.28945/3723

Ishikawa, T., Fujiwara, H., Imai, O., & Okabe, A. (2008). Wayfinding with a GPS-based mobile navigation system: A comparison with maps and direct experience. *Journal of Environmental Psychology*, *28*(1), 74–82. doi:10.1016/j.jenvp.2007.09.002

Kalita, B. (2017). Marketing Efficiency, Price Spread, Share of Farmers In Case Of Horticultural Markets of Assam. *International Journal of Advance Research and Development*, *2*(8), 65–72.

Karim, F., Karim, F., & Frihida, A. (2017). Monitoring system using web of things in precision agriculture. *Procedia Computer Science*, *110*, 402–409. doi:10.1016/j.procs.2017.06.083

Kearney, J. (2010). Food consumption trends and drivers. *Philosophical Transactions*, *365*(1554), 2793–2807. doi:10.1098/rstb.2010.0149 PMID:20713385

Klerkx, L., Jakku, E., & Labarthe, P. (2019). A review of social science on digital agriculture, smart farming and agriculture 4.0: New contributions and a future research agenda. *NJAS Wageningen Journal of Life Sciences*, *90-91*, 1–16. doi:10.1016/j.njas.2019.100315

Krikser, T., Zasada, I., & Piorr, A. (2019). Socio-Economic Viability of Urban Agriculture – A Comparative Analysis of Success Factors in Germany. *Sustainability*, *11*(7), 1–12. doi:10.3390u11071999

Krueger, R. A., & Casey, M. A. (2014). *Focus Groups: A Practical Guide for Applied Research*. SAGE Publications.

Kumaraswamy, A., Garud, R., & Ansari, S. (2018). Perspectives on Disruptive Innovations. *Journal of Management Studies*, *55*(7), 1025–1042. doi:10.1111/joms.12399

Lantzos, T., Koykoyris, G., & Salampasis, M. (2013). FarmManager: An Android application for the management of small farms. *Procedia Technology*, *8*, 587–592. doi:10.1016/j.protcy.2013.11.084

Li, C., & Niu, B. (2020). Design of smart agriculture based on big data and Internet of things. *International Journal of Distributed Sensor Networks*, *16*(5), 1–11. doi:10.1177/1550147720917065

Maldonado, E. A., Pogrebnyakov, N. A., & van Gorp, A. F. (2006). ICT Policies as a Means to Inhibit Social Exclusion: The South African Case. In E. M. Trauth, D. Howcroft, T. Butler, B. Fitzgerald, & J. I. DeGross (Eds.), *Social Inclusion: Societal and Organizational Implications for Information Systems*. Springer. doi:10.1007/0-387-34588-4_10

Meissner, D. (2019). Public-Private Partnership Models for Science, Technology, and Innovation Cooperation. *Journal of the Knowledge Economy*, *10*(4), 1341–1361. doi:10.100713132-015-0310-3

Miles, C. (2019). The combine will tell the truth: On precision agriculture and algorithmic rationality. *Big Data & Society*, *6*(1), 1–12. doi:10.1177/2053951719849444

Mirzaei, O., Micheels, E., & Boecker, A. (2016). Product and Marketing Innovation in Farm-Based Businesses: The Role of Entrepreneurial Orientation and Market Orientation. *The International Food and Agribusiness Management Review*, *19*(2), 99–130.

Mishra, N. (2008). New agro-technology and its impact on rural institutions in Orissa. *Social Change*, *38*(3), 411–433. doi:10.1177/004908570803800302

Mishra, P., & Homa, F. (2019). *Essentials of Statistics in Agricultural Sciences*. Apple Academic Press. doi:10.1201/9780429425769

Momaya, K. S. (2019). The Past and the Future of Competitiveness Research: A Review in an Emerging Context of Innovation and EMNEs. *International Journal of Global Business and Competitiveness*, *14*(1), 1–10. doi:10.100742943-019-00002-3

Mwangi, M., & Kariuki, S. (2015). Factors Determining Adoption of New Agricultural Technology by Smallholder Farmers in Developing Countries. *Journal of Economics and Sustainable Development*, *6*(5), 208–216.

Nagy, D., Schuessler, J., & Dubinsky, A. (2016). Defining and identifying disruptive innovations. *Industrial Marketing Management*, *57*, 119–126. doi:10.1016/j.indmarman.2015.11.017

Neuman, W. R. (2010). *Media, Technology, and Society: Theories of Media Evolution*. Digit Cult Books. doi:10.3998/dcbooks.8232214.0001.001

Njoroge, B. M., Fei, T. K., & Thiruchelvam, V. (2018). A Research Review of Precision Farming Techniques and Technology. *Journal of Applied Technology and Innovation*, *2*(1), 22–30.

Ommen, N. O., Blut, M., Backhaus, C., & Woisetschläger, D. M. (2016). Toward a better understanding of stakeholder participation in the service innovation process: More than one path to success. *Journal of Business Research*, *69*(7), 2409–2416. doi:10.1016/j.jbusres.2016.01.010

Østby, G. (2016). Rural–urban migration, inequality and urban social disorder: Evidence from African and Asian cities. *Conflict Management and Peace Science*, *33*(5), 491–515. doi:10.1177/0738894215581315

Peres, R., Muller, E., & Mahajan, V. (2010). Innovation diffusion and new product growth models: A critical review and research directions. *International Journal of Research in Marketing*, *27*(2), 91–106. doi:10.1016/j.ijresmar.2009.12.012

Qing-tao, Z., Qing, X., Liu, C., & Geng, S. (2013). Technologies for Efficient Use of Irrigation Water and Energy in China. *Journal of Integrative Agriculture*, *12*(8), 1363–1370. doi:10.1016/S2095-3119(13)60544-4

Queirós, A., Faria, D., & Almeida, F. (2017). Strengths and Limitation of Qualitative and Quantitative Research Methods. *European Journal of Education Studies*, *3*(9), 369–387.

Ra, S., Ahmed, M., & Teng, P. S. (2019). Creating high-tech 'agropreneurs' through education and skills development. *International Journal of Training Research*, *17*(sup1, s1), 41–53. doi:10.1080/14480220.2019.1629736

Rose, D., & Bruce, T. (2018). Finding the right connection: What makes a successful decision support system? *Food and Energy Security*, *7*(1), 1–2. doi:10.1002/fes3.123 PMID:29938109

Rose, D., & Chilvers, J. (2018). Agriculture 4.0: Broadening Responsible Innovation in an Era of Smart Farming. *Frontiers in Sustainable Food Systems*, *2*(87), 1–7. doi:10.3389/fsufs.2018.00087

Rosemary, I., Kraybill, D., & Erbaugh, M. (2014). Impact of integrated pest management technologies on farm revenues of rural households: The case of smallholder Arabica coffee farmers. *African Journal of Agricultural and Resource Economics*, *9*(2), 1–13.

Ruiz-Martínez, I., & Esparcia, J. (2020). Internet Access in Rural Areas: Brake or Stimulus as Post-Covid-19 Opportunity? *Sustainability*, *12*(22), 1–17. doi:10.3390u12229619

Rupnik, R., Kukar, M., Vracar, P., Kosir, D., Pevec, D., & Bosnic, Z. (2019). AgroDSS: A decision support system for agriculture and farming. *Computers and Electronics in Agriculture*, *161*, 260–271. doi:10.1016/j.compag.2018.04.001

Saad, M. M., Ibrahim, M. A., & Sayad, Z. M. (2017). Eco-City as Approach for Sustainable Development. *American Scientific Research Journal for Engineering, Technology, and Sciences*, *28*(1), 54–74.

Sachs, J., & Ki-moon, B. (2015). *The Age of Sustainable Development*. Columbia University Press. doi:10.7312ach17314

Samuel, K. S. (2014). The Role of Technology in Enhancing Corporate Performance. *International Journal of Research in Management & Business Studies*, *1*(1), 116–123.

Shackelford, S. J. (2020). *Governing New Frontiers in the Information Age: Toward Cyber Peace*. Cambridge University Press. doi:10.1017/9781108604000

Sommerville, I. (2015). *Software Engineering*. Pearson Education.

Spanaki, K., Sivarajah, U., Fakhimi, M., Despoudi, S., & Irani, Z. (2021). (in press). Disruptive technologies in agricultural operations: A systematic review of AI-driven AgriTech research. *Annals of Operations Research*. Advance online publication. doi:10.100710479-020-03922-z

Steiber, A., & Alänge, S. (2015). Organizational innovation: A comprehensive model for catalyzing organizational development and change in a rapidly changing world. *Triple Helix (Heidelberg)*, *2*(9), 1–25. doi:10.118640604-015-0021-6

Stelmach, B. L. (2011). A Synthesis of International Rural Education Issues and Responses. *Rural Educator*, *32*(2), 32–42.

Stone, M. (2015). The evolution of the telecommunications industry - What can we learn from it? *Journal of Direct, Data and Digital Marketing Practice*, *16*(3), 157–165. doi:10.1057/dddmp.2014.80

Tafidis, P., Sdoukopoulos, A., & Pitsiava-Latinopoulou, M. (2017). Sustainable urban mobility indicators: Policy versus practice in the case of Greek cities. *Transportation Research Procedia*, *24*, 304–312. doi:10.1016/j.trpro.2017.05.122

Tian, H., Grover, V., Zhao, J., & Jiang, Y. (2020). The differential impact of types of app innovation on customer evaluation. *Information & Management*, *57*(7), 1–17. doi:10.1016/j.im.2020.103358

Tilkov, S., & Vinoski, S. (2010). Node.js: Using JavaScript to Build High-Performance Network Programs. *IEEE Internet Computing*, *14*(6), 80–83. doi:10.1109/MIC.2010.145

Trindade, E. P., Hinnig, M. P., Costa, E., Marques, J. S., Bastos, R. C., & Vigitcanlar, T. (2017). Sustainable development of smart cities: A systematic review of the literature. *Journal of Open Innovation*, *3*(11), 1–14. doi:10.118640852-017-0063-2

Troshani, I., & Doolin, B. (2007). Innovation diffusion: A stakeholder and social network view. *European Journal of Innovation Management*, *10*(2), 176–200. doi:10.1108/14601060710745242

Urquhart, C., Liyanage, S., & Kah, M. (2008). ICTs and Poverty Reduction: A Social Capital and Knowledge Perspective. *Journal of Information Technology*, *24*(3), 203–213. doi:10.1057/palgrave.jit.2000121

World Bank. (2012). *Mobile applications for agriculture and rural development*. Retrieved from https://documents.worldbank.org/curated/en/167301467999716265/Mobile-applications-for-agriculture-and-rural-development/

Yakinni, O. T., Ladigbolu, T. A., Adeniyi, R. T., & Adebisi, G. L. (2019). Benefits Derived from the Use of Information and Communication Technologies among Rural Farmers in Northeast Nigeria. *Journal of Agricultural Extension*, *23*(3), 117–125. doi:10.4314/jae.v23i3.10

ADDITIONAL READING

Abobatta, W. F. (2021). Precision Agriculture: A New Tool for Development. In S. M. Abd El-Kader & B. M. Mohammad El-Basioni (Eds.), *Precision Agriculture Technologies for Food Security and Sustainability* (pp. 23–45). IGI Global. doi:10.4018/978-1-7998-5000-7.ch002

Bacco, M., Barsocchi, P., Ferro, E., Gotta, A., & Ruggeri, M. (2019). The Digitisation of Agriculture: A Survey of Research Activities on Smart Farming. *Array*, *3-4*, 1–11. doi:10.1016/j.array.2019.100009

Bhatnagar, V., & Poonia, R. C. (2019). Sustainable Development in Agriculture: Past and Present Scenario of Indian Agriculture. In R. C. Poonia, X. Gao, L. Raja, S. Sharma, & S. Vyas (Eds.), *Smart Farming Technologies for Sustainable Agricultural Development* (pp. 40–66). IGI Global. doi:10.4018/978-1-5225-5909-2.ch003

El-Gayar, O. F., & Ofori, M. Q. (2021). Disrupting Agriculture: The Status and Prospects for AI and Big Data in Smart Agriculture. In Management Association, I. (Ed.), Research Anthology on Food Waste Reduction and Alternative Diets for Food and Nutrition Security (pp. 771-812). IGI Global.

Goundar, S., & Rayani, P. K. (2021). *Applications of Big Data in Large-and Small-Scale Systems*. IGI Global. doi:10.4018/978-1-7998-6673-2

Gupta, A. K., Goyal, D., Singh, V., & Sharma, H. (2021). *Smart Agricultural Services Using Deep Learning, Big Data, and IoT*. IGI Global. doi:10.4018/978-1-7998-5003-8

Gupta, S., & Godavarti, R. (2020). IoT Data Management Using Cloud Computing and Big Data Technologies. *International Journal of Software Innovation*, *8*(4), 50–58. doi:10.4018/IJSI.2020100104

Jha, K., Doshi, A., Patel, P., & Shah, M. (2019). A comprehensive review on automation in agriculture using artificial intelligence. *Artificial Intelligence in Agriculture*, *2*, 1–12. doi:10.1016/j.aiia.2019.05.004

Kondaveeti, H. K., Priyatham Brahma, G., & Vijaya Sahithi, D. (2021). Deep Learning Applications in Agriculture: The Role of Deep Learning in Smart Agriculture. In P. Tomar & G. Kaur (Eds.), *Artificial Intelligence and IoT-Based Technologies for Sustainable Farming and Smart Agriculture* (pp. 325–345). IGI Global. doi:10.4018/978-1-7998-1722-2.ch020

Lohachab, A. (2020). Bootstrapping Urban Planning: Addressing Big Data Issues in Smart Cities. In R. C. Joshi & B. B. Gupta (Eds.), *Security, Privacy, and Forensics Issues in Big Data* (pp. 217–246). IGI Global. doi:10.4018/978-1-5225-9742-1.ch009

Maya Gopal, P. S., & Chintala, B. R. (2020). Big Data Challenges and Opportunities in Agriculture. *International Journal of Agricultural and Environmental Information Systems*, *11*(1), 48–66. doi:10.4018/IJAEIS.2020010103

Yang, J., Sharma, A., & Kumar, R. (2021). IoT-Based Framework for Smart Agriculture. *International Journal of Agricultural and Environmental Information Systems*, *12*(2), 1–14. doi:10.4018/IJAEIS.20210401.oa1

KEY TERMS AND DEFINITIONS

Agriculture 4.0: Set of state-of-the-art digital technologies integrated and connected by means of software, systems, and equipment capable of optimizing agricultural production, in all its stages.

Focus Group: Group interview-based research method that aims to obtain information and opinions quickly. It can involve different actors and the number of participants is typically small.

Human Capital: The capacity of a person's knowledge, skills, and personality attributes to perform work in order to produce economic value.

Internet of Things: Technology that enables inanimate objects to connect, store, and perform functions of all kinds. The core idea of IoT is to make things smarter and more connected.

Open Standard: Open access standards that allow for implementation and communication without the adoption of royalties or other fees that might inhibit their use.

Optical Fiber: Flexible filaments made of transparent materials such as glass or plastic fibers that are used as a light propagation medium. Optical fibers are generally very and have many applications, data transmission being one of the most common.

Unified Modelling Language (UML): Modeling language for specifying the construction of a software system.

XML: A markup language recommended by the W3C for creating documents with hierarchically organized data, such as text, databases, or vector drawings. The XML language is classified as extensible because it allows defining markup elements.

Chapter 8
A Comprehensive Entrepreneurship Model for the Internationalization of Green Innovation Businesses

José G. Vargas-Hernández
https://orcid.org/0000-0003-0938-4197
University Center for Economic and Managerial Sciences, University of Guadalajara, Mexico

ABSTRACT

The objective of this chapter is to analyze the functioning of an integral model of entrepreneurship in green innovation business (GIB) that is currently emerging and in the process of internationalization. Therefore, this work aims to study the central perspectives of technology that are based on the phenomenon of entrepreneurship and thus develop a strategy that adapts to companies with an ecological basis to achieve internationalization. This chapter analyzes a particular company that specializes in ecological biomineral organic fertilizer, where no chemical product is used to produce the composition; everything that is marketed is made up of a base of organic minerals and other organic compounds.

INTRODUCTION

Currently, there is a new business model which has as the objective to be eco-efficient. Eco-efficiency is defined as the production of products and services at competitive prices that meet human needs and provide quality of life, while the ecological consequences and the use of numerous resources during the life cycle are progressively reduced. level equivalent, at least, to the estimated capacity of the planet (World Business Council for Sustainable Development, 1991).

On the other hand, it is mentioned that eco-efficiency has the purpose of establishing production of manufactured products of high durability, reducing the intensity in the application of energy for the production of goods and services, maximizing the use of raw materials, managing and disposing of hazardous materials and waste in an efficient and environmentally acceptable manner, have management

DOI: 10.4018/978-1-7998-8900-7.ch008

systems and environmental quality, as well as procedures in occupational safety and health, among other provisions, that will bring them financial benefits and competitiveness (Cantú, 2008, page 78).

In both definitions, the authors agree that eco-efficient companies should have as their main objective, to develop quality products at competitive prices, as well as to reduce the environmental impact of producing or offering their products and services. Castro (1998) mentions that eco-efficiency aims to address three relevant aspects that correspond to 1) the total quality, which involves productivity and quality in the company, 2) the preservation of the environment, which is related to sustainable development; 3) occupational health and safety (Castro, 1998).

A COMPRENHENSIVE ENTREPRENEURSHIP MODEL OF GREEN BUSINESS INNOVATION (GBI)

Entrepreneurship, inclusiveness, civil and culture are variables that have complex but straightforward linkages and processes among them and much is still unknown. The last few years have witnessed the generation of extensive theoretical and empirical literature on entrepreneurship and its impact at the levels of regional and firm economic performance (Holmes and Schmitz, 1990; Evans and Leighton, 1989). However, when linking entrepreneurship to the variables of inclusiveness and civil culture, the literatures is inexistent. There is a relative void, despite some recent efforts to study the relationship between entrepreneurship in some other related fields beyond economic growth, which may be attributable to the lack of theoretical frameworks.

However, some studies conducted in economic development have shown reversed causality influencing entrepreneurial activities (Ortega-Anderezand, and Lai, 2017). Entrepreneurial activities tend to be overestimated in regions where there are start-ups playing a relevant role while entrepreneurship activities tend to be underestimated in regions where there is a new entrepreneurial formation and the startups are relatively few (Baptista, Escária, and Madruga, 2017).

The role of entrepreneurship in economic development has received attention from researchers showing the impact of entrepreneurship on employment and social development. Entrepreneurial intentions promote entrepreneurial behaviors (Krueger et al., 2000; Veciana et al., 2005; Souitaris et al., 2007). Entrepreneurship activity can be associated with human necessity providing an opportunity to earn money for living, until better alternative opportunities are found on the labor market (Carree and Thurik, 2010).

The concept of entrepreneurship is multidimensional and related to individual willingness, abilities, and activities on their own, in teams and in organizations to make decisions facing obstacles and uncertainty on the use of institutions and resources to create and develop new opportunities (Wennekers and Thurik, 1999). Entrepreneurial opportunities are the result of unemployment the influences start-up activity, the effect of a thriving economy and experience in entrepreneurial activities (Lin, Manser and Picot, 1998; Pfeiffer and Reize, 2000).

Entrepreneurship is the recognition and exploitation of opportunities leading to the creation and development of a firm (Aragon-Sanchez, Baixauli-Soler, Carrasco-Hernandez, 2017). Moreover, entrepreneurial intentions predict entrepreneurial behaviors, according to the theory of planned behavior (Ajzen, 1991). Entrepreneurial activity creates opportunities to influence economic performance (van Stel, Carree and Thurik, 2005) such as entering the markets with new production processes and products (Acs and Audretsch, 2003). Among the entrepreneurial models there is evidence that the entrepreneurial intentional models (Krueger et al., 2000) supports the theory of planned behavior.

The theory of planned behavior has been used in the analysis of entrepreneurial intentions (Shook et al. 2003). The determinants have significant positive relationships with behavior intention (Armitage and Conner 2001). Do Paço et al. (2011) found a positive and significant influence between entrepreneurial intentions, attitudes, and perceived behavior control. However, subjective norms have an indirect impact on entrepreneurial intentions.

The determinants of entrepreneurial intentions are linked with access to financial, natural, human, and cultural capital and resources through the influence of attitudes, individual subjective norms, perceived social control, and self-efficacy. The attitude is the individual favorable evaluation to start a new opportunity. Individual subjective norms are the perceived social pressure to start a new opportunity and perceived social control and self-efficacy is the perceived ease to start a new opportunity. Based on the Shapero and Sokol 1982) model, attitudes and subjective norms are linked to desirability and perceived behavioral control or self-efficacy with feasibility (Krueger et al., 2000).

METHODOLOGY OF RESEARCH

Background of the Problem and Assumption

Green innovation in urban areas is a neglected issue in terms of urban planning and policing. Still, more neglected is a concern for changes in urban green areas toward the implementation of green innovation initiatives to revitalize the cities, increase the economic growth, improve the social justice and inclusiveness, as well as the improvement of environmentally sustainable development, strengthen the biodiversity and socio-ecosystems. To achieve these goals, it is necessary to implement some actions following the design and implementation of a comprehensive entrepreneurship model.

This new eco-efficient business entrepreneurship model will be analyzed in the context of the integral model, analyzing in this way the tripod of the strategy, which integrates considerations based on industry, resources, and institutions. This paper begins with the assumption that the companies of ecological base present major difficulty at the time of wanting to internationalize, in comparison to the companies of the industrial base.

Instrument and Procedures

This analysis is based on the specific case study of an eco-efficient company using the methodological tools proposed for each one of the theoretical approaches.

For the analysis industry-based considerations are employed in the model of the five forces, also called the diamond model proposed by Porter (1980). The resource and capabilities analysis of the eco-efficient company is employed the model proposed by barney (2001) complemented with the SWOT analysis. Finally, the eco-efficient company is also analyzed in terms of the institutional considerations considered as the "rules of the game", formal and informal, that frame the entrepreneurial activities and the creation and development of companies according to the territorial environment.

Data Analysis: Use of Fertilizers in Mexico

The National Development Plan 2019-2024 establishes among the priority actions for food self-sufficiency and the rescue of the field the Fertilizer Program for the benefit of agricultural producers.

The Mexican Government's National Fertilizer Program aims to address the problem of low availability of national fertilizers at competitive prices for small producers. Includes chemical fertilizers and biofertilizers._This National Fertilizer Plan aims to reduce dependence on the import of these fertilizers. To achieve this, the Cosoleacaque Petrochemical Complex plants and the Pajaritos plant are reactivated to produce ammonia, an input to produce urea, which Mexico imports mainly from Ukraine. In the case of phosphates, in the Pacific, the Lázaro Cárdenas plant is operational and is the largest in Latin America.

The National Bio Fertilizer Program shows incipient progress. The fertilizer production in Mexico estimated for 2019 was 1.85 million tons, reflecting an annual reduction of 2%; while demand continues to rise, with a record estimate of around 5.5 million mt. However, by July 2020, the production volume of nitrogen fertilizers in Mexico almost reaches 48,800 metric tons, which represents a decrease of 32.7% compared to that reported during the same month in 2019. The production volume of phosphate fertilizers in Mexico exceeded 75,600 metric tons, which represents a decrease of 29.1% compared to that reported during the same month in 2019 ((Burgueño Salas, 2020).

Fertilizer Demand in Mexico

A recent analysis of the fertilizer market in Mexico pointed out that the consumption of fertilizers has changed the structure in favor of consumers with the highest concentration and diversification (UACH). This situation has contributed to a consumption drop of fertilizers since the farmer's real income has fallen.

On the change in the consumption pattern (sources), I have no accurate information about it. From experience in the field, there has been a growth in the use of physical fertilizer mixtures where the distributor makes certain formulations, mixing fertilizers, and thus the producer only buys one product. Regarding those of high concentration; it is precise because there has been significant growth in the protected agriculture area. As the strawberries and high-value vegetable area, where the drip irrigation system is used and through it, fertilizers are applied, as a result, highly soluble sources with minimal impurities are usually used, but these are not traditional sources.

However, with more recent data reported by Instituted Trusts concerning Agriculture (FIRA 2020), in the agricultural year, 2019, 71.8% of the sown area was fertilized, representing the seventh consecutive year with increases in the percentage of fertilized area. According to this report, the consumption of fertilizers in Mexico grew 5.8%.

In an interview with a FIRA fertilizer specialist analyst; Gallegos Cedillo (2021), asking about the trend in fertilizer consumption and asked about the demand and supply of fertilizers in Mexico, he warned that the lack of information with hard data does not always support trends in which the market moves. Regarding the consumption of fertilizers, the production information indicates that production has fallen: 15.4% from 2013 (2.06 million tons) to 2020 (1.75 million tons) from January-September. Even though, in 2014 and 2019, it increased compared to the previous year and the same period.

Now, if we consider that fertilized area has increased, as a percentage of the sown area. This proportion has increased from 2013 to 2019, the last reported year, going from 65.3% in 2012 to 71.8% in 2019. However, if we look at the apparent consumption, it increased from 5 million tons in 2013 to 6.7 in 2018, in 2019 (January-December), it does drop to 5.8, and in September 2020, it is 5.2 (Graph in the

Figure 1. Production and consumption of fertilizer in Mexico (Millions of tons)

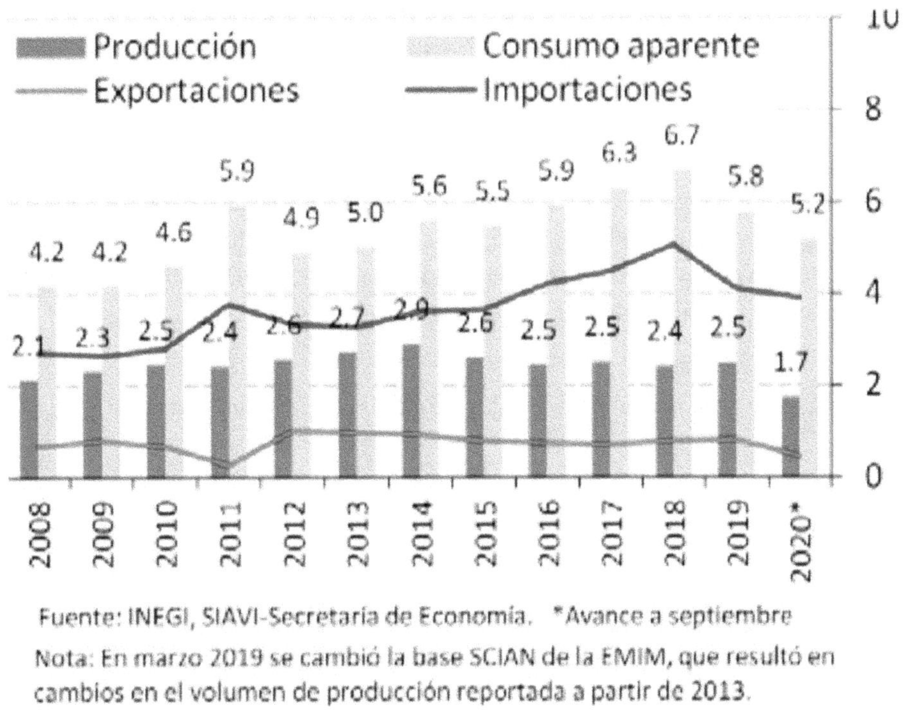

quarterly report). There are no elements to attribute the drop in consumption from 2018 to 2019 to the decrease in producers' income. But if someone affirms this, it is difficult to prove otherwise.

Fertilizer Production in Mexico

The increase in Mexican fertilizer production during the last years is attributed, in part, to the reforms made to the Pemex Law since 2008. These had the objective of promoting the production and productivity of the Mexican fertilizer industry, through the supply of raw materials, such as ammonia, at competitive prices for national manufacturers. In 2018, national production contributed 30.7% of apparent national consumption, which is estimated at 6.2 million dollars and represents an annual increase of 7.9%. The preceding shows the country's high dependence on fertilizer imports.

With data from 2019 the national production of fertilizers is mainly composed of Phosphate fertilizers (diammonium and others) with 55.7%. Nitrogen fertilizers (ammonium sulfate and nitrate and others) with 32.2% Acidic fertilizers (phosphoric, sulfuric, and nitric) with 10.9%. The estimated value of the fertilizer industry in Mexico in 2019 is 13,616.4 million pesos, 2% lower than that registered in 2018.

Fertilizer production processes in Mexico are not integrated, a situation that has an impact on domestic producers, especially urea and nitrate, being at a disadvantage compared to producers of international competition (UACH). According to data from the latest Quarterly Report on Fertilizers from the Directorate of Research and Economic Evaluation of the Instituted Trusts concerning Agriculture (FIRA, 2020), the production of fertilizers in Mexico decreased at an annual rate of 8.4% in the first nine months of the year compared to the same period in 2019.

Regarding the fact that the national producers of urea and ammonium nitrate are at a disadvantage with international competition, as a consequence of the lack of integration of their production processes, if the statement is well understood, especially the reason that is manifested by the lack of integration of their production processes, in general, it would be expected that greater integration greater profitability, but this is not considered to be the main reason for the disadvantage of domestic producers. It can affect but is not the main reason.

More weight would be given to the characteristics of the Mexican farmer, small areas and little specialized, farmer grow various products, as well as the dependence to a large extent on the import of fertilizers that expose him to have fertilizers without due quality control, lack of weight in the containers (49 kg bags instead of 50 for example), exchange rate, etc. Gallegos Cedillo, 2021).

*Figure 2. Acid fertilizer production in Mexico, 2013-2020 * (Thousands of tons)*

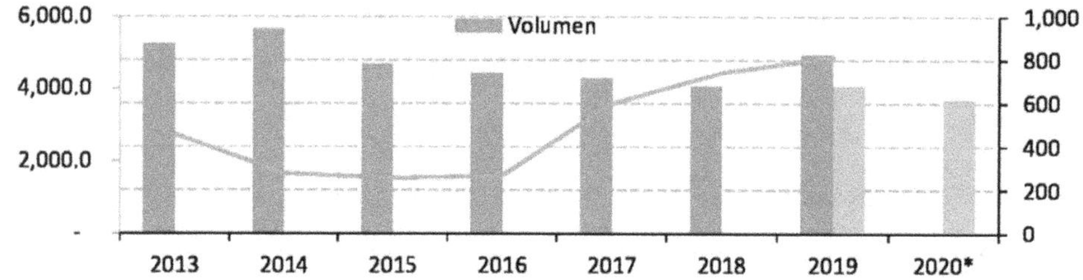

*Figure 3. Fertilizer production in Mexico, 2013-2020 * (Millions of tons)*

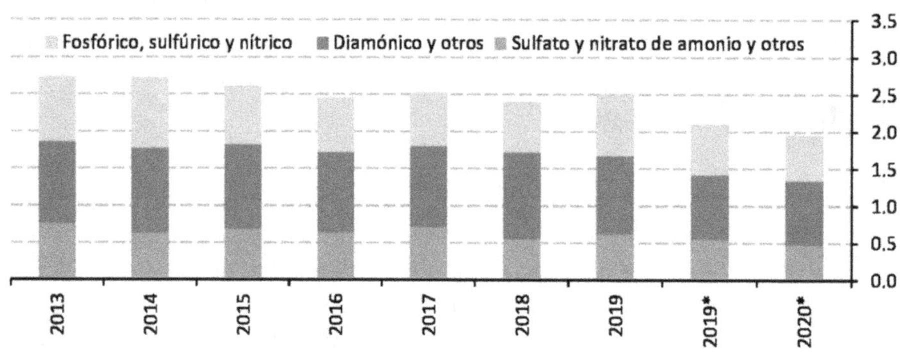

Fertilizer Exports and Imports in Mexico

In this regard, with information from the Ministry of Economy, imports of fertilizers in Mexico, during 2018, registered a volume of 5.06 million dollars and represented an annual increase of 12.6 percent. In said year, imports came from Russia (30.9%), the United States (14.3%), Norway (12.1%), and China (11.1%), mainly. In the last five years, these countries participated with 69.5% of the national fertilizer imports. In the case of exports, these stood at 0.79 million pesos and registered an annual growth of 10.5%. Between 2008 and 2018, consumption and imports grew on average at an annual rate of 6.3 percent.

Imports in 2019 are estimated at around 2.9 million tons, of which an advance of 92% was recorded as of November. Of the total imported, 67.2% corresponded to nitrogenous, 15.5% to phosphate, 12.5% to potash, and others with 4.8%. The commercial value of these imports was 807.8 million dollars, reflecting a reduction of 14.3% to the record of 943.1 million dollars registered in 2018. The main supplying countries were Russia 30.6%, China 16.9%, United States 11.3%, Algeria 7.9%, Malaysia 3.8%, Chile 2.9%, Canada 2.1%, Egypt 1.4%, and others 23.1%.

Between January and September 2020, Mexican fertilizer imports grew 9.9%, while exports decreased 16.6% at an annual rate, totaling 3.9 and 0.46 million dollars, respectively.

*Figure 4. Mexican Fertilizer Exports, 2003-2020 * (Millions of dollars)*

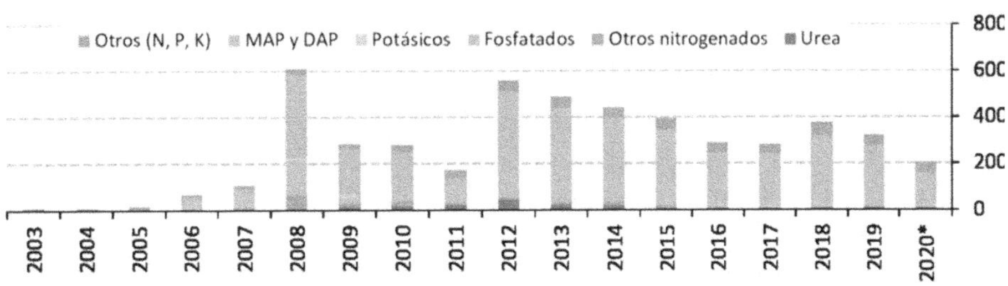

*Figure 5. Mexican fertilizer imports, 2003-2020 * (Millions of dollars)*

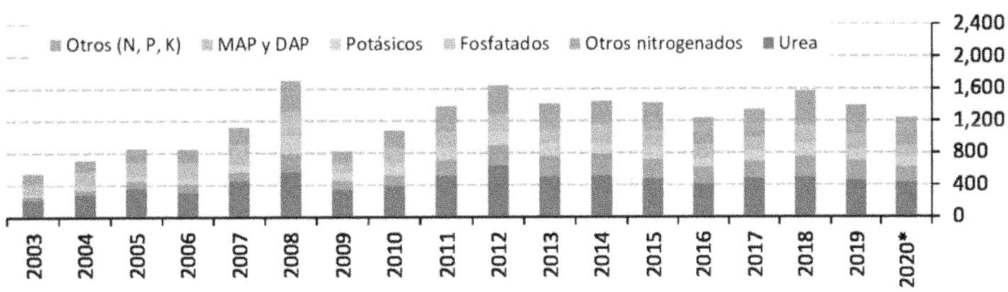

*Figure 6. Balance of the fertilizer trade balance, 2003-2020 * (Millions of dollars)*

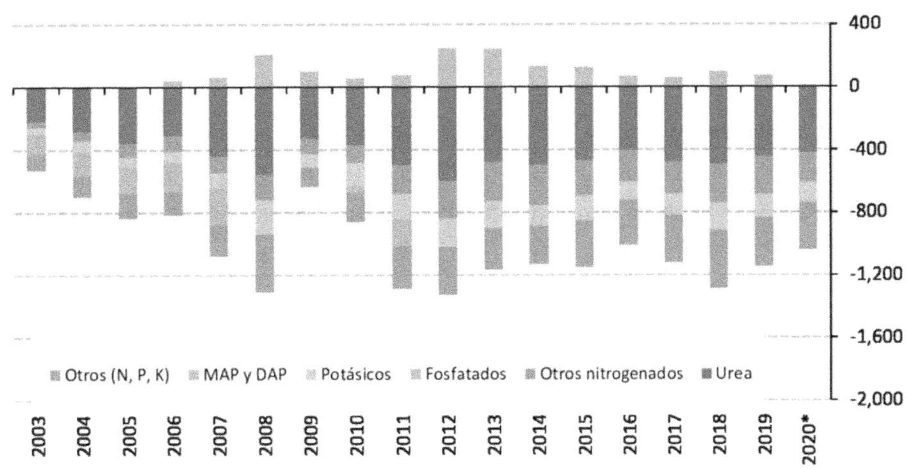

Figure 7. Trade balance (Millions of tons)

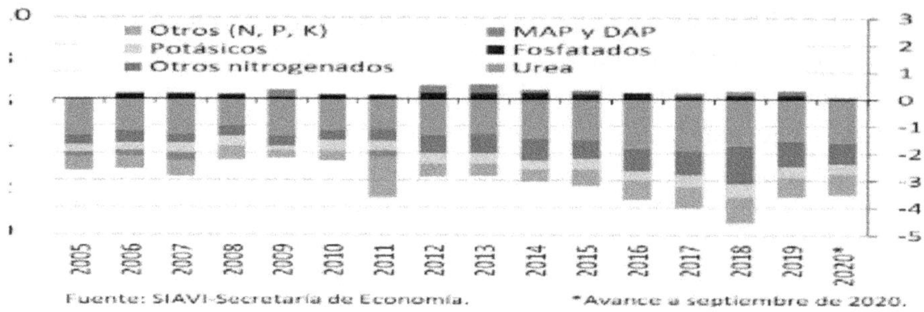

The Fertilizer Markets

The implicit price of fertilizer imports in Mexico in 2018 was $ 322.3 per ton and represented an annual increase of 6.7%. The implied fee of imports from Russia was 283.4 dollars per ton; 387.2 from China and 416.1 for those from the United States, and showed an annual growth of 10.8, 24.2, and 9.7%, respectively. Regarding the average price level of fertilizers in Mexico, since 2014, the behavior has been on the rise. In 2018, a ton of fertilizer in distribution centers was quoted on average, at 10,254 pesos, which represents an increase of 3.5% compared to 2017.

The fertilizers with the highest annual increases in prices were: diammonium phosphate (DAP) (10.4%), Triple 17 (10%), and potassium nitrate (9.4 percent). During January and February 2019, the average price was reduced by 2.0 and 1.3% compared to December 2018, so it would be expected that the average price in 2019 will be at similar levels to those registered during 2018. In the domestic market, urea prices increased 7.3% from 2018 to 2019; while those of diammonium phosphate rose 4.8%.

The average price of fertilizers in the country maintains an upward trend. In November 2020, it registered a growth of 2.7% at an annual rate and 3.3% from December 2019 to date. As of November, the price increases of triple 17 (12.2% annually), ammonia (9.9%), ammonium sulfate (8.3%), and ammonium nitrate (7.9%) stand out. The prices of simple superphosphate and potassium chloride were the ones that showed the highest annual decrease, at rates of 7.1 and 4.6%, respectively.

Bio Organic Fertilizers in México

The use of organic fertilizers in Mexico is not very common, this is because industrial-based companies need chemical products that make plants grow at a faster rate due to market demand. But these fertilizers damage the soil causing it to become unusable after a time for the harvest, as a result, the food absorbs these fertilizers that are harmful to the human being in the long run.

On the other hand, organic fertilizers bring many benefits to the soil along with plants and food. According to a study carried out by SAGARPA, organic fertilizers favorably influence the physical characteristics of the soil (physical fertility); These characteristics are structure porosity, air action, water retention capacity, infiltration, hydraulic conductivity, and stability of aggregates. The following table 1 shows a comparison made by the National Agricultural Survey (ENA) in which there is an increase in the use of chemical fertilizers in Mexico and a decrease in organic fertilizers.

Figure 8. Prices in the national marker (In Mexican pesos per ton)

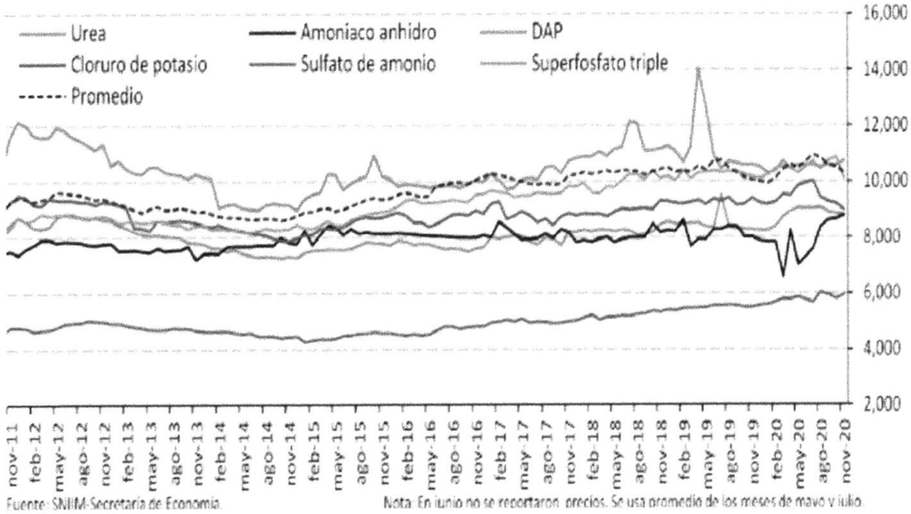

*Figure 9. Production, foreign trade, and apparent consumption of fertilizers in Mexico, 2013-2020 * (Millions of tons)*

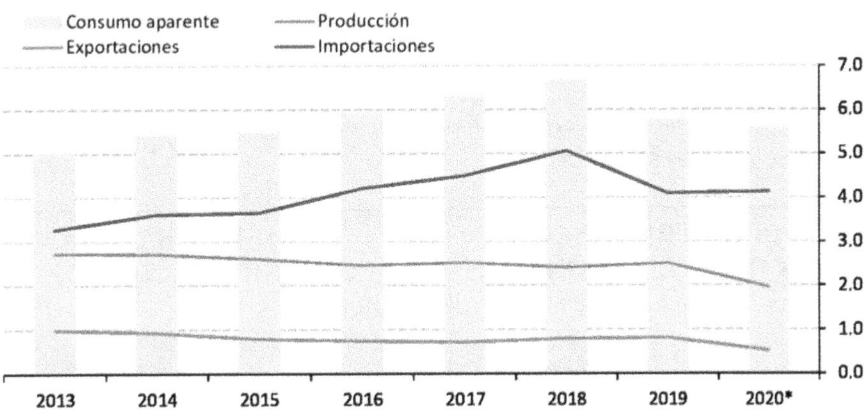

Table 1. Comparison of the use of chemical and organic fertilizers in Mexico

Employed technology	Percentage	
	ENA 2012	ENA 2014
Type of seed	60.9%	82.2%
Creole	29.7%	29.2%
Improved	Na	Na
Certified	Na	Na
Transgenic	Na	Na
Seedling	Na	21.0%
Chemical fertilizers	65.5%	68.8%
Natural fertilizers	40.4%	27.5%
Herbicides	61.7%	62.7%
Insecticides	45.3%	48.2%

THEORETICAL-CONCEPTUAL FRAMEWORK

The theoretical framework applied in this study is based on the analysis of the coefficient company using the main three theories of strategic design and implementation: The industry-based considerations, the resource and capabilities considerations, and the institutions-based considerations, as shown below in figures 10 and 11.

Figure 10. Model of strategic analysis

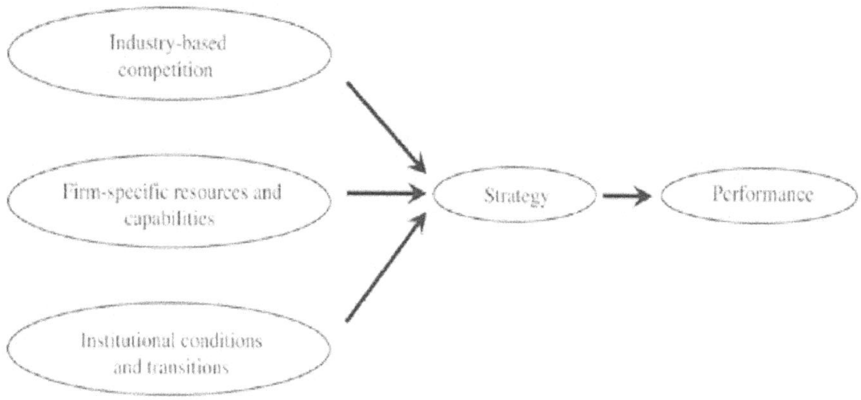

Figure 11. A comprehensive entrepreneurship model for the internationalization of green innovation business (GIB) and eco efficient companies

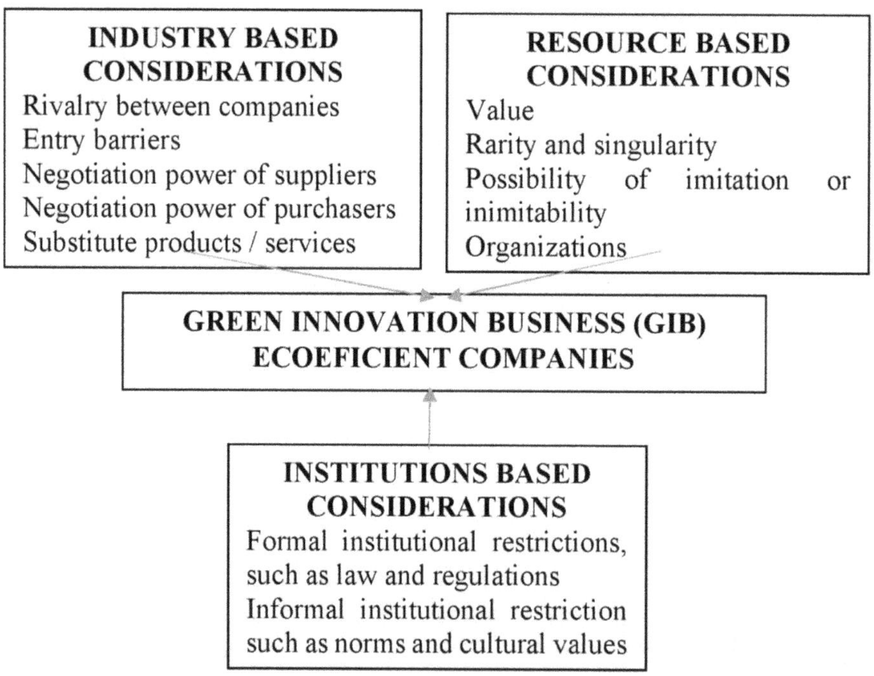

THE GREEN INNOVATION BUSINESS (GIB): BIO-FOM

BIO-FOM is a green innovation business (GIB) also characterized as an eco-efficient company to promote sustainable, profitable, and inclusive development through highly competitive Mexican seeds with fair prices. Using organic mineral bio-fertilizers healthily increased the profitability of the producers. It is in the metropolitan area of Guadalajara (Figure 12).

Figure 12. Localization of BIO-FOM a green innovation business (GIB)

BIO-FOM is the most complete Organic Mineral Bio-fertilizer available on the market, which is made from the interaction of elements: biological, organic, and mineral. It is a mineral organic bio-fertilizer for plant nutrition. The interaction of the BIO-FOM elements forms a Functional Plant Nutrition System; its results reflect: improving seed germination and initiation, healthy and adequate growth, larger roots, greater flowering, and tie, increase in the quality of the fruits, and provide resistance to pests and diseases. BIO-FOM increases soil fertility and contributes to the decontamination and regeneration of the soil.

BIO-FOM bases its functionality on the interaction of the biological, organic, and mineral elements that compose it: A Poly-Functional Consortium of Microorganisms, among them: antagonists to pests and diseases, nitrogen fixers, and; mycorrhizal fungi; that help release the nutrients provided by the BIO-FOM blend itself. The 17 Basic Mineral Elements of plant nutrition (nitrogen, phosphorus, potassium, calcium, magnesium, sulfur, iron, boron, copper, manganese, silicon, zinc, iodine, sodium, cobalt, molybdenum, selenium) Organic Matter, Amino Acids, Acids Humic, and Fulvi.

BIO-FOM contains live microorganisms: 8 bacteria and mycorrhizal fungi that help unblock the chemical components that exist in the soil allowing their absorption, in addition to an important contribution of organic matter, amino acids, humic and fulvic acids, this, together with its content of 16 mineral elements (nitrogen, phosphorus, potassium, magnesium, calcium, iron, boron, copper, manganese, silica, zinc, iodine, sulfur, sodium, cobalt and chlorine) together give rise to a molecular dialogue that results in a functional plant nutrition system, fertile soil, larger root size and helps improve pH, which is essential for proper and healthy plant growth.

Among other benefits, it potentiates the chemical fertilizers that complement the application of BIO FOM as:

a) It recovers the soil fertility, unlocking mineral nutrients present but not available, favoring their absorption due to their bio-organic profile.
b) Increases the specific weight of grains, resulting in higher production.
c) Increases the Br ix degrees of the fruit, having better sales possibilities.
d) Stimulates the immune system, so healthy, vigorous, and productive plants are achieved. Strengthens the root system of the plant, thereby improving the use of water and mineral nutrients available in the soil.
e) Respect the ecosystem, avoiding the contamination of the water table.
f) 100% assimilable, achieving a greater expression of the genetic potential of plants.
g) Retains and conserves moisture in soils.

EMPIRICAL RESULTS AND CONCLUSIONS

Industry-Based Considerations

When analyzing industry-based considerations, the framework of Porter's five forces is used, which considers factors such as rivalry between firms, barriers to entry, the power to negotiate with suppliers and buyers, as well as substitute products.

The Rivalry Between Companies

When talking about the rivalry of companies, it refers to the struggle that companies face every day to obtain a position before consumers. These battles can be for prices, quality in products, advertising battles, etc. On the other hand, the author Huyghebaert mentions that the rivalry between firms generates a direct impact on the likelihood that a business will be successfully undertaken, as well as having an impact on the existence of barriers to entry since fewer firms exist in one industry more complicated will be the entry of new firms (Huyghebaert, 2004).

The company to analyze is dedicated to organic fertilizers made from organic minerals. As it is an innovative product, it does not have much competition, nonetheless, there is competition with the substitute products that would be all chemical fertilizers, due to a great variety of substitute products that are available in the market. It is difficult to compete for price in the same way as being a medium-sized company is not even positioned in the consumer's mind when thinking of organic fertilizer.

Nowadays fertilizers and other chemical products for the treatment of the soil represent 15.7% of the income of the Nurseries and gardens industry in 2019. The industry has experienced a rapid growth in recent years, thanks to the constant increase in disposable income per capita and the consumer segment strengthened.

A decent part of the US market has been occupied by large companies such as Scotts Miracle-Gro Company, Walmart, and Lowe's. Because economies of scale, as well as large corporations can maintain a low cost while offering affordable prices. Price competition among large companies has led to slow growth throughout the sector and low profit margins. Some less profitable companies have been forced to leave the industry. Successful companies are those able to compensate the highest prices by offering a deeper knowledge of the product and superior customer service.

Over the next five years, the nursery and garden store industry are expected to continue to grow, as economic conditions are expected to remain stable. As the disposable income per capita increases, the consumer will be able to invest more time and money in the appearance and operation of their gardens. It is also expected that the industry will benefit from a growing demographic interest in small plants, gardening, and organic horticulture.

Below is a list of companies with which Mary's Poop competes in the United States.

Scotts Miracle-Gro Company (OH, USA)

Founded in 1868, Scotts Miracle-Gro Company is a multinational corporation and industry leader in the lawn and garden market. Its products include the infamous Miracle-Gro and a complete line of plant foods. The most popular Miracle-Gro fertilizers are inorganic, which are much less expensive than organic fertilizers.

PacaProGro (CA, USA)

The PacaProGro farm is in Somis, California, with an estimated 180 alpacas producing alpaca fiber and alpaca manure. Not only sell alpaca fertilizers, but also merchandise made of alpaca fiber, such as hats, gloves, scarves, and stuffed animals.

Alpaca Grow (VT, USA)

Alpaca Grow is a small family business operated in Vermont that began in 2014. Like Mary's Poop, Alpaca Grow sells only alpaca fertilizers.

Isolated wood farm (British Columbia, Canada)

Secluded Wood Farm is headquartered in British Columbia, Canada. They sell alpaca fiber and manure products.

Alpaca Island Company of Martha's Vineyard (MA, USA)

Island Alpaca Company has more than 15 acres and around 50 alpacas. They sell alpaca fiber and manure products, and they are charging higher prices.

Camelot Haven Alpacas (British Columbia, Canada)

Founded in 2002, Camelot Haven Alpacas produces and sells alpaca clothing, accessories, gifts, and alpaca fertilizers.

Entry Barriers

As already mentioned before, BIO-FOM faces competition from large chemical and organic fertilizer companies. Due to this, there are difficulties when it comes to wanting to enter the foreign markets, as the big brands are doing well positioned in the market, and it would be complicated to unseat them or compete for prices so that the company, in this case, could compete for product quality as well as innovation and for the benefits that its product provides.

It is also important to consider the transaction cost and the transportation cost if working with foreign currencies. If the costs become too high, BIO-FOM products would be less competitive in the foreign market. An alternative to export is the franchise, which allows local business units to produce BIO-FOM products for a fee.

One of the competing companies of Mary's poop is already positioned in 10 states of North America, and has large sales volumes, being this way, the products of Mary's poop and other small alpaca farms cannot compete by prices and volume, as already mentioned previously, which makes it be handled as an oligopoly market among large companies. These in turn set entry barriers when playing with prices since, if they see that a new product wants to enter the market, they lower their prices to continue to have an advantage over others or use promotions which makes them more attractive. the purchase of their products to the final consumer.

It is also important to consider the transaction cost and the transportation cost if working with foreign currencies. If the costs become too high, Mary's Poop products would be less competitive in the foreign market. An alternative to export is the franchise, which allows local alpaca farms to produce Mary's Poop products for a fee.

Substitute Products or Services

In the global marketplace and more specifically, North America, large, medium-sized, and small competitors produce similar products. Geographically, most of the production plants are in rural areas and seem to serve only local markets. The competitive advantage of BIO-FOM is that it is the company with the highest social awareness among all its competitors. And although its product is easy to imitate, it gives it a plus by mixing it with some other ingredients that make the compost a better quality. It is also easy to use as they pack the product inside disposable pod-made biodegradable materials that only must be deposited in a container with water waiting for it to dissolve and starts to water in a normal way.

But there is a high range of substitute products that, although they are not organic, the final use is the same (serve as fertilizer to the plants), sharing with substitute products is complicated since the raw material they are made with are very economical and this it makes the prices of the products accessible to all types of public. On the other hand, the organic products that Mary's poop handles are of a slightly higher price, this is due to the handmade process with which they are manufactured.

Bargaining Power with Buyers

According to Porter, at this point, it is defined as the ability of customers to impose prices and conditions of sale (Porter, 1980). This force can be established by customers directly, whether negotiating a discount or financing model, demanding delivery forms, or indirectly which is summarized with competitive purchases.

The bargaining power of buyers may depend on some variables such as a high supply of products and/or services and low demand for them. Another may be that the products offered have no differentiation among themselves, etc. At this point, the customer has the option to choose any product or service that is presented to them and that they consider to be the best and meet their expectations. As well as defining what is the maximum price customers are willing to pay for a product or service, as well as some other requirements that could be delivery times, product quality, etc. All this has an impact on the company's profits.

Mary Forte, the owner of Mary's Poop, manages and operates the company and interacts with customers and other interested parties through social networks. Nowadays, online / digital marketing is the most important marketing tool, especially in social networks. For example, use "Pinterest" to communicate with the company's audience. This social networking platform is popular with many users who share images that can be easily shared. Therefore, she can approach her audience and attract their attention and thus know what customers need and adapt to the needs of the market.

Bargaining Power with Buyers

In one of his writing papers, Peng mentions that when the bargaining power of suppliers becomes too great, business solutions must be found that can reduce it (Peng, 2012, page 127); this is since many options must be available to purchase inputs at reasonable prices.

The organic fertilizer based on organic minerals requires many inputs because of a lot of raw materials are used for its elaboration, so it is necessary to have many suppliers. But if it is necessary that with few suppliers that have contact make and establish the negotiation agreements and delivery times among many other things, in this way can reduce and to a certain extent eliminate transaction costs.

Considerations Based on Resources and Capabilities

The resources are any input in a productive system in which an output is generated. These can be classified as financial, physical, human, technological, organizational, knowledge, management team experience, and customer service, among others. The consideration based on resources groups some indispensable factors for entrepreneurship, some of them are creating value, being unique through the rarity factor among other aspects. For example, in an analysis Peng mentions that it is important to consider the resources of the company, they must be oriented to create value, they must have rarity, they must be difficult to imitate and finally business resources must be immersed in the organization (Peng, 2012).

On the other hand, Barney dogmatizes that the heterogeneity of organizations is due to the possession of resources: i) valuable, which must respond to environmental threats and take advantage of their opportunities; ii) rare or scarce, those that cannot allow obtaining competitive advantages with competitors; iii) difficult resources to imitate, without substitutes and organizational, which means that the company has aspects of order (Barney, 2001, page 41).

The company BIO-FOM offers the market an innovative product because it is taking the greatest benefit to a product that people see as a waste and transformed it into a product with added value; the strengths which this product provides are that they are organic, have a higher performance compared to other fertilizers, are favorable to the environment, are not expensive to produce. Something that gives a higher value is that the packaging is biodegradable by which makes the whole product itself is ecological and there is no waste of plastic or any other material that is difficult to decompose, the product is safe for children and pets so anyone can use it.

Therefore, for the considerations based on resources, a SWOT analysis was carried out to analyze both: the internal and external factors that provide added value to the firm, as the positive and negative factors that can cause the company not to grow as rapidly as expected.

Table 2. SWOT analysis

Strengths	Opportunities
Organic products / products without chemicals Excellent performance compared to other inorganic fertilizers environment friendly It is not expensive to produce The packaging is biodegradable No toxic, safe for children and pets It is not easy to imitate	Growing trend in organic products Market development Product development Product differentiation
Weaknesses	**Threats**
Lack of advertising + visibility Small production capacity-challenge for the company at scale The products have limited functions The products are not standardized Hard to forecast production Limited capital and strategic partners Do not go alone Expensive compared to inorganic fertilizers	Regulations that exist for exports Animal diseases Environmental factors may affect production Breach of demand levels Low market level

In this table, it is shown that the firm has many strengths that make its product different but also has many threats. This is due to the same rarity of the product that makes it somewhat complicated for sale, likewise, the product is easy to imitate, so at any time it could have a lot of competition, which can quickly get to the market firm. In his essay, Porter states that the competitive advantage comes from the value that a company manages to create for its customers, by offering special benefits that compensate a higher price of the product (Porter, 1980).

Considerations Based on Institutions

When talking about considerations based on institutions, we are talking about the rules of the game of these, in which the behavior of the company is determined and how they are developed around the world. Peng points out 5 strategies for an entrepreneurial company to be successful, which can be applied together, and they are growth, innovation, networks, financing/government, and harvest/output (Peng, 2012).

The BIO-FOM company integrates some of these strategies such as innovation in their products, as well as networks since they try to have a wide network to get known as well as to obtain advice and keep growing as the company participates in a program called X-culture where companies are assigned a group of people from different parts of the world, advising companies so they can expand or internationalize at the time the company uses all the networks that are possible for be able to expand. Peng mentions that there is evidence that networks, personal and organizational, represent significant resources and opportunities and that the successful creation of networks can lead to the efficient performance of firms (Peng, 2012, page 13).

Instead, McDougall notes that network analysis builds a very solid foundation and helps identify international opportunities, as well as establish credibility, provide access to critical resources, as well as knowledge, and lead strategic and cooperative partnerships (McDougall, 2000).

CHALLENGES AND OPPORTUNITIES

Green innovation business (GIB) is in the field of organic fertilizers are contributing with an input of fundamental importance in primary agricultural activity. The rehabilitation of organic fertilizer-producing plants in the country is not economically viable in the short term, because the reactivation of activities in the plants takes time. Besides, it is not common to find these kinds of producer units in urban areas.

It is necessary to invest more resources in the qualification and to reduce the import of organic inputs and other ingredients necessary in the production of fertilizers. The importation of chemicals as well as organic fertilizers will continue, since the national production of 2020 only covers 33.7% of the total demand, and the remaining 66.3% is brought from the foreign market.

CONCLUSION

When analyzing the green innovation business (GIB) and eco-efficient company through the comprehensive method of entrepreneurship, it is observed that just as the company has some advantages in the product also with many threats that are the large companies that are already positioned and that also compete with competitive prices. It is also determined that although the green innovation business (GIB) has a wide network that is using the best way possible, it is necessary to increase the international network of contacts so that the business can grow quickly.

Although BIO-FOM has been positioned in the local, regional, and national markets for a short time and has grown steadily, it has the challenge to enter the international markets in a very competitive position. The company Mary's poop has a competitive advantage that differentiates it from other firms, and that is that its product is innovative and ecological, as well as its processes are carried out in an ecological way which makes it an eco-efficient company.

Likewise, due to the culture of planting in Mexico, this company faces problems to become highly competitive in the country, since most farmers prefer chemical fertilizers and thus produce more in less time than caring for the land and having a product of a higher quality with organic fertilizers.

It is necessary to implement a culture of conscience in Mexico when we talk about caring for the land since the use of organic fertilizers would not only help the environment but also people, this is because the foods that are consumed will have better nutrients and they would not be contaminated with pesticides. Today some people started to make their gardens at home, so they grow their food, this is a good technique because they can be sure that the consumption of food is one hundred percent organic. Here is the importance of the topic for green innovation in urban areas.

REFERENCES

Acs, Z. J., & Audretsch, D. B. (2003). Innovation and Technological Change. In Z. J. Acs & D. B. Audretsch (Eds.), *Handbook of Entrepreneurship Research* (pp. 55–79). Kluwer Academic Publishers. doi:10.1007/b105789

Ajzen, I. (1991). The theory of planned behavior. *Organizational Behavior and Human Decision Processes*, *50*(2), 1–63. doi:10.1016/0749-5978(91)90020-T

Aragon-Sanchez, A., Baixauli-Soler, S., & Carrasco-Hernandez, A. J. (2017). A missing link: The behavioral mediators between resources and entrepreneurial intentions. *International Journal of Entrepreneurial Behaviour & Research*, *23*(5), 752–768. Advance online publication. doi:10.1108/IJEBR-06-2016-0172

Armitage, C. J., & Conner, M. (2001). Efficacy of the theory of planned behavior: A meta-analytic review. *British Journal of Social Psychology*, *40*(4), 471–499. doi:10.1348/014466601164939 PMID:11795063

Barney, J. (2001). Is the resource-based "view" a useful perspective for strategic management research? *Academy of Management Review*, *26*(1).

Burgueño Salas, E. (2020). *Volumen de producción mensual de fertilizantes fosfatados en México de enero de 2019 a julio de 2020 (en toneladas métricas).* Statista. Retrieve on the 2 of July 2021 from https://es.statista.com/estadisticas/594554/volumen-de-produccion-de-fertilizantes-superfosfatados-en-mexico/

Cantú, P. (2008). Desarrollo sustentable: conceptos y reflexiones. Universidad Autónoma de Nuevo León. Colección: Tendencias Científicas.

Carree, M. A., & Thurik, A. R. (2010). The impact of entrepreneurship on economic growth. In J. Zoltan & D. Audretsch (Eds.), *Handbook of entrepreneurship research.* Springer. doi:10.1007/978-1-4419-1191-9_20

Castro, A. (1998). *Hacia el desarrollo sostenible y la ecoeficiencia: integración de las normas ISO 9000, ISO 14000 e ISO 18000 diseño de un sistema de gestión ambiental certificable* (Tesis de maestría inédita). Obtenido de Facultad Regional de Buenos Aires: http://posgrado.frba.utn.edu.ar/investigacion/tesis/MIA-1998-Castro.pdf

Consejo Mundial Empresarial para el Desarrollo Sostenible. (1991). Ecoeficiencia y sustentabilidad. *Ciencia UANL, 35.*

Do Paço, F., Arminda, M., Ferreira, J. M., Raposo, M., Gouveia, R., & Dinis, A. (2011). Behaviours and entrepreneurial intention: Empirical findings about secondary students. *Journal of International Entrepreneurship*, *9*(1), 20–38. doi:10.100710843-010-0071-9

Evans, D. S., & Leighton, L. S. (1989). Some empirical aspects of entrepreneurship. *The American Economic Review*, *79*, 519–535.

FIRA. (2020). *Reporte trimestral de fertilizantes, Dirección de Investigación y Evaluación Económica del Fideicomisos Instituidos en relación con la Agricultura*. FIRA.

Gallegos Cedillo, G. (2021). *Working paper*. Agronegocios FIRA.

Holmes, T. J., & Schmitz, J. A. Jr. (1990). A theory of entrepreneurship and its application to the study of business transfers. *Journal of Political Economy*, *98*(2), 265–294. doi:10.1086/261678

Huyghebaert, N. (2004). *Comportamiento estratégico de los titulares en los mercados financieros y la salida de los arranques empresariales*. Academic Press.

Krueger, N. F. Jr, Reilly, M. D., & Carsrud, A. L. (2000). Competing models of entrepreneurial intentions. *Journal of Business Venturing*, *15*(5-6), 411–432. doi:10.1016/S0883-9026(98)00033-0

Lin, Z., Manser, M. E., & Picot, G. (1998). The Role of Self-Employment in Job Creation in Canada and the U.S. *OECD-CERF-CILN International Conference on Self-Employment*, Burlington, Ontario, Canada.

McDougall, P. (2000). *International entrepreneurship*. AMJ.

Ortega-Anderezand, D., & Lai, E. (2017, February). The Main Factors Influencing Entrepreneurial Activities and Intentions within a Country—A Case Study of Spain. *International Journal of Innovation, Management and Technology*, *8*(1).

Peng, M. (2012). *Global Strategy*. Thomson South-Western.

Pfeiffer, F., & Reize, F. (2000). Business Start-ups by the Unemployed - an Econometric Analysis Based on Firm Data. *Labour Economics*, *7*(5), 629–663. doi:10.1016/S0927-5371(00)00016-6

Porter, M. (1980). *Competitive Strategy*. Free Press.

Shapero, A., & Sokol, L. (1982). The social dimensions of entrepreneurship. In C. A. Kent, D. L. Sexton, & K. H. Vesper (Eds.), *Encyclopedia of Entrepreneurship* (pp. 72–90). Prentice Hall.

Shook, C. L., Priem, R. L., & McGee, J. E. (2003). Venture creation and the enterprising individual: A review and synthesis. *Journal of Management*, *29*(3), 379–399. doi:10.1016/S0149-2063(03)00016-3

Souitaris, V., Zerbinatib, S., & Al-Lahamc, A. (2007). Do entrepreneurship programmes raise entrepreneurial intention of science and engineering students? The effect of learning, inspiration and resources. *Journal of Business Venturing*, *22*(4), 566–591. doi:10.1016/j.jbusvent.2006.05.002

Van Stel, A., & Suddle, K. (2007). The Impact of New Firm Formation on Regional Development in the Netherlands. *Small Business Economics*.

Veciana, J. M., Aponte, M., & Urbano, D. (2005). University students' attitudes towards entrepreneurship: A two countries comparison. *The International Entrepreneurship and Management Journal*, *1*(2), 165–182. doi:10.100711365-005-1127-5

Wennekers, A. R. M., & Thurik, A. R. (1999). Linking entrepreneurship and economic growth. *Small Business Economics*, *13*(1), 27–55. doi:10.1023/A:1008063200484

Chapter 9
Regional Competitiveness:
Theoretical and Empirical Aspects

Miloš S. Krstić
Faculty of Sciences and Mathematics, University of Niš, Serbia

Vladimir Radivojević
https://orcid.org/0000-0002-3928-0623
Faculty of Economics, University of Priština in Kosovska Mitrovica, Serbia

ABSTRACT

The aim of the chapter was to model the impact of selected determinants (trade openness, human capital, entrepreneurship, and innovation) on regional competitiveness, as well as to propose future activities and measures required to be implemented to improve the competitive performance of the regions. The research was conducted on the sample of 18 regions in six European countries: Serbia, Croatia, Slovenia, Northern Macedonia, Montenegro, and Romania. The database was prepared, and the statistical processing was performed in SPSS. In this data analysis, the following methods were used: comparative analysis, correlation, and regression analysis. The results of the research showed that the impact of the determinants—import dependence, the number of pupils enrolled in secondary education, gross domestic expenditure on research and development, and the number of companies per 10,000 inhabitants on the competitiveness of the region—are (statistically) significant.

INTRODUCTION

What are the definitions of competitiveness, in the literature?. According to the definition of World Economic Forum, competitiveness is a set of institutions, policies and factors that determine the level of productivity of a country (Schwab & Porter, 2018; Krstić et al., 2020b; Krstić et al., 2016b; Radivojević et al., 2019b). At the micro level, competitiveness is defined as the ability of firms to compete, grow, and be profitable (Martin et al., 2006; Powell, 2001), or the ability of a firm to produce and sell products and services at a price lower than the competition (International Institute of Management Development, 2000). Between micro and macro level of competitiveness, regional competitiveness is defined.

DOI: 10.4018/978-1-7998-8900-7.ch009

One of the most frequently used definitions of regional competitiveness is the European Commission's definition, according to which, „the competitiveness of a region is its ability to produce goods and services that meet the requirements of the domestic and world market in terms of price, quality, etc., maintaining a high and sustainable level of income, or, more generally, the region's ability to generate, under external competitive pressures, a relatively high level of income and employment" (European Commission, 1999, p. 75).

What needs to be pointed out is that the concept of regional competitiveness has expanded over time, so that it includes the potential (or the strength) of the region or locality to create a sufficient level of export to achieve a sustainable level of income (and full employment) of the population. On the other hand, regional competitiveness is observed and analyzed as a result of the influence of several factors, and the most important are: (1) the business infrastructure; (2) availabilityand quality of human resources; (3) the production environment, etc. (European Commission, 2004).

Kitson and co-workers suggest that although theorists often use the term "regional competitiveness", it remains complex and controversial. „We are far from a consensus on what is meant by this term" (Kitson et al., 2004, p. 992). This is confirmed by numerous definitions of regional competitiveness that can be found in the literature. For example, Huggins believes that regional (or the local) competitiveness refers to conditions that allow companies to compete in selected markets and create value within a particular region (Huggins, 2003). Imre Lengyel and Mikosh Lukovic gave an overview of the competitiveness of Hungarian regions, using indicators, such as: GDP per capita, labor productivity, employment rate, etc. (Lengyel & Lukovic, 2006). Huggins and Davies created the European Competitiveness Index that measures the competitiveness of 27 European countries and 118 regions. In the report, the authors emphasize the role of knowledge, creativity and infrastructure for the analysis of regional competitiveness (Huggins & Davies, 2006). The Polish Regional Competitiveness index was calculated by Bronisz, Heiman and Miszczuk. They ranked 16 NUTS 2 Polish regions based on the weighting system used in the calculation of the final value of the Regional Competitiveness Index (Bronisz et al., 2008).

Due to the complexity of the concept of "regional or local competitiveness", often there are basics of determinants ((in terms of definitions)) within the regional competitiveness model. Modeling regional competitiveness using dual prices is found in Omoregie and Thomson's work (Omoregie & Thomson, 2001), and a similar approach is taken by Gardiner, Martin and Tyler (Gardiner et al., 2004) and Lengyel and Lukovics (Lengyel & Lukovics, 2006). Lukovics emphasizes the importance of the pyramid model and its application in many studies, because it follows the definition of competitiveness of the European Commission in a simple and logical way, and includes all the most important factors (Lukovics, 2007). The review of the conceptual model of regional competitiveness in the form of a "hat" was given by the European Commission (European Commission, 2004). It points out that GDP per capita as an indicator of competitiveness coincides well with the perspectives of EU regional policy, for example, with the aim of convergence.

There are two main goals of this research. One (goal) is to model the impact of selected determinants (trade openness, human resources, entrepreneurship and innovation) on the competitiveness of the regions of Serbia, Croatia, Slovenia, Montenegro, Northern Macedonia and Romania, by using the multiple linear regression model. The second goal is to propose activities that need to be implemented in order to improve the competitive performance of the region in the covered countries.

FACTORS OF REGIONAL COMPETITIVENESS

Although there are numerous theories and explanations of regional competitiveness factors, the most systematic way explains the factors (competitiveness) through "the prism" (or the pyramid) of the increasing quality of living. Factors, as sources of competitiveness, are defined at "the bottom" of the pyramid and include: (1) the economic structure; (2) the innovation; (3) the number and quality educational institutions; (4) the quality and availability of labor force, (5) the decision-making centers; and (6) the cultures. "The discovered regional competitiveness" is the central part of the pyramid and refers to: (1) the technological development; (2) the development of small and medium enterprises; (3) the volume of direct investments; (4) the infrastructure; (5) human capital; (6) institutions, etc. (Veselinović & Knežević, 2015, p. 144). Imre Lengyel gave the following explanation of the factors of regional competitiveness. According to him, there are the seven key factors of regional competitiveness:

(1) **The Economic Structure**. The workforce of more prosperous regions is usually concentrated in the branches of business services and/or the manufacturing industry.
(2) **The Innovation**. An adequately innovative environment allows the region to respond to any type of challenge.
(3) **The Regional Accessibility**. The proximity to large cities, the transport networks and the good geographical location can make the region more successful than others that do not have these characteristics.
(4) **The Trained Strength**. The share of educated workers in the total population is relatively high in competitive regions.
(5) **The Social Structure**. The knowledge-intensive economic activities and the growth of economic services affect regional competitiveness.
(6) **The Centering Decisions**. The presence of a corporate headquarters generates demand for highly qualified workers in the labor market, strengthens the local knowledge base, enhances the business environment, etc.
(7) **The Regional Identity**. Successful are those regions that are able to solve problems caused by structural changes in the economy and that achieve the rapid growth. It is important to encourage the regional identity of the population and to promote localism (Lengyel, 2004).

In the continuation of the paper, the following factors of regional competitiveness, which from our point of view, play an important role in the successful functioning of the economy, the initiating dynamic economic growth and the sustainable development will be considered: international trade/trade openness; human resources; dynamic entrepreneurship and networks; and innovation/the regional innovation systems.

Regional Competitiveness and International Trade/Trade Openness

Regional competitiveness can be defined as the competitive advantage of one region over another that can be seen through its share (the national and the international) in the export market (Kitson et al., 2004, p. 992). Porter, also, emphasized in his work the importance of export-oriented clusters as a basis for the achieving a high standard of living at the regional level (Porter, 1998, 2001).

However, the focus on regional exports as a measure of regional competitiveness is wrong. First of all, it uses the concept of competitiveness that was initially defined for national economies, without questioning whether this is the most useful or the most meaningful concept for application on a subnational scale (urban and regional scale). Second, the striving for regional export is characterized by all the problems and debates surrounding the notion of competitiveness. Even at the national level, there are significant disagreements over this notion. Thus, Nobel laureate Krugman considers competitiveness a a secondar product of political initiatives in the United States and the European Union since the early 1990s, where in both cases it was used as an excuse for unpopular economic policy measures (Kitanović & Krstić, 2010).

The metaphor of competitiveness, that implies that countries compete on the world market as companies do, for several reasons, is, according to Krugmann, wrong. First, states (and regions) in case they are uncompetitive cannot go bankrupt as is the case with firms; then, the efficiency of the country cannot be equated with the surplus in foreign trade; world trade is not a zero-sum game (regions can improve their positions at the same time without compromising the positions of other regions) and so on (Krugman, 1994; Krstić, 2012b).

There are also a kind of interpretation of the concept of competitiveness. If competitiveness is defined as the ability to create prosperity, then it has two components. One component is "outcome competitiveness" which measures the success of a country or region in achieving that goal. The second is "process competitiveness". It describes how the ability to "create well-being" is created (Aiginger, 2006, p. 174).

The Human Capital and Regional Competitiveness

Individuals of different characteristics, residents of municipalities and wider regions are the basis of human capital. In the long run, human capital enables technological progress and increases productivity. Investing in human capital raises the quality of the workforce with all positive outcomes (Tijanic, 2010; Radivojevic et al., 2019a).

The importance of human resources is emphasized in various models of regional competitiveness. Brooksbank and Pickernell ranked the regions of the United Kingdom according to selected indicators of education (that representing human capital) (Brooksbank & Pickernell, 1999). Gardiner considers various indicators of regional competitiveness related to knowledge and innovation and classifies them into a set of indicators of scientific capital, innovation capacity and products of the knowledge society in order to obtain a more detailed picture of these categories (Gardiner, 2003). The comparative analysis of the competitiveness of the eight Romanian regions based on human resource indicators was conducted by Constantin and Banica (Constantin & Banica, 2007). Human capital as a determinant of competitiveness and economic development at different levels is emphasized by Marija Bušelić in the book Knowledge and Competitiveness (Bušelić, 2007).

The Dynamic Entrepreneurship and Networks

Social milieu or local environment can be defined as a dynamic system which integrates know-how, rules and capital relations (capital relations that include the relationship of companies with the local environment) (Krstić et al., 2016a; Radivojević, 2019).

Networks are a key element in the creation and development of the local environment, because (they) interconnect different actors involved in the development of a particular locality (region). The quality and type of network affect the ability of regional actors to respond adequately to external stimuli (Krstić, 2012a).

Networks can be (of) different types. They can be either formal (business network structures, institutional networks) or informal. They can include the production of complete products, the exchange of information or the transfer of know-how, or a mixture of all three types. In any case, the network is an intermediary between the firms in the environment with which it is in contact (Wheeler et al., 2005). The exchange of information and know-how is extremely important, as it affects the ability of actors:

(1) to identify changes in the environment;
(2) to adjust their behavior to the identified changes;
(3) to improve the process of collective learning; and
(4) to create new products (Yusuf et al., 2016).

When we talk about institutional networks, we mean all the institutions/actors with which the company develops relationships/partnerships. These are: (1) public administration; (2) financial organizations; (3) education system; (4) other companies and the like (Garmise & Rees, 1997). Elements (or structures) of a business network can help businesses in some of the following ways:

(1) the creating new opportunities;
(2) the producting development and marketing;
(3) the combining individual products into a complete range of products, etc. (Håkansson & Olsen, 2015)

Innovation

It is generally accepted that the ability of regional economies to be competitive and to adapt to technological change is linked to their ability to innovate. In fact, innovation and flexibility are key to global success (Dasić et al., 2020; Topić & Lodorfos, 2021).

When we talk about innovation, we mean almost any improvement that could be introduced at the level of production, marketing, management and organizational systems of firms. When developed in the right environment (networking, availability of information, etc.), this know-how becomes a competitive advantage (or competitiveness) (Petrović & Nikolić, 2018; Krstić et al., 2016c).

The process of innovation takes place in two phases: conceptualization and realization. Innovative activity, in turn, triggers new opportunities for innovative projects. In local production systems, innovations are rarely the result of a single actor. In fact, it is more often the case that innovations are the result of a multifunctional innovation network. The innovation network consists of heterogeneous actors such as public laboratories, technical research centers, universities, companies, financial institutions, etc. The elements that characterize the innovative region are not given in advance, but may be induced by the authorities through regional policy measures. Some of the recommendations for developing innovative regions are:

(1) involve local actors and develop or renew intangible resources (know-how, specific technology, formal rules, principles of trust and reciprocity, etc.);
(2) stimulate synergies (interactions, networks, etc.) and the learning process between local actors; and
(3) permanent contact with advanced technology.

The types of regional innovation systems according to the management dimension are: basic, network and conductor regional innovation system. The grassroots regional innovation system is characterized by regional initiatives, diffuse funding, applied research, low level of technological specialization and local coordination. Network innovation systems can be established at several levels (local, regional, federal and federal levels). The financing of network innovation systems is mainly provided by banks, companies and state agencies. The research of the mentioned innovation systems is focused on the applied technology with flexible specialization. Conductor innovation systems are usually initiated and funded by the central government. The research is, to a large extent, basic (or fundamentally) with a high level of specialization, so that it can be used in larger firms/multinational companies in or outside the region (Braczyket al., 1998; European Commission, 2002).

The types of regional innovation systems according to the dimension of "business innovation" are: local, interactive and globalized innovation systems. The scope of research of an individual company in the local innovation system is large, but there is a relatively high degree of connection between entrepreneurs. Interactive innovation systems are characterized by a balance between large and small firms, as well as a combination of private and public research institutes that reflects involvement in the promotion of the innovation base of both large firms and local authorities. Third, globalized regional innovation systems are supported by supply chains (Sepic, 2005).

METHODOLOGICAL FRAMEWORK OF RESEARCH AND HYPOTHESIS

The analysis of regional competitiveness, which is presented in this paper, has two goals. One is to model the impact of trade openness, human resources, entrepreneurship and innovation on the competitiveness of the region by applying the multiple linear regression model. The second goal is to propose activities that need to be implemented in order to improve the competitive performance of the regions being studied.

In accordance with the set goals of the research, the following initial hypotheses were defined and tested:

Hypothesis One (H1): Import Dependence (IMD), as the approximation for trade openness, has the most negative impact on GDP per capita as an indicator of regional competitiveness.

In designing Hypothesis One, we were guided by the research titled *Competitiveness, Productivity and Economic Growth across the European Regions* by Ben Gardiner, Ron Martin and Peter Tyler. These authors analyzed the determinants of regional competitiveness in EU coutries and EU candidate countries. The results of the research showed that international trade (trade openness) is one of the factors influencing the convergence of regional productivity (or competitiveness) between "the center" (Western Europe) and "the periphery" of European Union (Central and Eastern European countries).

Hypothesis Two (H2): Human capital, that is quantified in this study using the variable The Number of Pupils Enrolled in Secondary Education per 10,000 Inhabitants significantly affects the level of GDP per capita.

Hypothesis Three (H3): Gross Domestic Expenditure on Research and Development, as an economic indicator most often used to measure the level of innovation in the country/region, has a positive and statistically significant impact on the competitiveness of the region in the analyzed countries in 2019.

To define hypotheses H2 and H3, we used the research titled *Regional Competitiveness, Economic Growth and Stages of Development* by Robert Huggins, Hiro Izushi, Daniel Proko and Piers Thompson. These authors analyzed the determinants of regional competitiveness and found that knowledge (education) and innovation are key factors that determine differences in the development of the region. These authors also came to the conclusion that regions at the same or similar level of economic development have similar opportunities for economic growth in the future (Huggins et al., 2014).

The key information basis for the implementation of the survey is the data from the database named *Regions*, which contains the main regional data within Eurostat. When data were not available within the *Regions*, we used other data sources such as: *Global Entrepreneurship Monitor Report*, *Smart Specialization Platform*, databases of the Statistical Office of Serbia, Croatia, Slovenia, Northern Macedonia, Montenegro and Romania etc.

In the study of regional competitiveness, we used the sample that includes NUTS2 regions in six countries, namely: Serbia, Croatia, Slovenia, Northern Macedonia, Montenegro and Romania. In the process of data processing, we used the SPSS program with the aim of calculating the coefficient of multiple linear correlation, regression constant, regression coefficients, t and p values, etc. Also, statistically significant differences were defined in relation to the rank of 95% of the confidence interval.

RESEARCH RESULTS

The data from 2019 are included in the regression model of regional competitiveness. GDP per capita is the dependent variable in the model. The independent variables (or the regressors) in the model are: Exports Per Capita (EXpc in euros), Export Dependence (EXD, in %); Import Dependence (IMD, in %); The Number of Students Enrolled in Tertiary Education per 10,000 Inhabitants (S10000); The Number of Pupils Enrolled in Secondary Education per 10,000 Inhabitants (P10000); Gross Domestic Expenditure on Research and Development (GERD); and The Number of Companies per 10,000 Inhabitants (E10000) (see Table 4 in the Appendix).

The relationship between the variables can be described by the model:

$$\ln Y_i = \beta_0 + \beta_1 \ln x_{1i} + \beta_2 \ln x_{2i} + \beta_3 \ln x_{3i} + \beta_4 \ln x_{4i} + \beta_5 \ln x_{5i} + \beta_6 \ln x_{6i} + \beta_7 \ln x_{7i} + u_i \quad (1)$$

where: Y – the dependent variable; X – the independent or the explanatory variable; β_0 – the regression constant, β_j (j = 1,2,… 7) the regression coefficient, u – the error.

The first part ($\beta_0 + \beta_1 \ln x_{1i} + \beta_2 \ln x_{2i} + \beta_3 \ln x_{3i} + \beta_4 \ln x_{4i} + \beta_5 \ln x_{5i} + \beta_6 \ln x_{6i} + \beta_7 \ln x_{7i}$) called "the systematic part" of the model is formed according to the assumed (average) theoretical relationship at which the value of Y for each observation (i = 1, 2,… 27) linearly dependent on the set of regressors X_j (j = 1,2,… 7). The second, "stochastic part" of the model (u_i) represents random variations Y which take into account the possible effect of other variables that are not explicitly included in the model, then errors of measuring functional shape and sample errors:

(1) E(u) = 0, for each i (mean error value is zero);

(2) E $(u^2) = \sigma^2$, for each i (constant common variance);
(3) E $(u_i u_j) = 0$, for each $i \neq j$ (mutual independence of errors);
(4) $u_i \sim N(0, \sigma^2)$, for each i (normal independent distribution);
(5) E $(u_i, X_j) = 0$, for all i, j (regressor independence);
(6) the independent variables or the regressors are not perfectly linearly interdependent (no perfect multicollinearity);
(7) the regressors are not stochastic variables and do not contain measurement errors; and
(8) there are more observations (sample units) than evaluation parameters ($n > k$).

The equation of the estimated (log-log) model is:

$$\ln\hat{Y}_i = \hat{\beta}_0 + \ln X_{1i} + \hat{\beta}_2 \ln X_{2i} + \hat{\beta}_3 \ln X_{3i} + \hat{\beta}_4 \ln X_{4i} + \hat{\beta}_5 \ln X_{5i} + \hat{\beta}_6 \ln X_{6i} + \hat{\beta}_7 \ln X_{7i} \quad (2)$$

To determine the parameter estimates ($\hat{\beta}_0$ and $\hat{\beta}_i$) we use the least squares method. With this method, parameter estimates are obtained by minimizing the sum of the square deviations of the observations from the estimated regression model or the sum of the squares of the residuals.

The sum of the squares of the residual

$$\sum_{i=1}^{n} \hat{u}_i^2 = \sum_{i=1}^{n} \left(Y_i - \hat{\beta}_0 - \hat{\beta}_1 \ln X_{1i} - \hat{\beta}_2 \ln X_{2i} - \hat{\beta}_3 \ln X_{3i} - \hat{\beta}_4 \ln X_{4i} - \hat{\beta}_5 \ln X_{5i} - \hat{\beta}_6 \ln X_{6i} - \hat{\beta}_7 \ln X_{7i}\right)^2 \quad (3)$$

is at its minimum when the first derivatives of that sum by $\beta 0$ and βj $_{(}j= 1,2,\ldots 7)$ are equal to zero, while the second derivatives are positive.

As explained earlier, one of the important assumptions of applying regression analysis is that the independent variables are not perfectly linearly interdependent, because, in case there is a perfect correlation between them, the estimates of parameters (the regression coefficients) become indeterminate. In practice, there are different ways of determining the serious multicollinearity problem. This paper uses the so-called “Klein's rule” to detect the multicollinearity of independent variables (Klein, 1962; see more in Jovčić, 2002, p. 156; Gujarati, 2003, p. 361).

In accordance with the “Klein's rule”, it is necessary to calculate the square of the multiple correlation coefficient (R_j^2) of each explanatory variable X_j from other (k-1) variables in the equation and compare it with the square of determination of the whole model (R^2), i.e. with the square the coefficient of multiple correlation between the dependent variable and all k independent variables in the equation. That independent variable (or more of them) that shows a higher level of linear dependence on other independent variables than the coefficient of multiple correlation for the whole equation is certainly the cause of harmful multicollinearity (see regression equations and Table 1).

$$\ln EXD = \hat{\gamma}_0 + \hat{\gamma}_1 \ln EXpc + \hat{\gamma}_2 \ln IMD_i + \hat{\gamma}_3 \ln S10000_i + \hat{\gamma}_4 \ln P10000_i + \hat{\gamma}_5 \ln GERD_i + \hat{\gamma}_6 \ln E10000_i \quad (4)$$

$$\ln \text{IMD submodel} = \hat{\delta}_0 + \hat{\delta}_1 \ln \text{EXpc}_i + \hat{\delta}_2 \ln \text{EXD}_i + \hat{\delta}_3 \ln \text{S10000}_i + \hat{\delta}_4 \ln \text{P10000}_i + \hat{\delta}_5 \ln \text{GERD}_i + \hat{\delta}_6 \ln \text{E10000} \quad (5)$$

$$\ln \text{S1000 submodel} = \hat{\varepsilon}_0 + \hat{\varepsilon}_1 \ln \text{EXpc} + \hat{\varepsilon}_2 \ln \text{EXD}_i + \hat{\varepsilon}_3 \ln \text{IMD} + \hat{\varepsilon}_4 \ln \text{P10000} + \hat{\varepsilon}_5 \ln \text{GERD} + \hat{\varepsilon}_6 \ln \text{E10000} \quad (6)$$

$$\ln \text{P10000} = \hat{\epsilon}_0 + \hat{\epsilon}_1 \ln \text{EXpc} + \hat{\epsilon}_2 \ln \text{EXD} + \hat{\epsilon}_3 \ln \text{IMD} + \hat{\epsilon}_4 \ln \text{S10000} + \hat{\epsilon}_5 \ln \text{GERD} + \hat{\epsilon}_6 \ln \text{E10000} \quad (7)$$

$$\ln \text{GERD submodel} = \hat{\theta}_0 + \hat{\theta}_1 \ln \text{EXpc} + \hat{\theta}_2 \ln \text{EXD} + \hat{\theta}_3 \ln \text{IMD} + \hat{\theta}_4 \ln \text{S10000} + \hat{\theta}_5 \ln \text{P10000} + \hat{\theta}_6 \ln \text{E10000} \quad (8)$$

$$\ln \text{E10000 submodel} = \hat{\vartheta}_0 + \hat{\vartheta}_1 \ln \text{EXpc}_i + \hat{\vartheta}_2 \ln \text{EXD}_i + \hat{\vartheta}_3 \ln \text{IMD}_i + \hat{\vartheta}_4 \ln \text{S10000}_i + \hat{\vartheta}_5 \ln \text{P10000} + \hat{\vartheta}_6 \ln \text{GERD} \quad (9)$$

Table 1. Klein's rule

Subregression Models	R_j^2	The Subregression Models and The log-log Model	R_j^2
EXpc	0.946684	P10000	0.696476
EXD	0.820298	GERD	0.903007
IMD	0.682159	E10000	0.889708
S10000	0.764768	The log-log model	$R^2 = 0.939162$

Source: Author's calculation – SPSS

Based on Table 1, we conclude that the variable Export per capita (EXpc) significantly contributes to the multicollinearity. Therefore, the regression model after removing the disputed variable from the model is listed below. The relationship between the variables can now be described by the model:

$$\ln Y_i = \beta_0 + \beta_1 \ln x_{1i} + \beta_2 \ln x_{2i} + \beta_3 \ln x_{3i} + \beta_4 \ln x_{4i} + \beta_5 \ln x_{5i} + \beta_6 \ln x_{6i} + u_i \quad (10)$$

After the dropping the variable that contributes to the problem of serious multicollinearity, the estimated regression model reads:

$$\ln \hat{Y}_i = \hat{\beta}_0 + \ln X_{1i} + \hat{\beta}_2 \ln X_{2i} + \hat{\beta}_3 \ln X_{3i} + \hat{\beta}_4 \ln X_{4i} + \hat{\beta}_5 \ln X_{5i} + \hat{\beta}_6 \ln X_{6i} \quad (11)$$

The results of the analysis of the regression model that does not contain the variable that contributed to the problem of serious multicollinearity are shown in Table 2.

Table 2. The results of regression analysis

Model	Unstandardized Coefficients B	Standardized Coefficient Beta	Sig.	Correlations Part
(Constant)	16113.032287		0.103181	
EXD	-63.258293	-0.162641	0.375685	-0.077489
IMD	-112.349085	-0.330733	0.040714	-0.194559
S10000	-3.462776	-0.107490	0.503994	-0.057981
P10000	26.691198	0.163559	0.083949	0.094544
GERD	0.000033	0.996138	0.000013	0.625152
E10000	16.450358	0.545312	0.011226	0.255236

Source: Author's calculation – SPSS

According to Table 2, the strongest impact on changes in GDP per capita has the score of the variable Gross Domestic Expenditure on Research and Development in 2019 (see column *Standardized Coefficient Beta*), followed by variables: The Number of Companies per 10,000 Inhabitants and Import Dependence, while other variables are many times smaller impact on GDP per capita change (e.g. The Number of Students Enrolled in Tertiary Education per 10,000 Inhabitants has a 9 times weaker impact on the region's GDP per capita than Gross Domestic Expenditure on Research and Development). Confirmation of this view can be found in the column *Correlations Part.* According to the data in this column, if we omit, for example, the variable Gross Domestic Expenditure on Research and Development, it would reduce the degree of explained variability (variability of the dependent variable GDP per capita) by 39% (0.625152 * 0.625152 * 100), while if we would omit the variable Number of Students Enrolled in Tertiary Education per 10,000 Inhabitants from the analysis, which would result in the reduction of the explained variability of the dependent variable by only 0.3% (0.057981 * 0.057981 * 100).

The values in the column *Unstandardized Coefficients B* show by how much the value of GDP per capita would change in the case of a unit change in the independent variables. Thus, if the score of Import Dependence increases by one, with unchanged values of other independent variables, it will result in a reduction of GDP per capita by 112.349085% and vice versa. If the score of The Number of Pupils Enrolled in Secondary Education per 10,000 Inhabitants per one, at unchanged values of other independent variables, it will lead to increase in GDP per capita by 26.691198%. If the score of the variable The Number of Companies per 10,000 Inhabitants is increased by one, with unchanged values of other independent variables, this will also result in an increase in GDP per capita by 16.450358% and vice versa.

When formulating the stated conclusions, the evidence on the statistical significance of the numerical influence of certain independent variables from the column Sig. was ignored. The statistical significance for the influence of the variables Export Dependence and The Number of Students Enrolled in Tertiary Education has not been confirmed, since it exceeds the marginal level of 0.1. This means that the (numerical) impact on regional GDP per capita in Serbia, Croatia, Slovenia, Northern Macedonia, Montenegro and Romania is strictly confirmed for Import Dependence, The Number of Pupils enrolled in secondary education, Gross Domestic Expenditure on Research and Development and The Number of Companies in 10,000 Inhabitants.

Considering the coefficient of determination R^2 as a descriptive statistic, it can be concluded following: the higher the coefficient of determination, the factors included in the model are of greater importance for explaining the variability of the analyzed phenomenon. In this study, the coefficient of determination R^2 2 is 92%. That means that the our model is very reliable, because it can explain 92% of all variations of the variable it describes. When we have determined the coefficient R^2, it is necessary to see whether the obtained value of the coefficient is statistically significant.

F test statistics show the statistical significance of estimating the parameters (regression coefficients) of all independent variables in the model. If the estimated regression model is 100% irrelevant, it means that independent variables in the model are not significant for explaining GDP per capita as an indicator of regional competitiveness. This further means that all regression coefficients (in the population and in the estimated log-log model or sample) are equal to zero.

We now formulate the null hypothesis to be tested:

$H_0 = \beta 1_{,} \beta 2, \ldots, \beta k = 0$

An alternative hypothesis is that at least one non-zero regression coefficient:

$H0_{=}$ at least one coefficient $\beta j \neq_0$

To test the null hypothesis that all regression coefficients are equal to zero, the Wald test was performed (which is one of the variants of the F test statistic). It was found that the p value (0.000) is less than any usual theoretical level of significance α, and therefore we can reject $H0_{,}$ which suggests that at least one variable in the model is significant (Table 3).

Table 3. Wald's test

Wald Test:				
Equation: Untitled				
Null Hypothesis:	C(2)=0			
	C(3)=0			
	C(4)=0			
	C(5)=0			
	C(6)=0			
	C(7)=0			
	C(8)=0			
F-statistic	21.826247		Probability	0.000000

Source: Author's calculation – SPSS

Finally, one can speculate about the reasons for "the deficiency" of the results of this research. The technical specification of the regression model, conditioned by the thematic setting of the research, which includes: a short observation period (one year), a relatively large number of independent variables (there are seven), a small number of observations, etc., can certainly be taken as the most acceptable. Problems related to the specification of indicators of regional competitiveness, trade openness, entrepreneurship and the like can also be taken into consideration. All the mentioned facts have a significant impact on the quality of research results.

CONCLUDING REMARKS

The presented analysis is part of complex researches on the issue of regional competitiveness (Joshi et al., 2016; de la Vegaet al., 2019; Libecap, 2011) which facing problems in defining, measuring, and then the analyzing the various determinants that determine it. The model proves the positive impact of trade openness, human resources and innovation on the competitiveness of the region in the analyzed countries, so future measures and activities at the regional, city and municipal level should be focused on: (1) the increasing exports and qualitative changes in export structure; (2) the increasing the quality of secondary education; and (3) the development and application of new technologies (or innovations).

(1) **The increasing exports and qualitative changes in the export structure**. The results of measuring the competitiveness factors of the region in Serbia, Croatia, Slovenia, Northern Macedonia, Montenegro and Romania show that the worst export results were achieved by the following regions: Montenegro, Nord-Est, Sud-Est, Southern and Eastern Serbia, Adriatic Croatia, and Sumadia and Western Serbia (see Table 5 in the Appendix). The poor export results of these regions show that they failed to build their competitive advantage ("competitiveness") or, better said, they lost a comparative advantage (cheap labor), and there was no finding a new source of production and export growth. Gone are the days when these regions could base their development only on comparative advantages and it is time to adopt a new approach and ask the question: how to build a competitive advantage?

Here are some suggestions to economic policy makers: reduce and/or abolish the tax burden on start-ups over a period of time, provide higher tax deduction for the taxes on personal income, make loans more accessible to entrepreneurs and reduce interest rates on loans to entrepreneurs, enable entrepreneurs to have a long-life learning process, technologically modernize existing firms and thus increase productivity, the attracting and realizing foreign export-oriented investments, and the implement appropriate trade policies that protect investors and more quickly adjust the analyzed countries and regional economic structures with world and EU standards and regulations so that they can be better integrated into global supply chains.

(2) **The increasing the quality of secondary education**. Economies in the analyzed areas need a proactive and innovative workforce with high productivity, with developed skills of teamwork, critical thinking, flexibility and adaptability. Therefore, it is necessary to invest a lot in the transformation of the educational system, then to apply effective tools and teaching methods that are adapted to the so-called "Generation Z." In our opinion, the education systems in Serbia, Croatia, Northern Macedonia, Montenegro and Romania are not sufficiently adapted to such persons, which results in reduced motivation for learning and unattainable outcomes (Krstić, 2021). In order to achieve the highest possible extent of the engagement of members of the "Generation Z", we believe that it is necessary for institutions in education (high schools) to use different types of games (on-line, off-line, rules games, and simulation) that they are used exclusively in teaching. Pupils benefit greatly from games in the classroom, which increases the motivation to learn, but also develop motor skills (which are still insufficiently encouraged in university teaching). By playing games, students develop the ability to think quickly and analyze the situation from different aspects with the help of coordination and concentration on visual details. The application of play in teaching in secondary education promotes: fun, active participation in teaching,

pleasant atmosphere in the classroom, the adoption of content in an interesting way, creativity, imagination of pupils and so on.

One of the key benefits of gamification in secondary education is that it introduces pupils to a mental state of ecstasy or "enchanting obsession state" in which the pupil is highly focused on the current task. In order to achieve enthusiasm, you need: clear goals that guide the student on "the right path", immediate feedback and clear information and balance between the required skill and the challenge presented.

(3) **The development and application of new technologies**. In order to create and direct new technologies, sectors and markets, the competent state institutions in the analyzed regions must be "armed" with the intelligence which is necessary for designing and making important decisions. This does not mean that state bodies, "armed" with experts in new technologies and economic sectors. will always succeed. In fact, the uncertainty inherent in the innovation process means that it will often be a failure. However, public policy makers must learn from unsuccessful investments and constantly improve their staff and practice (Krstić & Pavlvić, 2020a).

Firms, cities, regions or countries should be the promoters of the most radical, most innovative types of innovation, because they do not lead all innovations to the growth of the entire economy. The growth of the entire economy is most often stimulated by new products or processes that affect a wide range of economic sectors. Equally important is the establishment of public sector organizations (government agencies, laboratories, research centers, etc.) at the regional, city and municipal levels that imagine opportunities, engage in the most risky and uncertain early research and control the process of commercialization of innovations. The role of these organizations in the development of new technologies should not be limited to subsidizing the innovative activities of certain companies (private sector). They need to be able to spread new ideas quickly. They are able to shape the market and drive technological progress, thus acting as a catalyst for change - a "spark that ignites fire" (Matsukato, 2020).

Finally, it should be said that the development of technological innovation is not directly related only to spending on research and development. A necessary precondition for innovation is a networked economy with continuous feedback between different individuals and organizations in order to enable the exchange of knowledge and the shifting of borders.

ACKNOWLEDGMENT

This research was funded by the Ministry of Education, Science, and Technological Development of the Republic of Serbia (the number of the contract of the realization and financing of scientific research work: 451-03-9/2021-14/00124).

REFERENCES

Aiginger, K. (2006). Competitiveness: From a dangerous obsession to a welfare creating ability with positive externalities. *Journal of Industry, Competition and Trade*, *6*(2), 161–177. doi:10.100710842-006-9475-6

Armstrong, H., & Taylor, J. (2006). *Regional Economics and Policy*. Blackwell Publishing.

Braczyk, H.-J., Cooke, P., & Heidenreich, M. (1998). *Regional Innovation Systems*. UCL Press.

Bronisz, U., Heijman, W., & Miszczuk, A. (2008). Regional competitiveness in Poland: Creating an index. *Jahrbuch für Regionalwissenschaft*, *28*(2), 133–143. doi:10.100710037-008-0026-y

Brooksbank, D. J., & Pickernell, D. G. (1999). Regional competitiveness indicators: A reassessment of method. *Local Economy*, *13*(4), 310–326. doi:10.1080/02690949908726459

Bušelić, M. (2007). Knowledge and competitiveness. Pula, Croatia: University "Juraj Dobrila" in Pula, Department of Economics and Tourism "Dr. Mijo Mirkovic". (in Croatian)

Commission of the European Communities. (1999). *Regional Policy and Cohesion Sixth Periodic Report on the Social and Economic Situation and Development of the Regions of the European Union*. Commission of the European Communities.

Constantin, D. L., & Banica, G. C. (2007, August). Human Resources and Regional Competitiveness. *The Case of Romania*. Paper presented at the meeting of the European Regional Science Association, Paris, France.

Croatian Chamber of Commerce. (2019). *Foreign trade of counties with foreign countries in 2019 - provisional data*. Croatian Chamber of Commerce. (in Croatian)

Croatian Chamber of Commerce. (2019). *Foreign trade of counties in 2019 - provisional data*. Croatian Chamber of Commerce. (in Croatian)

Dasić, B., Dević, Z., Denić, N., Zlatković, D., Ilic, I. D., Cao, Y., Jermsittiparsert, K., & Le, H. V. (2020). Human development index in a context of human development: Review on the western Balkans countries. *Brain and Behavior*, *10*(9), 17–55. doi:10.1002/brb3.1755 PMID:33460259

De la Vega, J. C. S., Azorín, J. D. B., Segura, A. C. F., & Yago, M. E. (2019). A new measure of regional competitiveness. *Applied Economic Analysis*, *27*(80), 108–126. doi:10.1108/AEA-07-2019-0010

European Commission. (2002). *Regional Innovation Strategies under the European Regional Development Fund Innovative Actions 2000-2002*. European Commission.

European Commission. (2004). *A Study on the Factors of Regional Competitiveness*. European Commission.

Eurostat. (2021). *Regions database*. Retrieved from https://ec.europa.eu/eurostat/web/regions/

Gardiner, B. (2003, April). *Regional competitiveness indicators for Europe-audit, database construction and analysis*. Paper presented at the meeting of the Regional Studies Association, Pisa, Italy.

Gardiner, B., Martin, R., & Tyler, P. E. T. E. R. (2006). Competitiveness, productivity and economic growth across the European regions. *Regional Competitiveness, 30*, 55.

Garmise, S., & Rees, G. (1997). The role of institutional networks in local economic development: A new model of governance? *Local Economy*, *12*(2), 104–118. doi:10.1080/02690949708726379

Global Entrepreneurship Monitor Report. (2021). *Entrepreneurial behaviour and attitude*. Retrieved from https://www.gemconsortium.org/data

Gujarati, D. (2008). *Basic econometrics*. New York: McGraw-Hill.

Håkansson, H., & Olsen, P. I. (2015). The roles of money and business deals in network structures. *Industrial Marketing Management, 45*, 207–217. doi.org/10.1016/j.indmarman.2015.01.009

Huggins, R. (2003). Creating a UK Competitiveness Index: Regional and Local Benchmarking. *Regional Studies, 37*(1), 89–96. doi.org/10.1080/0034340022000033420

Huggins, R., & Davies, W. (2006). *European Competitiveness Index 2006-2007*. Retrieved from http://www.cforic.org/pages/european-competitiveness.php

Huggins, R., Izushi, H., Prokop, D., & Thompson, P. (2014). Regional competitiveness, economic growth and stages of development. *Zbornik radova Ekonomskog fakulteta u Rijeci: časopis za ekonomsku teoriju i praksu, 32*(2), 255-283.

International Institute of Management Development. (2000). *The World Competitiveness Yearbook 2000*. Lausanne, Switzerland: International institute of management development.

Joshi, S., Saxena, S., & Godbole, T. (2016). Developing smart cities: An integrated framework. *Procedia Computer Science, 93*, 902–909. doi.org/10.1016/j.procs.2016.07.258

Jovčić, M. (2002). *Econometric models*. CID. (in Serbian)

Kitanović, D., & Krstić, M. (2010). Uticaj deviznog kursa na privredni rast Srbije. *Ekonomske teme, 48*(1), 13-27.

Kitson, M., Martin, R., & Tyler, P. (2004). Regional competitiveness: An elusive yet key concept? *Regional Studies, 38*(9), 991–999. doi.org/10.1080/0034340042000320816

Klein, L. R. (1962). Introduction to econometrics. New York: Prentice Hall.

Krstić, B., Radivojević, V., & Stanišić, T. (2016a). Determinants of CEE countries' tourism competitiveness: A benchmarking study. *Management: Journal of Sustainable Business and Management Solutions in Emerging Economies, 21*(80), 11-22. doi:10.7595/management.fon.2016.0021

Krstić, B., Radivojević, V., & Stanišić, T. (2016b). Measuring and analysis of competition intensity in the sugar market in Serbia. *Economics of Agriculture, 63*(2), 389-406. doi:10.5937/ekoPolj1602389K

Krstić, B., Stanišić, T., & Radivojević, V. (2016c), The impact of innovativeness' factors on the EU countries competitiveness. *Industrija, 44*(2), 101-116. doi:10.5937/industrija41-4000

Krstić, M. (2012a). The role of rules in the evolution of the market system: Hayek's concept of evolutionary epistemology. *Economic Annals, 57*(194), 123-140. doi:10.2298/EKA1294123K

Krstić, M. (2012b). New theory of international trade and new economic geography-Volčkova N.: Novaja teorija meždunarodnoj torgovli i novaja ëkonomičeskaja geografija, Voprosy teorii, 2009. *Geographica Pannonica, 16*(1), 35–37.

Krstić, M. (2021). Higher Education as Determinant of Competitiveness and Sustainable Economic Development. In M. Topić & G. Lodorfos (Eds.), *The Sustainability Debate: policies, gender and the media* (pp. 15–35). Emerald Publishing.

Krstić, M., Filipe, J. A., & Chavaglia, J. (2020b). Higher Education as a Determinant of the Competitiveness and Sustainable Development of an Economy. *Sustainability, 12*(16), 6607. doi.org/10.3390/su12166607

Krstić, M., & Pavlović, N. (2020a). Behavioral Economics: New Dimension in Understanding the Real Economic Behavior. In U. Akkucuk (Ed.), *Handbook of Research on Sustainable Supply Chain Management for the Global Economy* (pp. 281–298). IGI Global.

Krugman, P. (1994). Competitiveness: A dangerous obsession. *Foreign Affairs, 73*(2), 28–44.

Lengyel, I. (2004). The pyramid model: enhancing regional competitiveness in Hungary. *Acta Oeconomica, 54*(3), 323-342. doi:10.1556/AOecon.54.2004.3.3

Lengyel, I., & Lukovics, M. (2006, August). *An attempt for the measurement of regional competitiveness in Hungary*. Paper presented at the meeting of the European Regional Science Association, Volos, Greece.

Libecap, G. D. (2011). Introduction: Entrepreneurship and global competitiveness in regional economies: Determinants and policy implications. In G. D. Libecap (Ed.), *Entrepreneurship and Global Competitiveness in Regional Economies: Determinants and Policy Implications* (pp. 9–12). Emerald Group Publishing Limited.

Lukovics, M. (2007, October). *Statistical Analysis on the Competitiveness of the Hungarian Sub-regions*. Paper presented at the meeting of the European Regional Science Association, Tatras, Slovak Republic.

MAKStat Database. (2021). *Nadvoresna Trgovija Indikatori*. Retrieved from http://makstat.stat.gov.mk

Martin, R., Kitson, M., & Tyler, P. (2006). *Regional Competitiveness*. Routledge.

Mazzukato, M. (2020). *Entrepreneurial state - dispelling myths about the conflict between the public and private sectors*. Novi Sad, Serbia: Academic Book.

Omoregie, E. M., & Thomson, K. J. (2001). Measuring regional competitiveness in oilseeds production and processing in Nigeria: A spatial equilibrium modelling approach. *Agricultural Economics, 26*(3), 281–294. doi.org/10.1111/j.1574-0862.2001.tb00070.x

Petrović, P., & Nikolić, G. (2018). Schumpeterian Growth Theory: empirical testing of barriers to competition effect. *Economic Annals, 63*(217), 7-37. doi:10.2298/EKA1817007P

Porter, M. E. (1998). Location, clusters and the new economics of competition. *Business Economics (Cleveland, Ohio), 33*(1), 7–17.

Porter, M. E. (2001). Regions and the New Economics of Competition. In A. J. Scott (Ed.), *Global City Regions: Trends, Theory, Policy* (pp. 139–156). Oxford University Press.

Powell, T. C. (2001). Competitive advantage: Logical and philosophical considerations. *Strategic Management Journal, 22*(9), 875–888. doi.org/10.1002/smj.173

Radivojević, V., Kahrović, E., & Krstić, M. (2019a). Population skills as an indicator of European countries' competitiveness in the modern economy. *Vojno delo, 71*(5), 105-116. doi:10.5937/vojdelo1905105R

Radivojević, V., Krstić, B., Krstić, M., & Petković, M. (2019b). Benchmarking agricultural and other environmental performances of Central and East European countries. *Economics of Agriculture, 66*(2), 471-484. doi:10.5937/ekoPolj1902471R

Republic Statistical Office (2019). *Municipalities and regions in the Republic of Serbia*. Belgrade, Serbia: Republic Statistical Office. (in Serbian)

Schwab, K., & Porter, M. E. (2018). *The Global Competitiveness Report 2018-2019*. Geneva, Switzerland: World Economic Forum.

Sepic, D. (2005). *The regional competitiveness: some notions*. Russian-European Centre for Economic Policy.

Statistical Office of Montenegro – MONSTAT. (2019). *External trade in goods of Montenegro January - December 2019*. Retrieved from https://www.monstat.org/cg/novosti.php?id=3195

Statistical Office of the Republic of Slovenia. (2021). *Exports, imports, trade balance and coverage of imports by exports, Slovenia, annually (cummulative data)*. Retrieved from https://pxweb.stat.si/SiStatData/pxweb/en/Data/-/2490002S.px/

The Joint Research Centre. (2021). *Smart Specialization Platform*. Retrieved from https://s3platform.jrc.ec.europa.eu/s3-trade-tool

The National Institute of Statistics. (2019). *Monthly statistical bulletin of counties no 11/2019*. The National Institute of Statistics.

Tijanić, L. (2010). Regional (non) competitiveness in the Republic of Croatia [in Croatian]. *Economic Review (Kansas City, Mo.), 61*(7-8), 419–454.

Topić, M., & Lodorfos, G. (2021). Introduction. In M. Topić & G. Lodorfos (Eds.), *The Sustainability Debate: policies, gender and the media* (pp. 15–30). Emerald Publishing.

Veselinović, P., & Knežević, M. (2015). Competitiveness enhancing as the incentive for the investment in the city of Kragujevac. *Ekonomski pogledi, 17*(3), 139-163. (in Serbian)

Wheeler, D., McKague, K., Thomson, J., Davies, R., Medalye, J., & Prada, M. (2005). Creating sustainable local enterprise networks. *MIT Sloan Management Review*, *47*(1), 33–41.

Yusuf, S. E., Ge, M., Hong, J. B., Kim, H. K., Kim, P., & Kim, D. S. (2016, December). Security modelling and analysis of dynamic enterprise networks. Paper presented at the IEEE International Conference on Computer and Information Technology, Nadi, Turkey.

ADDITIONAL READING

Hill, E. W., & Brennan, J. F. (2000). A methodology for identifying the drivers of industrial clusters: The foundation of regional competitive advantage. *Economic Development Quarterly*, *14*(1), 65–96. doi:10.1177/089124240001400109

Huggins, R., Morgan, B., & Williams, N. (2014). Regions as Enterprising Places: Governance, Policy and Development. In J. Liddle (Ed.), *Enterprising Places: Leadership and Governance Networks (Contemporary Issues in Entrepreneurship Research* (pp. 1–28). Emerald Group Publishing Limited. doi:10.1108/S2040-724620140000003007

Lechner, C., & Leyronas, C. (2012). The competitive advantage of cluster firms: The priority of regional network position over extra-regional networks–a study of a French high-tech cluster. *Entrepreneurship and Regional Development, 24*(5-6), 457–473. doi:10.1080/08985626.2011.617785

Lundberg, H. (2010). Strategic networks for increased regional competitiveness: Two Swedish cases. *Competitiveness Review, 20*(2), 152–165. doi:10.1108/10595421011029866

Mole, K., & Worrall, L. (2001). Innovation, business performance and regional competitiveness in the West Midlands: Evidence from the West Midlands Business Survey. *European Business Review, 13*(6), 353–364. doi:10.1108/EUM0000000006198

Sáez, L., Periáñez, I., & Heras-Saizarbitoria, I. (2017). Measuring urban competitiveness: Ranking European large urban zones. *Journal of Place Management and Development., 10*(5), 479–496. doi:10.1108/JPMD-07-2017-0066

Szczech-Pietkiewicz, E. (2017). Competitiveness of Cities and Their Regions in Poland–Changes in the Light of the Growth Pole Concept. In G.C. Pascariu, & M.A.P.D. S.Duarte, (Eds.) Core-Periphery Patterns Across the European Union, Emerald Publishing Limited (pp. 233-247), Bingley,UK: Emerald Group Publishing Limited. doi:10.1108/978-1-78714-495-820171008

Telfer, D. J. (2001). From a wine tourism village to a regional wine route: An investigation of the competitive advantage of embedded clusters in Niagara, Canada. *Tourism Recreation Research, 26*(2), 23–33. doi:10.1080/02508281.2001.11081340

KEY TERMS AND DEFINITIONS

Education: Is the process of transferring the knowledge and culture of humanity from generation to generation. The basic goal of education is to create a healthy society that excels high level of information. Education is responsible for changes in society and must be obey change more than any other segment of society.

Human Capital: Refers to the abilities, skills, and knowledge available to the total population, which are accumulated through formal education, experience in the labor market, additional education and the like.

Innovation: Is a change, a novelty or a process of making changes. In the humanities, the term refers to the process of modernization and positive change in services or their results. Innovation is the application of a new and improved idea, procedure, good, service, process that brings new benefits or quality in application. Innovations in a broader sense bring improvements in the field of product construction (technological innovations), process innovations, organization of work or business, marketing, service innovations, etc.

Regression Analysis: As a term is related to determining the mutual relations between two or more phenomena. For example, we may be interested in the relationship between the time spent preparing for the exam and the grade obtained on the exam, employees' salaries and their education, interest rates and money supply. In order to determine whether and to what extent these phenomena are dependent, we make regressions model. Regression analysis has a wide application in predicting and forecasting phenomena in various fields, such as economics, medicine, psychology, history.

Statistical Significance: Implies the decision whether the observed relationship between two or more variables was created by the action of a case or was created by the action of some experimental factor. In the social sciences, it is common to use a significance level of 0.05. This practically means that there is a 5% probability that the observed relationship between the variables is due to the action of the case. If 5% is considered to be a large value, a significance level of 0.01 (1%) can be selected. The level of significance is denoted by α, the Greek letter alpha.

APPENDIX

Table 4. The dependent and independent variables in the regression model of regional competitiveness

NUTS 2 regions	GDPpc	EXpc	EXD	IMD	S10000	P10000	GERD	E10000
Belgrade region	11319	2586	23	53	857	362	266911000	456
Vojvodina	6564	3061	47	53	308	342	99376000	370
Šumadia and Western Serbia	4340	1930	44	39	135	379	10457000	410
Southern and Eastern Serbia	4199	1739	41	33	187	358	17397000	324
Adriatic Croatia	12770	1780	14	19	371	374	77087000	503
Continental Croatia	13579	3989	29	53	421	395	424670000	392
Montenegro	7956	668	8	53	383	454	23490000	523
Southern Macedonia	5398	2934	54	72	289	346	39071000	744
Eastern Slovenia	19203	11503	60	53	204	407	296825000	807
Western Slovenia	27752	16590	60	53	549	486	595899000	1185
Nord-Vest	11576	2217	21	22	366	370	53742000	515
Centru	10897	2493	49	23	243	347	71646000	466
Nord-Est	7229	755	11	12	211	427	41368000	332
Sud-Est	9516	1461	15	19	162	384	17824000	372
Sud-Montenia	8897	2567	28	28	71	349	78196000	305
Bucuresti-Ilfov	26421	5074	18	41	746	334	644839000	272
Sud-Vest-Oltenia	8874	2217	21	15	148	414	36746000	404
Vest	11703	2493	43	37	314	365	80410000	412

Note: (1)EXD = (Export/GDP) * 100. (2) IMD = (Import/GDP) * 100. (3) S10000 = (The Number of Students Enrolled in Tertiary Education Population) * 10,000. (4) P10000 = (The Number of Pupils Enrolled in Secondary Education /Population)*10,000. (5)E10000 = (The Number of Companies /Population)*10,000

Source: Statistical Office of the Republic of Slovenia. (2021). *Exports, imports, trade balance and coverage of imports by exports, Slovenia, annually (cummulative data).* Retrieved from https://pxweb.stat.si/SiStatData/pxweb/en/Data/-/2490002S.px/.; The National Institute of Statistics. (2019). *Monthly statistical bulletin of counties no 11/2019.* Bucharest, Romania: The National Institute of Statistics; Croatian Chamber of Commerce. (2019). *Foreign trade of counties with foreign countries in 2019 - provisional data.* Zagreb, Croatia: Croatian Chamber of Commerce- (in Croatian); Republic Statistical Office (2019) Municipalities and regions in the Republic of Serbia. Belgrade, Serbia: Republic Statistical Office. (in Serbian); MAKStat Database. (2021). *Nadvoresna Trgovija Indikatori.* Retrieved from http://makstat.stat.gov.mk/PXWeb/; Statistical Office of Montenegro – MONSTAT (2019). *External trade in goods of Montenegro January - December 2019.* Retrieved from https://www.monstat.org/cg/novosti.php?id=3195.; Eurostat. (2021). *Regions database.* Retrieved from https://ec.europa.eu/eurostat/web/regions/data/database; Global Entrepreneurship Monitor Report (2021). *Entrepreneurial behaviour and attitude.* Retrieved from https://www.gemconsortium.org/data; The Joint Research Centre (2021). *Smart Specialization Platform.* Retrieved from https://s3platform.jrc.ec.europa.eu/s3-trade-tool

Table 5. Ranking of NUTS2 regions according to export per capita

NUTS 2 Regions	Export *per capita*	Rank
Western Slovenia	16590	1
Eastern Slovenia	11503	2
Bucuresti-llfov	5074	3
Continental Croatia	3989	4
Vojvodina	3061	5
Southern Macedonia	2934	6
Belgrade region	2586	7
Sud-Montenia	2567	8
Centru	2493	9
Vest	2493	10
Nord-Vest	2217	11
Sud-Vest-Oltenia	2217	12
Šumadia and Western Serbia	1930	13
Adriatic Croatia	1780	14
Southern and Eastern Serbia	1739	15
Sud-Est	1461	16
Nord-Est	755	17
Montenegro	668	18

Source: Statistical Office of the Republic of Slovenia. (2021). *Exports, imports, trade balance and coverage of imports by exports, Slovenia, annually (cummulative data).* Retrieved from https://pxweb.stat.si/SiStatData/pxweb/en/Data/-/2490002S.px/.; The National Institute of Statistics. (2019). *Monthly statistical bulletin of counties no 11/2019.* Bucharest, Romania: The National Institute of Statistics; Croatian Chamber of Commerce. (2019). *Foreign trade of counties with foreign countries in 2019 - provisional data.* Zagreb, Croatia: Croatian Chamber of Commerce- (in Croatian); Republic Statistical Office (2019) Municipalities and regions in the Republic of Serbia. Belgrade, Serbia: Republic Statistical Office. (in Serbian); MAKStat Database. (2021). *Nadvoresna Trgovija Indikatori.* Retrieved from http://makstat.stat.gov.mk/PXWeb/; Statistical Office of Montenegro – MONSTAT (2019). *External trade in goods of Montenegro January - December 2019.* Retrieved from https://www.monstat.org/cg/novosti.php?id=3195

Chapter 10
Eco-Innovation and IT Technologies for Sustainable Development of Health and Recreational Tourism of Serbia

Gordana Petar Djukic
https://orcid.org/0000-0001-5419-0725
Faculty of Economics, University of Belgrade, Serbia

Ilic S. Biljana
https://orcid.org/0000-0001-6137-8478
Faculty of Management, Megatrend University Belgrade, Zajecar, Serbia

Goran R. Milovanović
https://orcid.org/0000-0002-9758-3606
Faculty of Economics, The University of Nis, Serbia

ABSTRACT

The aim of the chapter is to point the importance of eco-innovation and IT technologies for the sustainable development of health and recreational tourism in Serbia. The subject of the research is the rehabilitation center in Eastern Serbia. The main idea of the chapter is to show how those hospital institutions use artificial intelligence-IT technologies for improving recovery services to patients in the post-COVID condition. The chapter will discuss the most common types of support and measures to facilitate the functioning of eco-tourism in Serbia with the aim to adopt good practices of developed countries (Hungary). Ecological tourism takes place in areas of pure and preserved nature. The contribution of the chapter is to point to new strategies in spa tourism, to shorten the time and reduce business costs. This would contribute to the sustainability of tourism.

DOI: 10.4018/978-1-7998-8900-7.ch010

INTRODUCTION

In today's time of modern business, the most important thing for all types of organizations is to accept changes. The sooner organizations adapt and introduce new innovative business techniques, the sooner they will be competitive in the market. However, with the globalization of the concept of sustainable development, every business has taken the premise of the sustainability.

Acting sustainably, as well as achieving the desired results without violating the right to natural resources, economy, and social dimension of future generations, is the basis of the concept of sustainable development. Tourism has also taken the premise of sustainability, becoming a new way of tourism business - ecotourism. Interest in ecotourism has never been more pronounced than it is nowadays. Based on the fact that the Coronavirus pandemic, as well as its suppression, has taken precedence on a global level, current situations have posed many challenges to modern mankind, primarily in healthcare. But first of all, the next question needs to be asked: What is the term or definition of sustainable tourism? Sustainable tourism represents an industry that has a minimal impact on the environment and local culture, while at the same time it earns money, creates new jobs, and protects local ecosystems (Ilic, Djukic, Radisavljevic, 2020). That is responsible tourism which is friendly to the natural and cultural heritage. Rural tourism is an inseparable part of ecological tourism because it is connected to the preservation and undisturbed nature. Health and recreational tourism belong to the category of rural tourism. Since the chapter will be based on health and recreational tourism, it is necessary to point to the fact that both types of tourism take place mostly in spas. Spas are destinations rich in sources of thermal and mineral waters, clean air, and favorable climatic conditions. It is not uncommon for specialized treatment institutions, i.e. rehabilitation centers, to be settled in spas, because of the providing patient accommodation services, using mineral healing water for medical purposes. Rehabilitation centers also provide many other medical services depending on the health problems that patients have. Considering the indications that are affected by the composition of mineral thermal water, specialized rehabilitation centers treat those diseases on which spas water has the best effect. Some spas treat that treat the skeletal system, air spas that are recommended for the treatment of lung diseases, diseases related to the nervous system, cardiovascular diseases, and more. In 2019, the COVID-19 virus crashed like a Tsunami on the Planet and attacked the entire human race (OECD,1-7). Mankind, as the youngest and most tender species on Earth, is trying to "save" life with all available knowledge - using modern technologies and scientific achievements, especially in medicine. The whole world is witnessing a race of mankind and viruses. Whether and how long, humanity can cope with the rates of mutation and adaptation of the virus (that is trying to survive) remains to be seen in the future. With the achievements of modern medicine, several types of vaccines against viruses have been found, and many of them are obtained with the most modern technologies - that have not been used before. Based on the fact that life always finds a way to survive, the entire humanity began to adapt to the situation caused by the pandemic. New technologies and new ways of business have been introduced in almost all branches of the economy. There are almost no economic sectors without huge business losses, but tourism as the branch of industry that promised the most income before the appearance of the COVID-19 virus, can be one of the branches with the biggest losses. Because the pandemic promotes social distance, it resulted in reducing travel to all world destinations. Income of the tourism sector (globally) dropped by 90 percent during 2020 and at the beginning of 2021 (Behsudi, 2020). Countries around the world, depending on their economic capabilities, responded by providing certain subsidies and assistance, but many organizations and employees in the tourism sector "unfortunately did not have that privilege." According to the forecasts of the World

Tourism Organization (UNWTO), the income from international tourism could fall by 60 to 80 percent in the coming period compared to the previous year. UNWTO estimates that the Coronavirus pandemic is the worst crisis that international tourism has faced since 1950. The worst affected areas in the world are Asia and the Pacific region with 33 million fewer tourist arrivals. Right behind them in Europe. If a better scenario is realized (a drop of 60 percent) and international borders will be opened, travel will be a little easier starting from July 2021, which would mean that losses in tourism will be approximately 910 billion dollars (Behsudi, 2020).

How much damage has tourism had in Serbia so far? According to some data, Serbia is approaching losses of 300 million euros (HORES). Although it seemed that Serbian tourism would have the most successful season "in recent history", based on the fact that this branch of the economy has been growing since 2013, accelerated its growth in 2017, and in 2019 set a record of 3.69 million arrivals (half domestic guests, half foreign) and 10.1 million overnight stays (about six million domestic and four million foreign tourists), at the beginning of 2020, more precisely in February, the fairy tale about the success of Serbian tourism is over. The nightmare in the form of the illness started - both domestically and globally - and there are still no indications when it will be finished (OECD, 1-24). Although there are no plans, people's desire for traveling still exists as there are exists the desires of the tourism sector to provide services. However, as the second year of the pandemic is still present, it is necessary to introduce new ways of business. Some of the tourist agencies in Serbia have replaced the offers of exotic travel with mountain tours in Serbia - introducing rural tourism. Travel agencies that introduced rural tourism stated that none of the passengers became infected on those trips, nor did they have any consequences! So, great news! In the chapter, the authors will try to answer (in some way) the following question "Should tourism organizations in the future base their business on rural tourism, as the only possible way out of the new pandemic situation?" But the answer to this question is not the only goal to which the authors aspire. Preference will be given to modern IT technologies and achievements in the field of health tourism, to emphasize the importance of their introduction for the treatment of patients. With the aim of applying good practice in Serbia when the use of new technologies is in focus, the authors will present the current situation in the health system of Serbia. Although less developed than other parts of Serbia, the Eastern part of the county has potential, especially for the rural, spa, and recreational tourism. The research will provide an insight into the sustainability of spa tourism, i.e. the entire chapter will indirectly investigate the degree of sustainability of health tourism – case study of Gamzigradska Banja spa (Eastern Serbia), which altogether represents the subject of this research.

BACKGROUND

Rural tourism represents tourism in rural households, but also certain vacations and places of residence like rural areas, as well as manifestations, festivals, recreations, production and sale of handicrafts, and agricultural products (Kosic, 2009). Rural tourism is therefore seen as a means of addressing the problem of weakening the agricultural potentials of agricultural areas to provide additional income (OECD, 1994; Davies & Gilbert, 1992; Alexander & McKenna, 1998). Serbia, especially its rural regions, is a country extremely rich in colorful flora and fauna, geothermal springs, healing mud, organic product production, and cultivation, as well as diverse folklore and cultural heritage. Therefore, it is necessary to implement a sustainable rural tourism policy in the future to maximize the positive and minimize the negative effects (Gajic, 2010). Ilic and colleagues have already researched the terms of rural tourism in Serbia. They concluded that the greatest potential of the Eastern part of Serbia lies in spas and recreational tourism, as a subfunction of rural tourism. In the research paper titled “Sustainable development directions of Rural Tourism of Timok Region“(2020), Ilic and colleagues considered the types of rural tourism such as: mountain, hunting, ecological, cultural, and recreational tourism, applying the mathematical methods of MCDM decision-making - ELECTRE and AHP, based on the taken criteria (economic, environmental and social). The authors opted for MCDM methods for several reasons (Ilic, Djukic, Balaban,2020).

1. The concept of MCDM refers to situations where there are several conflicting criteria (Cupic, Tummala, & Suknovic, 2001). MCDM methodology is based on the analysis of criteria and alternatives so that one alternative is better than the other (Durkalic et al., 2019).
2. The ELECTRE (Elimination and Choice Translating Reality) method was developed by Bernard Lee Roy (1968) as a response to the existing decision-making methods and as part of the Multicriteria Decision Theory. This method occupies a very significant place in the theory of Multi-Criteria Decision Making and in the papers of Vincke, Roy, Vanderpooten, and other authors who researched the field of MCDM (Vincke, P., 1992., & Roy, B., Vanderpooten, D.,1996). The ELECTRE method has 9 steps of calculating matrices based on certain criteria in scientific research of some phenomena. Because of the specific requirements and criteria - related to three fields: Economy, Sociology, and Ecology, by eliminating less favorable solutions, the ELECTRE method led to the most favorable project.
3. AHP method (Analytical Hierarchy Process) was proposed by Saaty, T. L. (1977,1980) to model subjective decision-making processes based on multiple criteria in a hierarchical system. This method is very convenient for determining the relative criteria weights (Saaty, 1990). For this reason, the authors used the method to determine the weight of project criteria.

The ELECTRE applied as the main, and AHP as an auxiliary method in choosing an adequate and optimal solution. According to these methods, one of the five proposed projects was the one that satisfied the choice of the optimal and best solution, and that was a project of health and sports-recreational tourism. This alternative was also very logical because the region of Eastern Serbia is rich in hydro-geothermal sources. With the chapter titled “Eco-Innovation and IT technologies for Sustainable Development of Health and Recreational Tourism of Serbia”, the authors continue to engage in scientific research on health and recreational tourism, wanting further contributing to business sustainability in modern conditions. Modern business means adapting to the situation in which the COVID -19 virus dictates new ways of life in all forms. During the COVID-19 crisis, many rehabilitation centers in Serbian spas redirected their health services to COVID hospitals, which treated and accommodated patients. However, as Serbia is one of the leading countries in Europe (but also in the world), as a country that has provided vaccines for its population (and for the region as well), as the number of COVID-19 patients has started to decrease, spa resorts have also begun to return to previous service delivery. However, based on the fact that there were not enough patients, for fear of contracting the virus, some rehabilitation centers have shifted their services, starting to treat and recover patients with post-COVID conditions, who need time to return to normal life. Although that the post-COVID condition is an insufficiently researched area, because such a condition is not yet treated as a special diagnosis in medicine, the authors will try to make pioneering steps in this research area. In the example of a spa resort in eastern Serbia, as much data as possible will be presented regarding the recovery of patients, as well as the introduction of new technology in the spa tourism sector. The main goal is to improve services, but above all to achieve sustainable business. Because of the fact that Rural tourism represents a form of sustainable tourism, the introduction of IT technologies contributes to business sustainability saving time and reducing costs. The introduction of an information system is a crucial step in improving any business. This is especially important for health facilities and hospitals, where this step is invaluable. The methods that the authors will use in the chapter are: descriptive, analytical, synthetic; and adequate literature in the field of health tourism, sustainability, and IT technology in health care.

IT Technologies in Healthcare and the Contribution of Sustainability

The worldwide use of computer technology in medicine began in the 1950s with the development of computers. In 1949, Gustav Wagner founded the first professional organization for health informatics in Germany. "Wagner's decision to start the journal came from his participation in the post-war German research on medical documentation (both paper and early computer-based) with its own professional society, the German Society for Medical Documentation and Statistics (GMDS.) Around this time, the documentation focus was reinforced by the consolidation of medical biometry and epidemiology research into academic departments" (Oberhoffer, 1962). Health informatics, called health information systems, is a discipline at the intersection of informatics, computer science, and health care. It consists of the resources, devices and methods, needed to optimize the collection, storage, retrieval, and using the information in health and biomedicine (Weiner and Wiley, 1948). Health Information Technology (HIT) is information processing that includes both computer hardware and software for a store, retrieve, share, and use health information, data, and knowledge for communication and making decisions (Bihari, 2021). HIT technology represents computers and communication systems that can be networked to create systems for moving health information. What is the background of information technology in medicine? Health informatics tools include computers, clinical guidelines, formal medical terminology, and information and communication systems(Hedman, Kalling, 2002). Knowledge of clinical protocols, their confirmations (or rejections) by screening in clinical practice should be widely available, as this increases the likelihood that better and more successful practices will be applied in an individual case - whether for diagnosis, treatment, or care. Specialized university departments and training programs in informatics began during the 1960s in France, Germany, Belgium, and the Netherlands (Kulikowski, 2017). During the research, medical informatics units began to appear, first in the 1970s in Poland and the USA. Since then, the development of high-quality health IT research, education, and infrastructure have been the goal of the United States, the European Union, and many developing countries. Education is faster with IT - with the development of IT, there have been significant changes in medical education around the world. The changes can be seen in the fact that most medical students these days are computer literate. New information on medical topics is easily available via the Internet - on computers and even mobile devices that have already become an indispensable part of medical equipment(Galbraith, 2012). Information technology can help in medical education in different ways, such as college networking and information sharing on the Internet. Computer-assisted learning (CAL), virtual reality (VR), human patient simulators as new digital products are just some of the options*(Figure 1).*

Medical students and their professors can stay in touch even when they are not in college and long after graduation, thanks to the "networking". It has never been easier to contact an authority for consultation than nowadays. Many medical schools use online programs such as Blackboard or a study center concept to coordinate their courses. Such programs provide quick access to information and a quick

Figure 1. New digital products at "Humana Studio H."
Source:https://images.app.goo.gl/kAMzhjBPs7mpRYoaA

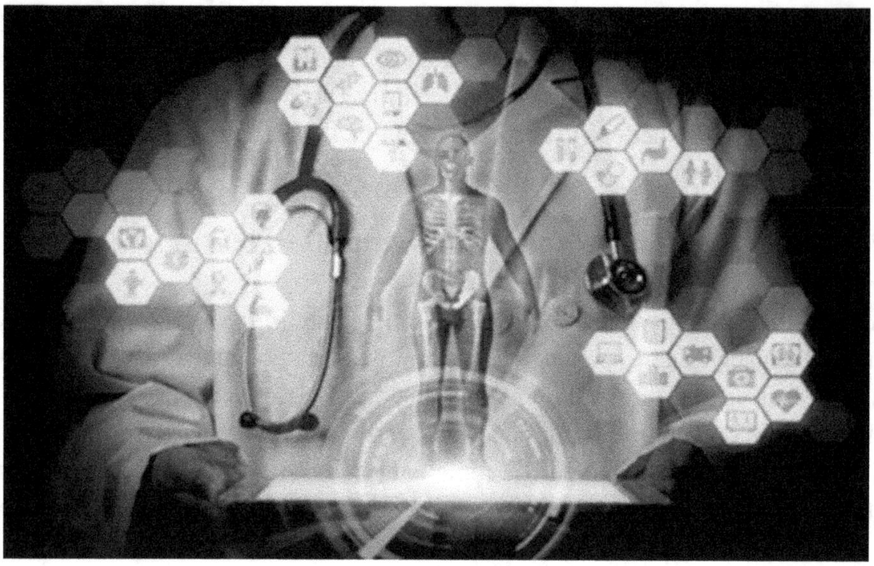

turnaround in evaluation and messaging. These teaching systems also enable all mentors, assessors, and students to look at their entire personal contribution and development (in the form of an automatic CV) anywhere. Similarly, the Internet provides opportunities to obtain up-date information on various aspects of health and disease talking to colleagues on different continents through online conferencing. Consul is unavoidable in more complicated cases, and the Internet provides instant consultation with the best diagnostic experts or treatment protocols (Burton, 2013).

Free access to various medical journals, online textbooks and the latest information on new developments in medicine also encourages and accelerates learning and research. Information technology is of great help to the entire health sector. The development and standardization of electronic medical records (EMR) is one example of significant advances that IT has provided to hospitals. This technology turns medical information into a single database.This not only reduces the cost of "paperwork", but also provides healthcare with access to relevant patient information, such as medical history, used medications and potential allergies or contraindications, insurance information, and more (Kates, Galbraith, 2007). This information can be accessed with just one click of the mouse, although there are large restrictions on the storage of this sensitive information and the issue of who may have access to this documentation. EMR, however, has great potential in the clinical arena. The ability to care for a patient with a medical record that is integrated with laboratory and pharmaceutical information, and provides service information related to preventive services, diagnosis, and treatment, represents a dramatic advance in the care and nursing of patients. Quality measurement would be immediately improved if all clinicians used EMR. In a situation where medical assistance is needed abroad or if the "chosen doctor" is simply not on duty when emergency intervention is necessary, the use of EMR solves these problems. With EMR, drug prescribing models of individual clinicians can be more carefully assessed and compared with established standards or even with doctors from completely different climates.The application of computer-based clinical support as part of EMR significantly improves the doctors efficiency, and thus the ability of patients to respond faster and to be correctly treated.

Reducing medical errors has become a priority all over the world. Potentially harmful drugs or inappropriate therapy are a common source of complications in hospitalized patients. In response to improve patient safety, computerized systems for entering a doctor's order (CPOE) are becoming more common. CPOE systems oblige doctors to write all medication orders online. These systems have the ability to check the accuracy of drugs - based on the patient's medication, automatically checking dose and contraindications. This way of working significantly reduces serious mistakes in treatment. Computerized orders can reduce treatment errors by as much as 80 percent, and negative (serious damage to patients) errors by 55%. In the United States, CPOE has become an almost mandatory tool in many clinics.Besides electronic medication, a standardized bar code drug dosing system could prevent a quarter of medication dosing errors - many of which are simple oversights that the system quickly detects. It is surprising how much information about patients - the risks of certain drugs, clear labels, avoidance of similar drug names, and appropriate instructions for use, can influence the appearance of unintentional mistakes in the therapy application. Arguments about the impact of informatics in medicine and medical education should not be disputed, but many aeas need to be improved, i.e. checked before IT can be harnessed.

And last but not the least, no matter how advanced the technology is, it can never replace the interaction that doctors or medical students have with the patient. The clinical assessment ultimately determines the excellent doctors. Researching modern technologies, it cannot be overlooked the importance of the relationship between doctor and patient. Today's computers are capable of performing far more complex tasks than quickly executing pre-programmed routines. Thanks to ML and AI techniques (Artificial Intelligence and Machine Learning), computers can "see" and "hear", and even analyze images and sounds. One of the goals is for computers to recognize, perceive, and solve problems as humans would (Genç,2019).

Figure 2 shows the evolution of AI. "The term "artificial intelligence" is often used to describe machines (or computers) that mimic "cognitive" functions that humans associate with the human mind, such as "learning" and "problem-solving". Machine learning is an application of Artificial Intelligence (AI) generating systems that can learn and improve without being programmed. Deep learning is a class of machine learning algorithms that uses multiple layers to progressively extract higher level features from the raw input" (Kedari, 2019). The undoubted advantage of computers is that they can work on a much larger set of data and faster than any person, even a group of people such as a medical council.

Figure 2. Evolution of AI
Source: Kedari, P. (2019) AI, ML and DL — Let's make it more clear!

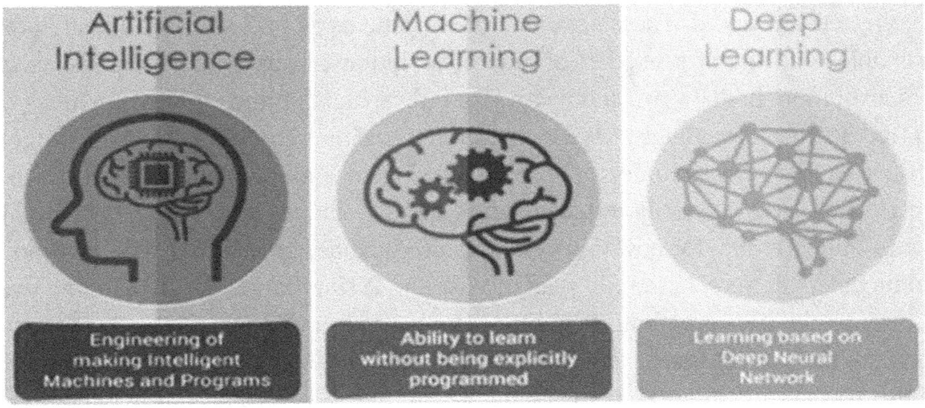

In some cases, computers have already taken precedence: Medical imaging analysis ¬ Computers can recognize elements in imaging, such as magnetic resonance imaging (MRI) or computed tomography (CT), and thus help determine diagnosis and therapy. ¬ Computers can analyze the relationships between data, people, and activities and help manage hospital admissions, optimize hospital bed occupancy, genomics, and many other areas. Computers can recognize and transcribe words spoken by doctors and nurses and make notes or reports at the same time. Because they understand the meaning of words and the nuances of medical terms, computers can translate in real-time, which is a great help to international medical teams (Digitalization in healthcare). IT technologies, web applications, service architecture of information systems for e-health, (for users and healthcare professionals) is a growing field in the public and private sectors (Radjenovic, Milovanovic and Milovanovic, 2017).

MAIN RESEARCH FOCUS

Due to the process of a long-term transition, Serbia "lags behind" in many advanced discoveries compared to the developed countries of Europe and the world. The social significance of the health information system in Serbia is based on the complex identification of the European health care systems, i.e. on their content, as well as on the possibility of integrating Serbian health care into European standards. Unfortunately, Serbia lags behind in the implementation of information systems comparing to middle-developed and developed countries. Inadequate use of information technology, as well as higher classes of information systems, is insufficient. No adequate and long-term information technology policy would gradually enable the IT application (Lutovac, 2008). For quality work, a doctor needs time and relief from administrative work, and this is exactly what ensures the use of IT and the transfer of certain administrative tasks to medical technicians (Petkovic, Lukic, 2013). In Serbia, the base of the Integrated Health System is used. The implementation of the Integrated Health Information System of the Republic of Serbia (ISIS - IZIS) began in 2016 in most health centers. This system, although not yet fully implemented, is available to both patients and healthcare professionals. Health care as one of the most important activities must be ready to respond to the challenges of rapid technological change, growing patient expectations, increasing equity, efficiency, and effectiveness. As a consequence, there was a need to integrate data on all health parameters as well as opportunities for international comparisons of national health-related data in all its sectors. With the emergence of the Integrated Health Information System, a significant technological and functional improvement has been achieved compared to the previous way of keeping data in health care. The health information system of the Republic of Serbia represents one central database in which all data are collected and stored. These data include all patients as well as all healthcare professionals. The system enables tracking of printed electronic prescriptions, instructions and all services provided by one health institution. In that way, an objective and transparent attitude towards patients is provided. Patients can find doctors through the "My Doctor" portal, i.e. they can see when there are free appointments with their chosen or desired doctor. The system includes all doctors employed in the Republic of Serbia, in General Hospitals or in tertiary institutions.The implementation of ISIS is still not fully and at all levels implemented, for now this system is used mainly by health workers within the Health Center. Health centers in Serbia are fully ready to use the ISIS program. By scheduling examinations for patients through the ISIS program, it enables their referral to Consultative-specialist examinations (ATD).In this way, patients schedule examinations faster - they are able to see the date and time when they need to call the Specialist Institution, ie the General Hospital. When ISIS is fully operational, patients will schedule appointments in tertiary institutions through a selected specialist doctor from the General Hospital. However, patients will still have to verify referrals through the Republic Health Insurance Fund. Information on free appointments with the selected doctor of the Health Center, patients can find on the official website of the system, and they can make an appointment by phone in general practice clinics every working day from 12 to 14 hours (www.mojdoktor.gov.rs).

As already written at the beginning of the chapter, Serbia is a country with unequally developed regions.„On the one hand, Belgrade, Novi Sad (big cities in Serbia) and the region of central Serbia represent the developed regions, while on the other hand the regions in the south and east of the country as less developed. Natural resources which can be used in a sustainable way, and which can improve the quality of life of citizens are more concentrated in the regions that are less developed. The concept of sustainable using natural resources can be considered as a substitution of renewable resources with non-renewable energy. Eastern region in Serbia is abundant with renewable energy resources, and it is advisable to use them“ (Ilić,2019). „The region of Eastern Serbia includes cities such as Zajecar, Bor and Negotin which are among the larger centers of this part of Serbia.This part of the county are rich in geothermal energy. The tourism potential of the Eastern Serbia, and Timok region is also unevenly developed. From relatively developed - Sokobanja, archeological sites "Felix Romuliana" and Lepenski Vir, Stara Planina Nature Park (Babin zub), Djerdap National Park, Gamzigrad Spa, Brestovac Spa, to underdeveloped cities - Zajecar and Bor and other municipal centers, with over 50 cultural, artistic and sporting events, natural and cultural-historical sights, hunting grounds, tourist spots in the neighborhood and the beginnings of rural tourism in a small number of surrounding villages. The existing tourists and recreational offer of the whole Timok region are not sufficiently recognized and developed“(Ilic, Djukic, Balaban, 2020). Uneven regional development in Serbia also refers to the situation of the health system. In Eastern Serbia, the real state of health is not very satisfactory. Patients expect a well-organized system, almost perfect, but in Serbia, especially in some parts of the country, the lack of „essence" is covered by „excess of form“, and instead of investing in the development and implementation of innovations in health, it is investing in some less important things (such as health uniforms, etc.). According to some authors, i.e. doctors, health care in Serbia is dehumanizing because doctors are often“surprised“ by the arising problems during the treatment of patients. Because of the situation with the COVID-19 virus, 2020 was especially difficult.

The health system was not functioning within the organization. The doctors who did not want to „make“ (give) a special effort, found the „Ideal Excuses“, while on the other hand, many doctors worked too hard, exposing themselves to the risk of contracting the virus. These doctors, who worked hard, were objectively unable to do everything necessary because, in less developed parts of the country, there was a lack of a well-organized health system in emergency conditions, as well as the lack of necessary and modern medical equipment. However, the state system, especially the health system, is expected to respond and provide appropriate support in such situations. Unfortunately, it remains to analyze the past unfortunate events as the pandemic passes. Because, if we start from the assumption that Serbia has an organized health system in all its parts, then the factor of "good luck" should not have any influence in such relations. This means that all patients should have equal status, i.e. the same treatment (Jović et al., 2016). Nature has given the Spa several sources of thermal-mineral waters, which are considered to be alkaline-earth - hyperthermia, i.e. Acrotherms.

The construction and arrangement of the spa bath began in 1920. In 1978, a modern Special Hospital for Rehabilitation "Gamzigrad" was built and started to work on the very shore of Crni Timok, on natural springs of thermo-mineral water whose temperature from a depth of 303 m, is 42° C. The hospital for

Figure 3. Network of health institutions ("as-is") in the region of Eastern Serbia
Source: Current situation - Public health institutions in the region of Eastern Serbia - Zajecar district https://optimizacijazdravstva.rs/lat/region-istocne-srbije

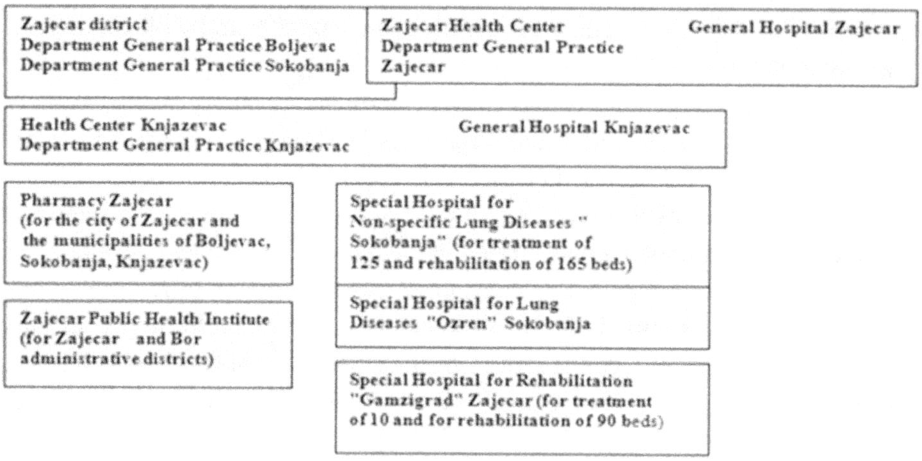

rehabilitation "Gamzigrad" has special importance for this place. Rehabilitation Center "Gamzigrad" is a specialized hospital for the treatment of peripheral blood vessel diseases (organic disorders of arterial blood vessels, functional disorders of arterial blood vessels, diseases of peripheral venous blood vessels, diseases of the lymphatic system) and vibration disease. The hospital also performs general rehabilitation of: connective tissue diseases, articular forms of rheumatism, extra-articular forms of rheumatism, orthopedic diseases and post-traumatic conditions, childhood anomalies, neurological diseases, gynecological diseases (Gamzigrad spa). There are two indoor pools with thermo-mineral water in the Special Hospital - a large one for adults and a small pool for children, which are surrounded by wall mosaics, impressing visitors with both their artistic expression and their size (the work of academician Prof. Srbinović). Thermal Mineral water is used in mineral and galvanic baths, local baths (four-cell baths, Haufe baths, alternating baths), then for underwater massage and gynecological spraying. The special hospital also applies electrotherapy with relatively modern devices. When the technical equipment of the hospital is in focus, from modern devices, i.e. technology, the following types of currents are used:

1. galvanic,
2. diadynamic,
3. interference,
4. exponential and
5. high-frequency currents.

Microwave and shortwave diathermy, as well as ultrasound, are also part of the rehabilitation therapy in Gamzigradska Banja, in the hospital. Special devices that work on the principle of changing the pressure: "Vasculator" and "Vacusac" are used for the treatment of blood vessels. The hospital also uses phototherapy, ultraviolet, and infrared lamps. Chinese therapy and occupational therapy are conducted in special halls. Thermo-therapy is applied with paraffin therapy. Hyperbaric oxygenation therapy is administered in a single hyperbaric chamber. Diagnosis and examination of the function of peripheral blood flow, heart, lungs, and peripheral nerves and muscles are especially represented. According to the data of the Statistical Office of the Republic of Serbia, in the Republic of Serbia in June 2020, compared to June 2019, the number of tourist arrivals decreased by 52.5%, and the number of overnight stays by 43.3%. In June 2020, compared to June 2019, the number of overnight stays of domestic tourists also decreased by 17.2%, and the number of overnight stays of foreign tourists by 84.8% (Statistical Office of the Republic of Serbia). This trend has been noticed in many spas in Serbia, especially in smaller ones, such as Gamzigradska Banja. To be able to return to the previous level of business, as well as to help patients who are recovering from the consequences of the COVID-19 virus, the Specialized Hospital "Gamzigrad" is included in the rehabilitation of these patients in the Spa's program. The trend of adjusting to the new business conditions began in the previous 2020 when the World Physiotherapist's Day was marked in September at the Special Hospital for Rehabilitation "Gamzigrad".

World Physiotherapist's Day is celebrated every year in over 100 countries, since 1951, when the World Confederation of Physical Therapy was founded. In 2020, the organization changed its name to "World Physiotherapy", but with the same goals - to point out the important role that this profession (physiatry) plays in preserving human health, to raise awareness of the physiotherapy profession, and to represent the profession before legislators and ministries, so that the position of the physiotherapist in the health system would be at the highest possible level. The topic of the World Physiotherapist's Day was related to the rehabilitation of patients after the virus COVID-19 (Postcovid) and also to the importance of physiotherapists in the treatment of people suffering from this disease (Gamzigrad Spa).

The meeting highlighted the fact that physiotherapists are crucial for the early and permanent rehabilitation of patients, who are recovering from severe forms of COVID-19. Recovery, as well as exercises, i.e. therapy, have a great effect on both the physical recovery and the mental recovery of the patient.

What is the situation in the Special Hospital "Gamzigrad", in terms of using IT? According to the conducted research, interview technique, i.e. conversation with the hospital director and management team, the authors learned that the special hospital "Gamzigrad" has not yet become an integral part of the ISIS base, although the spa management hopes that this will happen in the near future. Information technologies, especially modern achievements in the field of healthcare are not represented in Gamzigradska Banja, as well as in other spas in Serbia.The fact that the state of Serbia needs to allocate certain funds for the purchase and purchase of modern devices for diagnostics and treatment, as well as for the introduction of the necessary equipment for the implementation of information technologies within the health system of Special Hospitals, is emphasized. The authors emphasized the fact that the state of Serbia needs to allocate certain funds for the purchase of modern devices for diagnostics and treatment, as well as for the introduction of the necessary equipment for the implementation of information tech-

Figure 4. Illustrated representation - Rehabilitation after COVID-19
Source: https://www.gamzigradskabanja.org.rs/index.php/vesti

nologies within the health system of Special Hospitals. It is necessary for the state to invest the money in the education of health care employees, to be educated for using networked computer systems, as well as for working on them. From all the above, the authors conclude that Serbia, i.e. Serbian spas, lags far behind spas in other parts of Europe, especially for regulated and developed countries, such as Switzerland, Austria, and Germany.

According to Csapó and Marton (2017), the product of spa and wellness medical tourism is widespread throughout the country, except hilly areas and mountains, and represents a priority in Hungary's tourism policy. There are 224 tourist places in Hungary with medical and thermal spas, spread over 1000m^2. 1763 swimming pools on an area of 401,876m^2. and 202,967 maximum number of persons allowed. Schwertner, Jurayand Martyn (2014), considered that spa tourism in Hungary is a good example for Serbia because the experiences of Hungary can be successfully applied, in terms of infrastructure and facilities. Namely, spa facilities in Hungary have been expanded and adapted for health resorts that can be used not only by visitors but also by social security users. Additionally, for disease prevention and health recovery, they can use wellness services for physical and mental rejuvenation (Schwertner, Juray and Martyn, 2014). Agnes (2017), pointed that for every spa tourist destination, the tourist offer is important - diverse in terms of quality in health and wellness services. It is also important how to establish cooperation in the region and the surrounding area i.e. the cooperation with service units "which may even be competitors".Mórahalom spa in Hungary is a representative example of good cooperation between individual tourist service units (Ágnes (2017). According to Martyn (2013), in Mórahalom, located in the southern part of the state, which was an agricultural place, decision-makers recognized how to invest in thermal springs - building a health center.). As a country, Hungary has done much to promote own spa tourism based on the natural resources of their spa destinations. The fact is that in the Hungarian spas there is not enough data about the use of modern IT technology.

Issues, Controversies, Problems

Based on the official definition in Serbia, a spa is an area where there are healing springs such as: thermal and mineral water, air, gas, and healing mud (peloid), whose healing properties have been scientifically tested. Spas are natural resources for the common good, managed by the state based on thelaw (RS, 2018). According to Topalovic (2013), there are numerous problems in the spa tourism of Serbia that affect its development and expansion, such as: an inadequate strategic approach to spa tourism; the tourist offer is not satisfactory; insufficient stable financial resources; domestic tourists dominate over foreign ones; legal acts related to tourism have shortcomings, they are not applied to a sufficient extent; inappropriate tourist animation, potential tourists; lack of tourism market researching; obsolescence of medical equipment; unsatisfactory general and tourist infrastructure. Sustainable development is not fully accepted in practice, which has resulted in several negative effects, such as endangering natural geothermal springs and the ambiance of spas (Topalovic, 2013). According to Djukic (2015), for the sustainable and successful development of health tourism, it is necessary for local communities to make maximum efforts to use their existing resources adequately and to pay attention to spas as general socio-economic priorities. In this regard, the motivation of tourists to visit the spa is a quality, attractive, and "well-designed" integral health tourism product. This includes tourist programs with medical treatment services and other activities: sports and recreation, wellness programs, etc. to gain the comparative advantages of the spa, as a tourist destination (Djukic, 2015). According to (Belij, 2016), in his research, the most important factors influencing the attractiveness of spas as a tourist destination of spa and wellness tourism are:

the Attractiveness of health and wellness treatments in the spa (0.349), Natural conditions, and relaxing atmosphere of spas (0.206) and Price competitiveness (0.172), followed by the offer of sports activities (0.041). In terms of accommodation capacity, it is not satisfactory in terms of infrastructure because in some spas there are old hotels and special hospitals that have not been renovated for many years. Although spas have attractive natural potentials, which are competitive with European spas, they are not sufficiently used (Belij, 2016; Milicevic, 2015). Manic (2018), emphasizes the importance of tourism relations management (CRM) as a powerful marketing tool with appropriate technology because it is important for innovative changes. In addition, new technologies pose a challenge to tourism organizations, as the application of modern information technology could attract target tourists. For the key solution of problems in health and spa tourism of Serbia is important that policy-makers consider the great impact of information technology and digitization, media, and social networks; it is also necessary to provide additional education and training of staff and qualification structure, in line with trends in the tourism market and the requirements for professionalism and skills of health workers. It is the responsibility of governments to face the potential opportunities for the development of underdeveloped spas in rural regions and their natural resources and the application of IT and digitalization in all spas and health centers.

SOLUTIONS AND RECOMMENDATIONS

According to Radivojevic (2020), for recovery of the tourism sector in Serbia, state assistance is important, with a selective strategy in the coming period that is primarily focused on the categories in which it would have the greatest impact on recovery. In this regard, the recommendations for recovery also refer to spas as health and recreation centers that are important in the era of COVID-19 for recovery and having in mind that Serbia is a spa region. The secondary attitude towards tourists in the pandemic era should be based on the standards and prescribed health protocols of the government, domestic and foreign health organizations. Besides, for a unique spa tourist offer, Serbia should diversify its tourist product more, to attract foreign tourists; given that in Serbia there are inexhaustible opportunities for spas as a primary tourism industry, highly educated and professional staff is one of the important factors for quality medical and wellness services. Medical and non-medical staff employed in spas and recreation and health centers should be continuously professionally educated in the field of medicine (and modern information technology). Based on a study by the WB6 CIF expert team (2020), which deals with research on opportunities and tourist travel in the era of the COVID-19 pandemic, in the region of 6 Western Balkan countries (WB6), and based on the Report – The COVID-19 crisis in the Western Balkans of OECD), the following recommendations were made: Effective (and continuous) cooperation is needed between the government; Promote domestic tourism;Prominent medical measures of protection and prevention of health and safety to restore the trust of tourists; Introduce targeted marketing campaigns promoting ecotourism, and Consider the current situation in tourism policy and define following modern policy, innovative tourism products. Besides, it is recommended to the governments of the countries in the region to "reduce taxes for tourism and travel industries until coronavirus pandemic is put under control" and to provide medical resources to public health stakeholders (WB6 CIF, 2020).

FUTURE RESEARCH DIRECTIONS

According toWong and Sa'aid Hazley (2020), the future directions of health tourism are in the application of advanced and currently available technology, especially IR 4.0, because today the modern, virtual way of health care is increasingly used. This was even more pronounced during the Corona pandemic, which caused a stalemate in global tourist travel. Technological innovations are very important for future travel and mobile healthcare. Medical diagnoses can be made based on digitized medical devices, such as portable medical devices, digital sensors, biotelemetry. The results of the digital diagnosis can be linked to the user's smart mobile phone (health-travel). Wong and Hazley further explained that knowledge and technologies in health and tourism are very important in the coming period in all parts of the world. Technological innovations provide opportunities for healthcare professionals to become familiar with them, which in turn provides users with proactive measures for better healthcare services and controls, digitally and remotely. Applied virtual rehabilitation would contribute to the health tourism industry, as health services are integrated with digital sensors carried by users and the range of their movement is recorded. Therapists and healthcare professionals analyze digital data and continuously monitor users 'medical parameters via smartphones or tablets. The application and availability of information technology virtually would increase the mobility of professionals (doctors, nurses, professors, and nursing staff). This creates the conditions to improve health tourism with IR 4.0, in developing countries (and Serbia) to be ready to welcome tourists. The traditional way of medical examinations of tourists would be technologically innovative in health care facilities and in the time and post COVID - 19 era, therefore, a large amount of physical movement will not be necessary (Wong and Sa'aid Hazley, 2020). In the context of the application of technology in health tourism, it would be the key to rapid tourism and recovery. Governments of countries (and Serbia) should, as decision-makers, engage as a matter of priority in the implementation of technological innovations in health care institutions and health tourism, "providing a timely platform for knowledge exchange and development of strategic partnerships" (Wyman, 2020).

CONCLUSION

Serbia has no sea, that is true. However, Serbia has something else - a lot of groundwater that springs in spas. Those sources should be valued much more than the state of Serbia does. Considering that spas are small oases of health because of the abundance of healing mud, mineral, and thermo-mineral water of various compositions and quality chemical structure, it is necessary for the development of spa centers, to give these places a phenomenal and complete atmosphere. Namely, spas are not exclusively rich in healing thermal water, but also the natural environment and healthy air, due to the lush vegetation and the altitude at which they are located.

Although the emphasis is often placed on their tourism potential, spas are centers for medical care - each of the spas is specialized to help a certain group of patients. Another characteristic is a very good price for accommodation and travel. There are large hotels and accommodation complexes in the spas that besides delicious traditional foods also have swimming pools (even though spas have their sources of mineral water). Many cafes and restaurants leave an even stronger impression in such an environment, and what is most important is that in spas vicinity, tourists can visit natural beauties - cultural and historical monuments and monasteries that attract both domestic and foreign tourists. For the younger population of Serbia, going to spas is a rather foreign thing, except in the period when they were children, so they went there with their parents and family. However, in the present situation that is related to the COVID-19 virus, spas can provide even more than going to sea. It is necessary to emphasize the fact that the high sophistication of tourist demands, the desire for recreation and relaxation in Serbian spas, indicates the necessity of adjusting the holders of tourist offers and health - to take a competitive position. The introduction of new technology and innovation is the future of any business, especially businesses related to health care and the treatment of patients. The benefits of IT technologies in modern and turbulent business conditions are immeasurable. The chapter presents the achievements in IT technologies, especially their application and importance in the health sector, on a global scale. However, based on the facts that are given in the text, the conclusion is that the introduction of IT technologies in the Serbian health system is at the very beginning, while in Serbian spas, or specialized institutions for the treatment of patients, these technologies are still not represented. In the short-term future period, related to the development and use of modern technologies, which could refer to the spas of Serbia, it is necessary to promote the spa facilities of Serbia through modern Internet technology. One of the ways to make the Serbian spas recognizable and competitive on the tourist market is the application of an appropriate combination of communication technologies with potential tourists or patients who need a stay in the spas. Because of the growing demands, based on the fact that tourism and health demand for spa services is quite "physical" away, it is necessary to inform the target group about the content and services of spas in Serbia. The application of modern information technology in the promotion of the spa tourism offer would enable significant benefits for tourists and carriers of the tourist offer. Creators of marketing activities in Serbian spas would realize the benefit in the form of simpler, more economical, and more meaningful contact with potential tourists and patients. From the aspect of visitors, the benefits of information technology would be reflected in simpler information about the spas offer as well as in the possibility of sharing pictures and other contents about the spa, in which they stayed or possibly plan to stay. Of course, it is always needed to see the best practices of the developed countries of Europe and the world.

ACKNOWLEDGMENT

This research received no specific grant from any funding agency in the public, commercial, or not-for profit sectors.

REFERENCES

Ágnes, U. T. (2017). The Examination of the Importance of the Thermál Panzió in Mórahalom. *EBEEC Conference Proceedings The Economies of Balkan and Eastern Europe Countries in the Changed World*, 226. 10.18502/kss.v1i2.659

Alexander, N., & McKenna, A. (1998). Rural tourism in the heart of England. *International Journal of Contemporary Hospitality Management*, *10*(5), 203–207. doi:10.1108/09596119810227820

Behsudi, A. (2020). Wish you were here - tourism-dependent economies are among those harmed the most by the pandemic. *Finance & Development*. https://www.imf.org/external/pubs/ft/fandd/2020/12/pdf/impact-of-the-pandemic-on-tourism-behsudi.pdf

Belij, M. S. (2016). *Transformation of Serbian spas into modern centers of spa and wellness tourism: dissertation*. Faculty of Geography, University of Belgrade.

Bihari, M. (2021). *Using Health Information Technology to information management*. https://bs.approby.com/koriscenje-zdravstvene-informacione-tehnologije-za-upravljanje-vasim-informacijama/

Bodeker, G., & Cohen, M. (2008). Understanding the Global Spa Industry. Elsevier.

Burton, R. M. (2013). An Interpretative Synthesis in Three Themes. *Journal of Organization Design*, *1*(2), 42–44. doi:10.7146/jod.7363

COVID-19 Crisis in Serbia. (2020). *OECD*. https://www.oecd.org/south-east-europe/COVID-19-Crisis-in-Serbia.pdf

COVID-19 Crisis in the Western Balkans: economic impact, policy responses and short-term sustainable solutions. (2020). *OECD*. https://www.oecd.org/south-east-europe/COVID-19-Crisis-Response-Western-Balkans.pdf

Csapó, J., & Marton, G. (2017). The Role and Importance of Spa and Wellness Tourism in Hungary's Tourism Industry. *Czech Journal of Tourism*, *6*(1), 55–68. doi:10.1515/cjot-2017–0003

Cupic, M., Tummala, R., & Suknovic, M. (2001). *Decision Making: Formal Approach*. SZR Zuhra.

Davies, E. T., & Gilbert, D. C. (1992). A case study of the development of farm tourism in Wales. *Tourism Management*, *13*(1), 56–63.

Digitalizacija u zdravstvu [Digitalization in healthcare]. (2019). PC Press.https://pcpress.rs/digitalizacija-u-zdravstvu-i-medicini/

Djukic, N. (2015). *Economic indicators of sustainable tourism in the leading spas in Serbia: master thesis*. Faculty of Mathematics and Natural Sciences, University in Nis, Serbia, p. 3.

Durkalic, D., Furtula, S., & Borisavljevic, K. (2019). Ranking tourism market performance in EMU countries: Results of PROMETHEE-GAIA approach. *Menadžment u hotelijerstvu i turizmu*, *7*(2), 67-76.

Gajic, T. (2010). Tourism in the Intensification of the Economic Development of Receptive and Emitting Areas - An Example of the South Backa District. *Industry, 38*(3), 139-155.

Galbraith, J. (2012). The evolution of enterprise organization designs. *Journal of Organization Design*, *2*(1), 1–13.

Gamzigradska banja [Gamzigrad Spa]. (n.d.). https://www.gamzigradskabanja.org.rs/index.php/vesti

Gamzigradska banja – Specijalna bolnica [Gamzigrad Spa – Special hospital]. (n.d.). https://euroturs.rs/bamjezszrbija-banja/gamzigradska-banja-specijalna-bolnica

Genç, Ö. (2019). *Notes on Artificial Intelligence, Machine Learning and Deep Learning for curious people*. https://towardsdatascience.com/notes-on-artificial-intelligence-ai-machine-learning-ml-and-deep-learning-dl-for-56e51a2071c2

Hedman, J., & Kalling, T. (2002). *IT and Business Models: Concepts and Theories*. Copenhagen Business School Press.

HORES - Business Association of Hoteliers and Restaurateurs of Serbia. (n.d.). http://www.hores.rs/odrzivost.html

Ilić, B. (2019). Social Component of Sustainable Development and Quality of Life: Region of the Balkans, Eastern Serbia: chapter. In Handbook of Research on Creating Sustainable Value in the Global Economy. IGI Global.

Ilic, B., Djukic, G., & Balaban, M. (2020). Sustainable development directions of rural tourism of Timok region. *Economics of Agriculture*, *67*(1), 157–174.

Ilic, B., Djukic, G., & Radisavljevic, G. (2020). Understanding ecotourism and protection of natural values - a comparison of America and Serbia. *Conference Proceedings Innovation as a driver of development, Innovation to business success*, 253-262.

Jovic, M., Jovic, L. J., Milosevic, J., Manic, S., Rajkovic, T., & Djordjevic, P. (2016). Pravni i etički problemi prinudne hospitalizacije – prehospitalni aspekt [Legal and ethical problems of involuntary hospitalization - prehospital aspect]. *HALO*, *22*(1), 60–89.

Kates, A., & Galbraith, J. (2007). *Designing your organization using the star model to solve 5 critical design challenges*. Jossey-Bass.

Kedari, P. (2019). *AI, ML and DL — Let's make it more clear!* https://community.letsupgrade.in/post/ai-ml-and-dl----let-s-make-it-more-clear-5eda6110d5a8b90b708491c2

Kosic, K. (2009). *Vojvodina Rural Tourism and Sustainable Development: Ph.D. Thesis*. The University of Novi Sad, Faculty of Science, Department of Geography, Tourism and Hotel Management, Novi Sad.

Kulikowski, C. A. (2017). Historical Roots of International Biomedical and Health Informatics: The Road to IFIP-TC4 and IMIA through Cybernetic Medicine and the Elsinore Meetings. *Yearbook of Medical Informatics*, *26*(1), 257–262.

Lutovac, D. M. *Informacioni sistem i informacione tehnologije u zdravstvu* [Information system and information technologies in healthcare]. Škola biznisa. http://www.vps.ns.ac.rs/SB/2008/1.15.pdf

Manic, M. (2018). Marketing approach to specific forms of tourism in the context of revitalization of underdeveloped areas of Eastern Serbia. *Marketing*, *49*(4), 305–306.

Martyn, Z. (2013). Tourism and local development – the case of Mórahalom. *Geographica Timisiensis*, *22*(2), 49–64.

Milicevic, S., Milovanovic, V., & Milutinovic, S. (2012). Wellness tourism in the function of change management in the tourism market. *Economics: Theory and Practice,* *5*(2), 29-41.

Oberhoffer, G. (1962). Report on an international seminar for medical documentation and statistics. *Methods of Information in Medicine*, *01*, 27–31.

OECD. (1994). Tourism strategies and rural development. Organization for Economic Co-Operation and Development.

OECD. (2020). https://www.oecd.org/south-east-europe/COVID-19-Crisis-in-Serbia.pdf

OECD. (2020). https://www.oecd.org/south-east-europe/COVID-19-Crisis-Response-Western-Balkans.pdf

Optimizacija zdravstva - Zaječarski okrug [Optimization healthcare, Zajecar district]. (n.d.). https://optimizacijazdravstva.rs/lat/region-istocne-srbije

Petkovic, M.& Lazarevic, S. (2012). Upravljanje interorganizacionim odnosima: dizajn centra zajedničkih usluga [Inter-organizational relations management: the design of a shared services center]. *Management, 64*.

Petkovic, M., & Lukic, J. (2013). Uticaj informacione tehnologije na dizajn organizacije: Primer organizacije u zdravstvu [The influence of information technology on the design of the organization: the example of the organization in health care]. *Sociologija, 55*(3), 439–460.

Radivojevic, A. (2020). *The Tourism sector in the Republic of Serbia during the pandemic. Quarterly Monitor, 61.*

Radivojevic, A. (2020). *The Tourism sector in the Republic of Serbia during the pandemic. Quarterly Monitor, 61.*

Radjenovic, Z., Milovanovic, S., & Milovanovic. (2017). Heliant Health Information System as a Support to Electronic Business of Healthcare Organizations in Serbia. *Facta Universitatis Series: Economics and Organization, 14*(3), 267. doi:10.22190/FUEO1703265R

Republic of Serbia. (2018). *Spa Law*. Official Gazette of RS, 95/2018, Belgrade, the Republic of Serbia.

Republic of Serbia. (2018). *Spa Law*. Official Gazette of RS, 95/2018. https://www.paragraf.rs/propisi/zakon-o-banjama-republike-srbije.html

Roy, B., & Vanderpooten, D. (1996). The European School of MCDA, Emergence, Basic Features, and Current Works. *Journal of Multi-Criteria Decision Analysis, 5*(1), 22–37.

Saaty, T. L. (1990). How to make a decision: The analytic hierarchy process. *European Journal of Operational Research, 48*(1).

Schwertner, J., Juray, T., & Martyn, Z. (2014). *Integrated spa development strategy for spas situated in the area along the main highway traffic route crossing the Hungarian-Serbian border region. In Cooling Cubes Project.* The University of Szeged, Department of Economic and Social Geography.

Statistical Office of the Republic of Serbia. (n.d.). https://publikacije.stat.gov.rs/G2020/Pdf/G20201205.pdf

Topalovic, S. (2013). *Status and perspectives of spa tourism in Serbia. Globe, No. 37-38.* Serbian Geographical Society. http://sgd.org.rs/publikacije/globus/3738/7-banjski%20turizam.pdf

Ursache, M. (2015). *Niche Tourism Markets – Means of Enhancing Sustainable Economic Development in EU's Eastern Periphery.* CES Working Papers, ISSN 2067-7693. Centre for European Studies, Alexandru Ioan Cuza University, 7(2a), p. 654.

Vincke, P. (1992). *Multicriteria Decision - Aid.* John Wiley and Sons.

WB6 CIF (Western Balkans 6). (2020). Prospects for Travelling and Tourism Sector in the Western Balkans in 2020: Study no. 1. *Studies on obstacles and opportunities for doing business in the region: study no. 1*, 1-52.

Weiner, N. (1948). *Cybernetics or Control and Communication in the Animal and the Machine*. Wiley.

Wong & Hazley (2020). The Future of Health Tourism in the Industrial Revolution 4.0. Era. *Journal of Tourism Futures*. DOI doi:10.1108/JTF-01-2020-0006

WTO (World Tourism Organization) &European Travel Commission (ETC). (2018). *Exploring Health Tourism – Executive Summary*. UNWTO. doi:10.18111/978928442030.8

Wyman. (2020). *The Future of Travel &Tourism in the Wake of COVID-19*. World Travel & Tourism Council.

ADDITIONAL READING

Almeida-Dias, J., Figueira, J. R., & Roy, B. (2020). Electre Tri-C: A multiple criteria sorting method based on characteristic reference actions. *European Journal of Operational Research*, *204*(3), 565–580. doi:10.1016/j.ejor.2009.10.018

Esfendiari, H., & Choobchian, S. (2020). *Designing a Wellness-Based Tourism Model for Sustainable Rural Development*. Research Square., doi:10.21203/rs.3.rs-63947/v1

Figueira, J., & Roy, B. (2002). Determining the weights of criteria in the ELECTRE type methods with a revised Simos' procedure. *European Journal of Operational Research*, *139*(2), 317–326. doi:10.1016/S0377-2217(01)00370-8

Karadayi-Usta, S., & Asan, S. S. (2020). A Conceptual Model of Medical tourism Service Supply Chain. *Journal of Industrial Engineering and Management*, *2*(13), 246–265. doi:10.3926/jiem.3008

KEY TERMS AND DEFINITIONS

AHP (Analytical Hierarchy Process): Is a structured technique of MCDM for organizing and analyzing complex decisions, based on mathematics and psychology. Individual experts' experiences are utilized to estimate the relative magnitudes of factors through pair-wise comparisons. Also, AHP enables the evaluation of alternatives with unclear and very subjective criteria. An inconsistency index is calculated for each estimation matrix, which represents how much is expert well filled in the assessment matrix, i.e. how much he respected the principle of transitivity.

Competitiveness of Tourism Destination: Is the ability to increase tourism expenditure, to increasingly attract visitors while providing them with satisfying, memorable experiences, and to do so in a profitable way, while enhancing the well-being of destination residents and preserving the natural capital of the destination for future generations.

ELECTRE (ELimination Et Choix Traduisant la REalité - Elimination and Choice Expressing the REality): Family methods, designed for multiple criteria decision aiding. These methods use as a preference model an outranking relation on the set of actions - it is constructed as a result of concordance and non-discordance tests involving a specific input preference information.

MCDM (Multi-Criteria Decision-Making Methods): Decision-making at all levels in the organization should be based on stimulating competitive proposals and choosing the best possible alternative for business. It sets three levels of decision-making: strategic decisions - which have a long-term effect, tactical decisions - which represent decisions for the next business year, and operational decisions that include daily organizational activities. The various methods of MCDM help managers to choose the one that is best for achieving the set goals from all the offered alternatives.

Post-COVID Condition: Condition after suffering from COVID-19 virus, which requires a serious approach to patients, i.e. serious treatment. The most common problems in post-COVID conditions are psychological.

Spa Tourism: Spa tourism as the use of thermal, mineral, and thermal-mineral waters, for treatment, rehabilitation, and recreation where numerous recreational, sports, and cultural activities are held.

Tourist Attraction: Is a particularly attractive tourist attraction destinations, of a natural or social character within the tourist space: natural attractions, cultural attractions, museums, historical heritage, events of the different spectrum, entertainment, etc.

Wellness: Originates as a coin of two terms, well-being, and fitness, denoting the orientation towards the maximum use of the potential of individuals. This primarily includes activities that affect the improvement of the physical and mental structure of the personality, such as a healthy diet, regular exercise, active lifestyle, quality time spent with friends and family, etc.

Chapter 11
What Drives Eco-Design Innovations in European SMEs?

Gamze Ozturk Danisman
https://orcid.org/0000-0003-3684-6692
Kadir Has University, Turkey

ABSTRACT

Building on the natural-resource-based view, and using a sample of 7,165 European SMEs, this chapter investigates the drivers of eco-design innovations among SMEs under three categories: (1) sustainability-oriented firm capabilities, (2) technological capabilities, and (3) access to finance. The findings reveal that sustainability-oriented capabilities achieved through investments into circular economy are the strongest driver of SMEs' eco-design innovations. Firms' technological capabilities are also found to boost their ability to adopt eco-design innovations. While equity finance increases the possibilities for SMEs to devote resources to eco-design, grant finance is interestingly observed to decrease such possibilities. The more traditionally used form of debt finance remains detached from eco-design implementations. The study contributes to a better understanding of how eco-design practices can be broadened within SMEs and highlights policy recommendations in this regard.

1. INTRODUCTION

The social and regulatory pressures for companies to take into account their environmental impact are on the rise and penetrate the activities of small businesses as well as their large counterparts. Hence, enterprises of all sizes are expected to introduce environmental innovations (i.e., eco-innovations) to ensure their environmental impact can be curbed whilst corporate performance is not affected (Battaglia et al., 2020). Despite some recent contributions, the eco-innovation (EI) efforts and potential of Small and Medium-Sized Enterprises (SMEs) remain less explored in the eco-innovation literature, with most efforts concentrating on large firms (Hoogendoorn et al., 2015). Given their substantial impact on the environment, with 60-70% of the total pollution being caused by SMEs alone, exploring SMEs' eco-innovations and their environmental impact stands as a crucial issue (European Commission, 2010).

DOI: 10.4018/978-1-7998-8900-7.ch011

Understanding the product eco-innovations are particularly key for SMEs that focus the majority of their innovation efforts on products (Bossle et al. 2016). Recent literature states that it is crucial to intervene during the design stage of the product/service innovation through "eco-design," evaluating the environmental consequences of products through the product life cycle (Boks, 2006; Deutz et al., 2013). The design stage of the innovation process is a key intervention point in the life cycle of a product for any environmental considerations (Deutz et al. 2013) and determines 80% of the environmental impact any product or service is likely to incur through its life cycle (McAloone and Bey, 2009). Recent policy efforts such as the Waste Electrical and Electronic Equipment (WEEE) Directive and Restriction of the use of Hazardous Substances (RoHS) Directive and Eco-design Directive (European Commission, 2009) aim to systematically integrate environmental considerations in product design. While eco-design is accepted as an important element in new product development, more detailed empirical evidence is required for a better understanding of the factors that affect the introduction of such practices within firms in general and SMEs in particular.

Building on the natural-resource-based view of the firm (NRBV) and eco-innovation literatures, we aim to shed light on this under-researched area by exploring the role of (1) sustainability-oriented firm capabilities, (2) technological capabilities, and (3) access to finance, on eco-design innovations of SMEs. We use a database of 7,165 SMEs from 28 EU countries acquired by the Flash EuroBarometer 441 survey in 2016. Given the crucial role of SMEs in the economy and businesses (European Commission, 2010), the eco-design practices of SMEs deserve close attention (Hoogendoorn et al., 2015). Our results indicate that sustainability-oriented capabilities are the strongest driver of eco-design innovations. Broader technological capabilities of SMEs are also observed to increase SMEs' ability to adopt eco-design practices. Access to equity finance is positively and significantly associated with adopting eco-design activities, increasing the willingness of SMEs to devote resources to such projects. However, access to grant finance decreases the possibilities of SMEs to conduct eco-design. Debt finance, on the other hand, does not have a significant impact on eco-design activities.

The following sections outline the theoretical foundations and empirical evidence, as well as the methodology and data that were used to empirically investigate the relationships focal to the study. Finally, the results are presented, and implications are discussed. The study concludes with avenues for further research.

2. THEORETICAL BACKGROUND AND HYPOTHESES DEVELOPMENT

The *resource-based view* (RBV) and the *natural resource-based view* (NRBV) provide the theoretical foundations for discussing eco-innovations in general and eco-design innovations in particular. The RBV from the strategic management literature assumes that unique, non-substitutable, and difficult to imitate resources and capabilities are the sources of competitive advantage (Barney, 1991). The NRBV, building on RBV and dynamic capabilities literature, states that environmental constraints lead firms to adjust and renew their capabilities and resources in a way to obtain new sources of competitive advantage through achieving sustainable technologies and products (Hart and Dowell, 2011; Hart, 1995). In this paper, we use the NRBV as a foundation to understand the role of a firm's resources and capabilities that facilitate its ability to introduce eco-design innovations.

2.1 Eco-Design Innovations

Eco-innovation is defined as "the development of products (goods and services), processes, marketing methods, organizational structure, and new or improved institutional arrangements, which, intentionally or not, contribute to a reduction of environmental impact in comparison with alternative practices" (see OECD, 2009, p. 2). *Eco-design* is a typical example of product eco-innovation that aims to redesign products by undertaking environmental assessments to reduce any harmful effects. The design stage of the innovation process is crucial because it determines 80% of the environmental impact any product or service is likely to incur through its life cycle (McAloone and Bey, 2009). The design stage enables proactive planning by either improving existing systems or creating an entirely new system that focuses on minimizing environmental impacts (Carrillo-Hermosilla et al. 2010). It focuses on the entire life-cycle of a product from raw materials to final disposal, which helps to determine the stages with the highest environmental impact and set design interventions accordingly. Many tools and methodologies are available to facilitate eco-design practices, such as the life-cycle analysis (LCA), nature-inspired design strategies (NIDS) such as biomimicry, the cradle-to-cradle (C2C) approach, and Design for the Environment (DFE), among others. Many of these tools can be classified under the umbrella term, Design for X (DFX), which incorporates several features of environmental design strategies such as Design for Recycle, Design for Disassembly, Design out Waste, Design for Modularity, etc. (Rashid et al., 2013; Prieto-Sandaval et al., 2018). While eco-design aims to minimize natural resources and energy consumption and reduce the harmful impacts on the environment, it further aims to increase benefits for consumers. Products with improved environmental attributes lead firms to benefit from cost reductions, improved product life cycles, enhanced competitive advantage to firms, and increased customer satisfaction (Boks, 2006; Plouffe et al., 2011).

The eco-innovation literature suggests that most eco-innovations are stimulated by external factors such as *regulatory requirements* (Bossle et al., 2016) and *market-demand* (Horbach et al., 2012; Lin et al., 2013) as well as firm-specific capabilities, including *technological capabilities* (Ghisetti & Pontoni, 2015) and *sustainability-oriented firm capabilities* embodied in sustainable corporate practices and commitment (Demirel and Kesidou, 2019; De Marchi, 2012; Dangelico et al., 2017; Melander, 2018). While a range of different eco-innovations (e.g., product, process, and organizational eco-innovations) are considered in these studies, no large-scale study, to date, examines the factors that boost eco-design innovations. Previous literature on eco-design innovations mainly focuses on case studies with a relatively small number of companies that either excel with best practices or have implemented a unique eco-design approach (Deutz et al., 2013). In this paper, we aim to shed light on the drivers of eco-design innovations under three categories that are central to the NRBV and focus on the firm's internal resources and capabilities: (1) sustainability-oriented firm capabilities, (2) technological capabilities, and (3) access to finance.

2.2 Sustainability-Oriented Firm Capabilities

In addition to external drivers of eco-innovations (i.e., regulatory requirements and market demand) (Bossle et al., 2016; Ghisetti & Pontoni, 2015); recent literature, motivated by the NRBV, points out that sustainability-oriented firm capabilities are essential for eco-innovations (Demirel and Kesidou, 2019; Dangelico et al., 2017). Sustainability-oriented capabilities refer to the set of resources, abilities, and competencies that allow firms to integrate environmental sustainability into new product development through eco-innovations (Demirel and Kesidou, 2019; Dangelico et al., 2017;).

Strategic management literature has emphasized the importance of environmentally-specific capabilities -such as a green corporate culture, environmental research and development, human resource and organizational capabilities, as well as green marketing capabilities for achieving eco-innovations (Melander, 2018; Kabongo and Boiral, 2017). In this paper, we argue that the most relevant sustainability-oriented capabilities for eco-design innovations come from a deep understanding of the principles of the circular economy. Circular economy (CE) envisions a shift away from the traditional industrial model of take-make-dispose towards a circular system with a waste of all kinds eliminated (Ellen MacArthur Foundation, 2012). It aims to maximize the competitive advantage of circular firms through the superior design of products and systems (De los Rios and Charnley, 2017). CE proposes a range of circular strategies such as reduce, reuse, repair, recycle, restore, and cascading (Blomsma et al., 2019). The holistic and transformative changes envisioned in CE inherently require changes from the very early stages of the creation of the product or service. Circular thinking needs to be carefully integrated through the design process because resources, infrastructure, and life-cycle of a product are determined at the design product specification stage (Bocken et al., 2016; De los Rios and Charnley, 2017; Demirel and Danisman, 2019; Henry et al., 2020). In this paper, we argue that firms' strong commitment and investments in circular thinking and circular activities will develop sustainability-oriented capabilities that can help to foster eco-design innovations.

H1: Sustainability-oriented capabilities influence eco-design innovations positively.

2.3. Technological Capabilities

Eco-innovation literature highlights the role of technology push and the technological capabilities acquired through R&D as positively influencing firms' ability to introduce eco-innovations (Costantini et al., 2015; Ghisetti and Pontoni, 2015). The enhancements in technical knowledge by R&D help firms to improve their abilities in meeting new environmental regulations, become more adept in terms of their ecological impact and innovations, and accumulate knowledge in favor of eco-innovation (Horbach, 2016; Jové-Llopis & Segarra-Blasco, 2018). Technological capabilities help the advancement of internal resources and capabilities to reduce the firm's environmental impact, leading to improvements in the firm's competitive advantage (Lee and Min, 2015).

Therefore, in line with the aforementioned discussions, we expect to find that SMEs that focus on technological capabilities through improvements in R&D investments and process eco-innovations have an increased tendency of adapting eco-design.

H2: Technological capabilities influence eco-design innovations positively.

2.3 Financial Resources

Although the innovation literature investigates the relationship between firms' financial resources and their innovative ability in great depth, the eco-innovation literature is less focused on the role that financial resources affect firms' ability to engage with environmental innovations (Migendt et al., 2017). Firms that lack internal financial resources struggle to find external finance for their green innovations because the uncertainties are even larger compared to regular innovation projects (Revell et al. 2010; Cecere et al. 2018). SMEs typically need external finance to introduce eco-innovations. Yet, they face further difficulties when searching for eco-innovation funding because (1) the payback period for eco-innovations is longer due to higher capital costs and patient capital is only a fraction of externally available finance, (2) eco-innovations rely a lot on regulatory pressures, which brings another uncertainty for financiers and (3) returns to eco-innovation projects are uncertain and extremely skewed, and they often need collaterals where small firms face difficulty to obtain, especially in the case of bank loans (Cecere et al., 2018; Demirel and Parris, 2015).

The empirical SME literature has found that when applying for bank loans, SMEs have decreased chances to find financial support mainly due to uncertainties and lack of collaterals (Revest and Sapio, 2012). However, despite these difficulties, bank loans still constitute the main source of external finance for innovative SMEs, especially when venture capital markets are less developed (Cosh et al., 2009). Equity and venture capital can help mitigate the negative consequences of credit rationing for such SMEs and mitigate informational asymmetries because VC investors, with their specialized knowledge, can help SMEs by providing active management support and advice (Mrkajic et al., 2017; Hall, 2005). However, equity and venture capital funding are limited for SMEs as investors favor short-term gains and focus on firms with innovations closer to the commercialization stage. In those cases, governments and public funding complement private funding and help to promote and push the development of eco-innovations of SMEs (Olmos et al., 2012). Government grants serve as vehicles to relieve these market imperfections and support innovative SMEs. Since firms do not need to pay back these government grants, they constitute very valuable support and act as signals showing their ability to translate such innovations from conceptualization into commercialization.

Although the importance of access to finance is well-recognized in the literature, the design of an appropriate setting for financial support for green innovations is still at the development stage in the literature, and little attention has been paid to the different types of funding that would boost their advance. Considering the specific case of eco-design innovations, the literature lacks much exploration of the financial ecosystem surrounding SMEs. We hypothesize as follows:

H3: Access to external finance (equity finance, grant finance, and debt finance) affects eco-design innovations positively.

3. DATA AND METHODOLOGY

Data

This paper utilizes data from the "Flash EuroBarometer 441: European SMEs and the Circular Economy" survey coordinated by the European Commission Directorate General Environment in April 2016. The survey aims to understand the barriers and drivers of SME participation in the circular economy through eco-innovation activities. It covers 10,618 SMEs with less than 250 employees in 28 European Union member countries, with an approximately equal number of SMEs from each country. For the analysis, we considered the subsample covering only the firms where the information on revenue growth was available, consisting of 7,165 SMEs. According to their NACE categories, these SMEs operate in various industries, the breakdown of which is presented in Table 1.

Table 1. Industry breakdown

NACE Industries	Number of SMEs	% of total
B- Mining and quarrying	16	0%
C- Manufacturing	1005	14%
D- Electricity, gas, steam, and air conditioning supply	35	0%
E- Water supply; sewerage; waste management and remediation activities	62	1%
F- Construction	829	12%
G- Wholesale and retail trade	2450	34%
H- Transporting and storage	437	6%
I- Accommodation and food service activities	507	7%
J- Information and communication	314	4%
K- Financial and insurance activities	239	3%
M- Professional, scientific and technical activities	927	13%
N- Administrative and support service activities	344	5%
Grand Total	**7165**	

Variables and Methodology

The brief explanations for the variables are shown in Table 2, and the descriptive statistics are displayed in Table 3. The dependent variable in the study is the activity of "eco-designing products and services (ECO-DESIGN)" to minimize material use or to use recycled materials. It is constructed dichotomously and takes the value of 1 if a firm conducts eco-design activities and 0 otherwise. Table 3 indicates that 35% of the SMEs in our sample conduct eco-design activities.

We investigate the drivers of eco-design innovations under three categories that reflect the internal firm resources and capabilities: (1) sustainability-oriented firm capabilities, (2) technological capabilities, and (3) access to finance. Sustainability-oriented firm capabilities are measured by the investments into the circular economy (CIRCULARITY), which is measured as a percentage of turnover invested in circular economy activities over the last 3 years (2013–2016) and coded as a categorical variable that is classified into four categories: (1) 0%, (2) 1–5%, (3) 6–10%, and (4) 11% or more. Table 3 indicates that 49.21% of the firms invest in circular economy with shares of 39.02% of the firms investing 1–5% of their total turnover, 7.01% of the firms investing 6–10% of their turnover, and only 3.18% of the firms investing 11% or more of their turnover. As a robustness check, we use an alternative variable for circularity as being circular investments (CIRCULARINV) which takes a value of 1 if more than 1% of the company's turnover is invested on average per year to undertake circular economy activities over the last three years; 0 otherwise. It is observed from Table 3 that 70% of the SMEs invest more than 1% of the turnover into the circular economy. Technological capabilities are measured by process eco-innovations (PROCESS) which is constructed with a value of 1 if the firm conducts process eco-innovations such as re-planning of water and energy usage, using renewable energy, and minimizing waste; 0 otherwise. It is evident in Table 3 that 65% of the firms in our sample conduct process EI. As an additional variable for technological capabilities, we consider R&D intensity (RD) that is measured as a percentage of firm revenue in 2015. Table 3 indicates that, on average, SMEs invest 3.7% of their

Table 2. Description of the variables

	Description
Dependent variable	
Eco-designing products and services (ECO-DESIGN)	Shows whether the firm is redesigning or in the process of redesigning products and services to minimize the use of materials or use recycled materials; takes a value of 1 if so, 0 otherwise.
Independent variables	
CIRCULARITY (0%)	Takes a value of 1 if 0% of the company's turnover is invested on average per year to undertake these circular economy activities over the last 3 years; 0 otherwise.
CIRCULARITY (1-5%)	Takes a value of 1 if 1-5% of the company's turnover is invested on average per year to undertake these circular economy activities over the last 3 years; 0 otherwise.
CIRCULARITY (6-10%)	Takes a value of 1 if 6-10% of the company's turnover is invested on average per year to undertake these circular economy activities over the last 3 years; 0 otherwise.
CIRCULARITY (11% OR MORE)	Takes a value of 1 if more than 11% of the company's turnover is invested on average per year to undertake these circular economy activities over the last 3 years; 0 otherwise.
CIRCULAR_INV	Takes a value of 1 if more than 1% of the company's turnover is invested on average per year to undertake these circular economy activities over the last 3 years; 0 otherwise.
Process eco-innovations (PROCESS)	Takes the value of 1 if the firm conducts process eco-innovations such as re-planning of water and energy usage, using renewable energy, and minimizing waste; 0 otherwise
Grant finance (GRANTFIN)	Equals 1 if the firms had access to external finance through EU funds and government grants; 0 otherwise
Equity finance (EQUITYFIN)	Equals 1 if equity financing, such as venture capital, business angels, and capital markets, are available to the company; 0 otherwise
Debt Finance (DEBTFIN)	Equals 1 if the firms had access to finance through bank loans, green banks, and peer-to-peer lending; 0 otherwise
R&D expenditures (RD)	R&D expenditures measured as a percentage of company revenue in 2015
GROWTH	Annual revenue growth measured as a percentage between 2015 and 2016
Size (SIZE)	The natural logarithm of the number of employees in 2016
Age (AGE)	The number of years since establishment. It is comprised of dummy variables out of three categories: (i) age=1 year (AGE1) (ii) 1year<age<=5 years (AGE1-5) (iii) age>5 years (AGE_OLDER)
B2C	Equals 1 if the firm sells products/services directly to consumers ; 0 otherwise
B2B	Equals 1 if the firm sells products/services to companies or other organizations; 0 otherwise
REGULATIONS_COMPLEX	Equals 1 if the firm had encountered complexity in administrative and legal procedures; 0 otherwise
REGULATIONS_COST	Takes a value of 1 if the firm had encountered cost of meeting regulations and standards; 0 otherwise
Note: This table shows the list of variables used in the analysis and their brief descriptions.	

incomes in R&D. Considering the access to finance variables of interest, we measure those through three indicator variables: grant finance (GRANTFIN), equity finance (EQUITYFIN), and debt Finance (DEBTFIN). GRANTFIN is a dummy variable equals one if the firms had access to external finance through EU funds and government grants; 0 otherwise. EQUITYFIN equals one if equity financing, such as venture capital, business angels, and capital markets, are available to the company; 0 otherwise. Finally, DEBTFIN equals one if the firms had access to finance through bank loans, green banks, and peer-to-peer lending; 0 otherwise. We observe that 9% of the firms had access to GRANTFIN, 23% had access to EQUITYFIN, and 30% had access to DEBTFIN.

Table 3. Descriptive statistics

	Obs.	Mean	Min	Max	Median	Stand. Dev.
ECO-DESIGN	7165	0,35	0	1	0	0,48
CIRCULARITY	7165	1,88	1	4	2,00	0,76
CIRCULARITY (0%)	3639 (50.78%)					
CIRCULARITY (1-5%)	2796 (39.02%)					
CIRCULARITY (6-10%)	502 (7.01%)					
CIRCULARITY (11% OR MORE)	228 (3.18%)					
CIRCULAR_INV	5069	0,70	0	1	1,00	0,46
PROCESS	7165	0,65	0	1	1	0,48
GRANTFIN	7165	0,09	0	1	0,00	0,29
EQUITYFIN	7165	0,23	0	1	0,00	0,42
DEBTFIN	7165	0,30	0	1	0	0,46
RD	6554	3,66	0	100	0	11,46
GROWTH	7165	2,83	-100	100	0,00	22,43
SIZE	7165	2,12	0	5,53	1,79	1,27
AGE						
AGE1	7165	0,01	0	1	0	0,12
AGE1-5	7165	0,14	0	1	0	0,35
AGE_OLDER	7165	0,84	0	1	1	0,37
B2B	7165	0,37	0	1	0	0,48
B2C	7165	0,22	0	1	0	0,42
REGULATIONS_COMPLEX	7165	0,30	0	1	0	0,46
REGULATIONS_COST	7165	0,27	0	1	0	0,44

Note: The table shows summary statistics for the variables.

We use other firm-specific controls in the regressions such as growth, size, age, dummies for business to business (B2B), and business to consumer enterprises (B2C) and regulations. Growth (GROWTH) is measured as annual revenue growth as a percentage between 2015 and 2016. Table 3 shows that SMEs in our sample have an average annual growth rate of 2.83%. Size (SIZE) is calculated as the natural logarithm of the number of employees in 2016, and age (AGE) is comprised of dummy variables out of three categories: (i) age=1year (AGE1) (ii) 1year<age<=5 years (AGE1-5) (iii) age>5 years (AGE_OLDER). While most SMEs in our sample (84%) are older than five years, 14% of the firms are between 1 to 5 years old. B2B equals one if the firm sells products/services to companies or other organizations, and B2C equals one if the firm sells products/services directly to consumers; 0 otherwise. While 37% of the firms in the sample sell products/services to companies or other organizations, 22% sell directly to consumers. To control for the impact of regulatory requirements, we include two variables: REGULATIONS_COMPLEX and REGULATIONS_COST. They are dummy variables that equal one if SMEs had encountered complexity in administrative and legal procedures and encountered cost of meeting regulations and standards, respectively; and 0 otherwise. It is apparent from Table 3 that 30% of SMEs

Table 4. Correlation table

	(1)	(2)	(3)	(4)	(5)	(6)	(7)	(8)	(9)	(10)	(11)	(12)	(13)	(14)	(15)
(1) RD	1														
(2) GROWTH	0.1274*	1													
(3) SIZE	-0,0025	0.0342*	1												
(4) AGE1	-0,0094	0.0687*	-0.0550*	1											
(5) AGE1-5	0.0583*	0.1151*	-0.1865*	-0.0567*	1										
(6) B2B	0.0504*	0.0593*	0.0860*	-0.0229*	-0.0197*	1									
(7) B2C	-0.0457*	-0.0802*	-0.1106*	0.0222*	0.0280*	-0.4064*	1								
(8) PROCESS	0.0321*	0.0608*	0.1528*	-0,0174	-0.0665*	-0.0627*	-0,0121	1							
(9) CIRCULAR_ INV	0.0881*	0.0302*	0.1103*	-0.0483*	-0,0209	-0,0217	-0,0153	0.1046*	1						
(10) CIRCULARITY	0.1594*	0,0261	0.0676*	-0.0306*	0,0023	-0,0204	-0,0221	0.1019*	0.7742*	1					
(11) REGULATIONS_ COMPLEX	0.0553*	0,0051	0.0677*	-0,0113	-0,0158	-0.0403*	-0.0328*	0.1432*	0.1678*	0.1654*	1				
(12) REGULATIONS_ COST	0.0240*	-0,0048	0.0453*	0,0026	-0,0116	-0.0569*	-0,0085	0.1331*	0.1460*	0.1349*	0.4711*	1			
(13) GRANTFIN	0,0115	-0,0121	-0,0082	0.0207*	0.0393*	-0,0047	0,0117	-0.2848*	0.0626*	0.0812*	0,0096	0,0025	1		
(14) EQUITYFIN	0.0618*	0.0686*	0.0517*	0,0066	0,0118	0.0605*	-0.0890*	0.0602*	0.0622*	0.0674*	0.0814*	0.0943*	-0,0052	1	
(15) DEBTFIN	0.0224*	0,013	0,0059	-0,0059	0,0137	0,0136	-0,0188	0.0533*	0.1023*	0.1158*	0.1039*	0.1195*	-0.0542*	0.3227*	1

report that they faced complexity when dealing with administrative and legal procedures, and 27% of them incurred high costs of meeting regulations and standards.

Table 4 presents the correlation coefficients between the independent variables, and there is no indication of multicollinearity.

The study employs a logistic regression model (logit model) with cross-sectional estimation techniques to account for the dichotomous nature of the dependent variable in the analysis, aiming to investigate the sources of eco-design innovations in SMEs. We employ country and industry fixed effects to account for the heterogeneity between the countries and the industries. In the reported regressions, we show marginal effects, which help the coefficients to be interpreted as predicted probabilities.

4. RESULTS

Table 5 displays the findings on the influence of sustainability-oriented firm capabilities and technological capabilities as drivers of eco-design innovations. Columns 1 &2 consider the impact of sustainability-oriented capabilities by employing CIRCULARITY and CIRCULAR_INV as independent variables, respectively. The findings indicate that circularity investments significantly enhance the probability of eco-design practices, supporting Hypothesis 1. Specifically, considering Column 1, when investment in circularity is between 1-5%, the firm's likelihood of introducing eco-design innovations increases by 19.8

Table 5. The influence of sustainability-oriented firm capabilities and technology push

	(1) (Sustainability-oriented capabilities)	(2) (Sustainability-oriented capabilities)	(3) (Technology-push)	(4) (Technology-push)	(5) (Technology-push)	(6) (Technology-push)
CIRCULARITY (1-5%)	0.198*** (0,016)					
CIRCULARITY (6-10%)	0.327*** (0,026)					
CIRCULARITY (11% OR MORE)	0.351*** (0,037)					
CIRCULAR_ INV		0.223*** (0,015)				
PROCESS			0.312*** (0,012)			
RD	0.002** (0,001)	0.002*** (0,001)		0.002*** (0,000)	0.002*** (0,000)	0.002*** (0,000)
REGULATIONS_ COMPLEX					0.127*** (0,012)	
REGULATIONS_ COST						0.092*** (0,013)
GROWTH	0,000 (0,000)	0,000 (0,000)	0.000* (0,000)	0.001** (0,000)	0.001* (0,000)	0.001** (0,000)
SIZE	0,001 (0,006)	0,000 (0,006)	0.013*** (0,004)	0.027*** (0,005)	0.025*** (0,005)	0.026*** (0,005)
AGE1	-0,057 (0,066)	-0,059 (0,067)	-0,044 (0,046)	-0.091* (0,050)	-0.086* (0,050)	-0.093* (0,050)
AGE1-5	0,017 (0,021)	0,018 (0,021)	0,017 (0,016)	0,001 (0,017)	0,003 (0,017)	0,003 (0,017)
B2B	0,006 (0,017)	0,005 (0,017)	-0,017 (0,013)	-0.037*** (0,013)	-0.029** (0,013)	-0.031** (0,013)
B2C	-0,011 (0,020)	-0,01 (0,020)	-0,024 (0,015)	-0.034** (0,016)	-0,024 (0,016)	-0.029* (0,016)
Pseudo R2	0,061	0,055	0,106	0,041	0,054	0,047
LR Chi2	390,19	351,7	983,76	348,75	450,95	398,97
p value-Chi2	0,000	0,000	0,000	0,000	0,000	0,000
Country FE	YES	YES	YES	YES	YES	YES
Industry FE	YES	YES	YES	YES	YES	YES
Observations	4640	4640	7165	6554	6554	6554

Note: This table displays the logit regression results regarding the impact of sustainability-oriented firm capabilities and technology push on eco-design innovations. Country and industry fixed effects are included in the analysis. Standard errors are robust and given in parentheses. *p<0.10,**p<0.05,***p<0.010.

percentage points as compared to the case when there is no investment. When the investment increases to 6-10% and 11% or more, the probability of eco-design implementation further increases by 32.7% and 35.1%, respectively, being significantly and economically high. We further consider CIRCULAR_INV as an alternative circular investment variable in Column 2, and our results are robust with this alternative

Table 6. The impact of access to finance on eco-design

	(1) (Access to Finance)	(2) (Access to Finance)	(3) (Access to Finance)	(4) (Access to Finance)
GRANTFIN	-0.259*** (0,027)			-0.260*** (0,026)
EQUITYFIN		0.048*** (0,014)		0.049*** (0,014)
DEBTFIN			0.025** (0,013)	0,007 (0,013)
RD	0.002*** (0,000)	0.002*** (0,000)	0.002*** (0,000)	0.002*** (0,000)
GROWTH	0.001** (0,000)	0.001** (0,000)	0.001** (0,000)	0.001** (0,000)
SIZE	0.027*** (0,005)	0.026*** (0,005)	0.027*** (0,005)	0.026*** (0,005)
AGE1	-0,077 (0,050)	-0.090* (0,050)	-0.090* (0,050)	-0,075 (0,050)
AGE1-5	0,009 (0,017)	0,001 (0,017)	0,002 (0,017)	0,009 (0,017)
B2B	-0.036*** (0,013)	-0.038*** (0,013)	-0.036*** (0,013)	-0.037*** (0,013)
B2C	-0.029* (0,016)	-0.030* (0,016)	-0.032** (0,016)	-0,026 (0,016)
Pseudo R2	0,055	0,043	0,042	0,057
LR Chi2	463,7	360,27	352,75	477,22
p value-Chi2	0,000	0,000	0,000	0,000
Country FE	YES	YES	YES	YES
Industry FE	YES	YES	YES	YES
Observations	6554	6554	6554	6554

Note: This table displays the logit regression results regarding the impact of access to finance on eco-design innovations. Country and industry fixed effects are included in the analysis. Standard errors are robust and given in parentheses. *p<0.10,**p<0.05,***p<0.010.

specification, indicating that investments of more than 1% of turnover on circularity significantly improve the probability of eco-design innovation adoption for SMEs. Therefore, we observe that sustainability-oriented firm-level capabilities have a positive effect on eco-design practices. Our findings are in line with the strategic management literature that highlights the importance of environmentally-specific capabilities for achieving eco-innovations (Melander, 2018; Kabongo and Boiral, 2017).

Columns 3-6 in Table 5 display the impact of technological capabilities on the SME's propensity of eco-design innovation implementations. Technological capabilities considered in these specifications are PROCESS in Column 3 and RD in Columns 4-6, respectively. Columns 5&6 further include the impact of regulatory requirements as control variables. Due to concerns of multicollinearity, all variables are not entered into the regression simultaneously. Our findings reveal that implementing process eco-innovations such as re-planning of water and energy usage, using renewable energy, and minimizing waste significantly improves the probability of involving in eco-design by 31% compared to the case when SMEs do not implement any process eco-innovations. While the sustainability-oriented capabilities constitute the strongest driver of eco-design innovations, the impact of broader technological capabilities of SMEs on eco-design practices is also significant. Columns 4-6 use RD as an alternative technological capability variable and confirm the previous findings. This is in line with Hypothesis 2 that technological capabilities are likely to improve eco-design activities. This confirms the findings in the eco-innovation literature that highlights the role of technology-push and technological capabilities as positively influencing the introduction of eco-innovations (Costantini et al., 2015, Demirel and Kesidou, 2019; Ghisetti and Pontoni, 2015).

Considering the access to finance variables, Table 6 reports our findings and implies that access to equity finance through venture capital, business angels, and capital markets is positively and significantly associated with greater willingness to conduct eco-design innovations. This is in line with the literature stating that equity and venture capital mitigate information asymmetries and help to lessen the negative consequences of credit rationing for SMEs. Equity finance can help SMEs by providing active management support and advice (Mrkajic et al., 2017; Hall 2005). Access to grant finance in terms of EU funds and government grants, on the other hand, is interestingly observed to significantly decrease the likelihood of eco-design innovations. Access to debt finance through bank loans, green banks, and peer-to-peer lending is not significantly related to eco-design adoption. Therefore, grant finance and debt finance are ineffective for SMEs in promoting eco-design activities.

Considering the control variables that might potentially influence eco-design innovations, we observe in Tables 5&6 that the likelihood of eco-design increases for larger and high growth SMEs and under more stringent regulatory requirements, which is in line with the findings in the literature (Bossle et al., 2016; De Marchi, 2012). On the other hand, younger SMEs which are less than one year old and SMEs that focus on B2B are likely to conduct fewer eco-design innovations.

5. DISCUSSION AND CONCLUSIONS

This paper investigates the role of firm capabilities and resources for driving eco-design innovations among European SMEs. The findings of the paper indicate that sustainability-oriented capabilities that are measured by the investments into the circular economy significantly improve the practice of eco-design innovations. Likewise, technological capabilities have an important impact on the adoption of eco-design innovations. These two findings argue for the importance of building SMEs' innovation capabilities with a particular emphasis on their relationship with the natural environment. While external factors such as regulatory pressures and demand from stakeholders are essential to moving firms in the right direction to control their impact on the environment, firms' internal innovation and sustainability capabilities are what determine firms' ability to effectively respond to these stimuli.

In line with this finding, it is important to evaluate how SMEs can grow their technological and sustainability-oriented capabilities in order to maximize the industrial efforts towards low-carbon transitions. Most SMEs present limited resources to invest in innovation and sustainability as a natural outcome of their size. Hence, the business case for sustainability needs to be clear in order to incentivize SMEs' decision to invest in their sustainability-oriented capabilities. While basic regulations exist in all EU countries to control the environmental impact of corporations, they are often not aligned with the underlying principles of the circular economy. SMEs typically comply with environmental regulations through end-of-pipeline solutions that contain the pollution created. This presents little space for innovation and high costs. Instead, regulations should focus on incentivizing firms to avoid pollution and waste of all types throughout all stages of operations, rather than containing it at the end of operations. Industrial and environmental policies should be carefully considered with SMEs in mind to ensure their participation in the CE. For instance, recently designed tax breaks for repairs in Sweden are an excellent policy intervention in support of the circular economy. It is intended to boost eco-design with repairs in mind as well the emergence of a repairs industry that is likely to accommodate many SMEs.

The final finding of the paper highlights that only equity finance promotes eco-design innovations. While access to equity finance is positively and significantly associated with eco-design innovations, the availability of equity finance is limited to an extremely small number of SMEs. It only applies to those young firms with the potential to scale fast, often using digital technologies. Indeed, we have seen the emergence of environmental start-up firms with strong eco-design innovations to tackle ocean pollution, single-use plastic pollution, etc., in recent years, and these firms have been able to secure investment capital. Yet, they remain to be the exception in a financial climate where cleantech investments remain small. Additionally, it is not reasonable to expect that the majority of SMEs will be attractive to investors for most sectors. Hence, equity finance cannot be relied on to boost the eco-design innovations of SMEs. Grant finance and debt finance, on the other hand, are ineffective at best for promoting eco-design activities. Grant finance serves a particularly important role in supporting technology-oriented SMEs, and there is a strong incentive for these funds to be channeled into circular economy activities more effectively as governments work toward environmental targets. Likewise, even though alternative debt finance options such as peer-to-peer lending complement traditional bank loans as potential funding sources for eco-innovations of SMEs, our findings suggest that there are not significant synergies between debt finance and eco-design innovations. A changing attitude for banks, promoted by favorable regulations towards sustainability in lending, is necessary to ensure that SMEs can be integrated effectively into the circular economy through their design activities.

The paper has some limitations to be mentioned, which mostly originate from the nature of the dataset used in the empirical analysis and thereby bring forth the possible avenues for future research. We use a cross-sectional dataset that does not allow considering the SME's prior eco-design practices, which is a common shortcoming in EI literature. Nevertheless, we tried to overcome this limitation by using lagged independent variables where possible (e.g., investments in EI in the last three years).

REFERENCES

Barney, J. (1991). Firm Resources and Sustained Competitive Advantage. *Journal of Management*, *17*(1), 99–120. doi:10.1177/014920639101700108

Battaglia, M., Gragnani, P., & Annesi, N. (2020). Moving Businesses toward Sustainable Development Goals (SDGs): Evidence from an Italian "Benefit-For-Nature" Corporation. *Entrepreneurship Research Journal*, *20190305*(4). Advance online publication. doi:10.1515/erj-2019-0305

Blomsma, F., Pieroni, M., Kravchenko, M., Pigosso, D. C., Hildenbrand, J., Kristinsdottir, A. R., ... Li, J. (2019). Developing a circular strategies framework for manufacturing companies to support circular economy-oriented innovation. *Journal of Cleaner Production*, *241*, 118271. doi:10.1016/j.jclepro.2019.118271

Boks, C. (2006). The soft side of ecodesign. *Journal of Cleaner Production, 14*(15-16), 1346–1356. doi:10.1016/j.jclepro.2005.11.015

Bossle, M. B., de Barcellos, M. D., Vieira, L. M., & Sauvée, L. (2016). The drivers for adoption of eco-innovation. *Journal of Cleaner Production, 113*, 861–872. doi:10.1016/j.jclepro.2015.11.033

Carrillo-Hermosilla, J., del Río, P., & Könnölä, T. (2010). Diversity of eco-innovations: Reflections from selected case studies. *Journal of Cleaner Production, 18*(10), 1073–1083. doi:10.1016/j.jclepro.2010.02.014

Cecere, G., Corrocher, N., & Mancusi, M. L. (2018). Financial constraints and public funding of eco-innovation: Empirical evidence from European SMEs. *Small Business Economics*, •••, 1–18.

Cosh, A., Cumming, D., & Hughes, A. (2009). Outside enterpreneurial capital. *Economic Journal (London), 119*(540), 1494–1533. doi:10.1111/j.1468-0297.2009.02270.x

Costantini, V., Crespi, F., Martini, C., & Pennacchio, L. (2015). Demand-pull and technology-push public support for eco-innovation: The case of the biofuels sector. *Research Policy, 44*(3), 577–595. doi:10.1016/j.respol.2014.12.011

Dangelico, R. M., Pujari, D., & Pontrandolfo, P. (2017). Green product innovation in manufacturing firms: A sustainability-oriented dynamic capability perspective. *Business Strategy and the Environment, 26*(4), 490–506. doi:10.1002/bse.1932

De los Rios, I. C., & Charnley, F. J. (2017). Skills and capabilities for a sustainable and circular economy: The changing role of design. *Journal of Cleaner Production, 160*, 109–122. doi:10.1016/j.jclepro.2016.10.130

De Marchi, V. (2012). Environmental innovation and R&D cooperation: Empirical evidence from Spanish manufacturing firms. *Research Policy, 41*(3), 614–623. doi:10.1016/j.respol.2011.10.002

Demirel, P., & Danisman, G. O. (2019). Eco-innovation and firm growth in the circular economy: Evidence from European small-and medium-sized enterprises. *Business Strategy and the Environment, 28*(8), 1608–1618. doi:10.1002/bse.2336

Demirel, P., & Kesidou, E. (2019). Sustainability-oriented capabilities for eco-innovation: Meeting the regulatory, technology, and market demands. *Business Strategy and the Environment, 28*(5), 847–857. doi:10.1002/bse.2286

Deutz, P., McGuire, M., & Neighbour, G. (2013). Eco-design practice in the context of a structured design process: An interdisciplinary empirical study of UK manufacturers. *Journal of Cleaner Production, 39*, 117–128. doi:10.1016/j.jclepro.2012.08.035

Ellen MacArthur Foundation. (2012). *Towards the Circular Economy: Economic and Business Rationale for an Accelerated Transition. Isle of Wight*. Ellen MacArthur Foundation.

European Commission. (2010). *SMEs and the Environment in the European Union*. Publications Office of the European Union.

European Commission. (2009). Directive 2009/125/EC of the European Parliament and of the Council of 21 October 2009 Establishing a Framework for the Setting of Ecodesign Requirements for energy-related products (recast).

Ghisetti, C., & Pontoni, F. (2015). Investigating policy and R&D effects on environmental innovation: A meta-analysis. *Ecological Economics*, *118*, 57–66. doi:10.1016/j.ecolecon.2015.07.009

Hall, B. H. (2005). The Financing of Innovation. In S. Shane (Ed.), *The Handbook of Technology and Innovation Management* (pp. 409–430). Wiley.

Hart, S. L., & Dowell, G. (2011). Invited editorial: A natural-resource-based view of the firm: Fifteen years after. *Journal of Management*, *37*(5), 1464–1479. doi:10.1177/0149206310390219

Henry, M., Bauwens, T., Hekkert, M., & Kirchherr, J. (2020). A typology of circular start-ups: Analysis of 128 circular business models. *Journal of Cleaner Production*, *245*, 118528. doi:10.1016/j.jclepro.2019.118528

Hoogendoorn, B., Guerra, D., & van der Zwan, P. (2015). What drives environmental practices of SMEs? *Small Business Economics*, *44*(4), 759–781. doi:10.100711187-014-9618-9

Horbach, J. (2016). Empirical determinants of eco-innovation in European countries using the community innovation survey. *Environmental Innovation and Societal Transitions*, *19*, 1–14. doi:10.1016/j.eist.2015.09.005

Horbach, J., Rammer, C., & Rennings, K. (2012). Determinants of eco-innovations by type of environmental impact- the role of regulatory push/pull, technology push and market pull. *Ecological Economics*, *78*, 112–122. doi:10.1016/j.ecolecon.2012.04.005

Jové-Llopis, E., & Segarra-Blasco, A. (2018). Eco-innovation strategies: A panel data analysis of Spanish manufacturing firms. *Business Strategy and the Environment*, *27*(8), 1209–1220. doi:10.1002/bse.2063

Kabongo, J. D., & Boiral, O. (2017). Doing more with less: Building dynamic capabilities for eco-efficiency. *Business Strategy and the Environment*, *26*(7), 956–971. doi:10.1002/bse.1958

Lee, K. H., & Min, B. (2015). Green R&D for eco-innovation and its impact on carbon emissions and firm performance. *Journal of Cleaner Production*, *108*, 534–542. doi:10.1016/j.jclepro.2015.05.114

Lin, R. J., Tan, K. H., & Geng, Y. (2013). Market demand, green product innovation, and firm performance: Evidence from Vietnam motorcycle industry. *Journal of Cleaner Production*, *40*, 101–107. doi:10.1016/j.jclepro.2012.01.001

McAloone, T. C., & Bey, N. (2009). *Environmental Improvement through Product Development: A Guide*. Danish Environmental Protection Agency.

Melander, L. (2018). Customer and supplier collaboration in green product innovation: External and internal capabilities. *Business Strategy and the Environment*, *27*(6), 677–693. doi:10.1002/bse.2024

Migendt, M., Polzin, F., Schock, F., Täube, F. A., & von Flotow, P. (2017). Beyond venture capital: An exploratory study of the finance-innovation-policy nexus in cleantech. *Industrial and Corporate Change*, *26*(6), 973–996. doi:10.1093/icc/dtx014

Mrkajic, B., Murtinu, S., & Scalera, V. G. (2019). Is green the new gold? Venture capital and green entrepreneurship. *Small Business Economics, 52*(4), 929–950. doi:10.100711187-017-9943-x

OECD. (2009). *Sustainable Manufacturing and Eco-innovation: Framework, Practices and Measurement*. Organisation for Economic Cooperation and Development.

Olmos, L., Ruester, S., & Liong, S.-J. (2012). On the selection of financing instruments to push the development of new technologies: Application to clean energy technologies. *Energy Policy, 43*, 252–266. doi:10.1016/j.enpol.2012.01.001

Plouffe, S., Lanoie, P., Berneman, C., & Vernier, M.-F. (2011). Economic benefits tied to ecodesign. *Journal of Cleaner Production, 19*(6), 573–579. doi:10.1016/j.jclepro.2010.12.003

Rashid, A., Asif, F. M. A., Krajnik, P., & Nicolescu, C. M. (2013). Resource Conservative Manufacturing: An essential change in business and technology paradigm for sustainable manufacturing. *Journal of Cleaner Production, 57*, 166–177. doi:10.1016/j.jclepro.2013.06.012

Revell, A., Stokes, D., & Chen, H. (2010). Small businesses and the environment: Turning over a new leaf? *Business Strategy and the Environment, 19*(5), 273–288.

Revest, V., & Sapio, A. (2012). Financing technology-based small firms in Europe: What do we know? *Small Business Economics, 39*(1), 179–205. doi:10.100711187-010-9291-6

Chapter 12
Investigation of Green Criteria With Clustering Analysis in Green Supplier Selection

Mehmet Ali Taş
Turkish-German University, Turkey

Serap Akcan
Tarsus University, Turkey

ABSTRACT

Green supplier selection has a crucial importance for businesses. In the past, the selection of suppliers was solely based on conventional criteria such as cost, quality, and flexibility whereas expectations of businesses transformed in today's world on grounds of raised environmental awareness, public pressure, and regulations. Alternatives called green suppliers sensitive to the environment, preserving the ecological balance, managing wastes, and preventing pollution increased in value. This study analyzes practices on the selection of green suppliers. The articles between 2014 and 2021 were analyzed from the perspective of green criteria. The green criteria in the 50 articles determined are divided into 28 groups. With the k-means algorithm, these criteria groups are divided into four clusters, which was aimed to analyze the usage frequency of green criteria. This study is intended to contribute to green supplier selection practices in academia and industry in the future.

INTRODUCTION

Businesses establish supply chains for products and services they provide. One of the most important steps in any supply chain is the selection of a supplier that fits for demands and expectations of the business. Suppliers that are part of a green supply chain are expected to be green suppliers. Suppliers that adopt green practices, manage their waste, comply with societal and legal norms about the environment and run environmental management systems are preferable for green supply chains (Büyüközkan &

DOI: 10.4018/978-1-7998-8900-7.ch012

Vardaloğlu, 2008). Multi-criteria decision-making (MCDM) methods are adopted often for the selection of a green supplier.

These methods are based on comparison and grading of alternatives, which provides selection of the best one or ranking the alternatives. It enables to offer rational solutions for complicated problems in the light of considerations that contradict one another (Triantaphyllou, 2000). There are a variety of MCDM methods in the literature. Among the most popular ones are Analytic Hierarchy Process (AHP), Technique for Order Preference by Similarity to Ideal Solution (TOPSIS), Elimination and Choice Translating Reality English (ELECTRE), and Preference Ranking Organization Method for Enrichment Evaluations (PROMETHEE). Freeman and Chen (2015) implemented Entropy and TOPSIS along with integrated AHP. Espinilla et al. (2015) adopted ELECTRE, Gupta and Barua (2017) used BWM (Best-Worst Method) and fuzzy TOPSIS, and Kumar et al. (2017) implemented fuzzy ELECTRE. Similarly, literature also offers studies over the selection of green suppliers through goal programming and optimization. A thorough literature review was performed in this study. It was intended that observing current trends by choosing a time period covering recent dates produces more accurate and consistent results. Thus, it was deemed appropriate to include the studies between 2014-2021. The Web of Science Core Collection database was selected for literature review. The reason for choosing this database is to access articles published in quality journals with well-known indexes. An online search was performed on this database using these keywords: "green supplier" or "green supplier selection" or "green supplier evaluation" or "green supplier multi-criteria decision making" in the topic section. Only articles were chosen and books, book chapters, conference papers, theses, and other academic publications were excluded from the search. In addition, the timespan was set from 2014 to 2021, and only articles written in English were considered. It was desired to select green supplier selection articles that included at least one MCDM method, one case study, and considered appropriate for the scope of the study. All of the articles reached from the database were first subjected to abstract scanning and those that were not suitable for the study were excluded. 50 articles that were eligible after full-text scanning were included in the study.

BACKGROUND

Recently, many literature reviews of green supplier selection and green supply chain have been applied. Srivastava (2007) divided studies over green supply chain management into three categories: Importance of a green supply chain, green design and green operations. As a part of the study where the focus was on reverse logistics, criteria were tabulated using multi-criteria decision-making and other methods that were previously adopted. Genovese et al. (2010) compiled a review on the selection of environmental suppliers and presented a list of criteria employed. Sarkis et al. (2011) classified the studies over green supply chain management within the framework of organizational theories. Igarashi et al. (2013) carried out a compilation study that includes 60 articles over the selection of green suppliers from 1991 to 2011. The articles were reviewed and analyzed from the normative, empirical, and conceptual perspectives. Govindan et al. (2015) reviewed the evaluation and selection of green suppliers between 1997 and 2011. The study, which analyzed the selection perspectives and common criteria for selection, suggested that the most common criterion is environmental management systems. Fahimnia et al. (2015) performed a wide-ranging literature review about green supply chain management, and divided the articles into clusters by journal, country, year and citations. My Dung et al. (2016) reviewed the studies where AHP and Analytical Network Analysis (ANP) were adopted for the selection of green suppliers. Soda et al.

Table 1. Green supplier selection review studies

Source	Year	Number of articles reviewed	Years gap	Scope
Genovese et al. (2010)	2010	15	2003-2008	Review on awareness of sustainability for selection of green suppliers.
Igarashi et al. (2013)	2013	60	1991-2011	Review on theoretical, empirical and analytical models of studies over the selection of green suppliers.
Villanueva-Ponce et al. (2015)	2015	34	2007-2013	Multi-criteria decision-making methods adopted for the selection of green suppliers and analysis of criteria.
Govindan et al. (2015)	2015	33	1997-2011	Selection methods, the most common criteria and limitations for the evaluation and selection of green suppliers.
My Dung et al. (2016)	2016	36	2002-2014	Compilation of studies where AHP and ANP multi-criteria decision-making methods are adopted for the selection of green suppliers.
Ghadimi et al. (2016)	2016	61	2008-2014	Classification by response to seven questions such as criteria reviewed for the selection of green/sustainable suppliers, methods adopted and impact of a green supplier on environmental performance.
Konys (2019)	2019	190	2007-2019	Ontology of criteria for selection of green suppliers.

(2016) performed a literature review on green supply chain management (GSCM). The performance of GSCM practices, functional barriers and practices on selection of green suppliers were reviewed. Konys (2019) performed a literature review that covers green criteria. Konys employed PRISMA methodology to confirm the data of green criteria which are used before. It is a far-reaching study in which a database for green criteria was established. Published between 2002 and 2014, 36 articles serve as the basis of the review. The Table I displays some of review studies analyzed.

MAIN FOCUS OF THE CHAPTER

There are several ways to set criteria for selecting a supplier. The most common one is to review studies in the literature and determine criteria appropriate for the scope of implementation. The purpose of benefiting from previous studies is to use similar criteria in studies conducted in similar industries. The criteria to be considered in the selection of green suppliers are expected to be green criteria. This study is intended to be of use for further studies over the selection of green suppliers. The green criteria, which were previously adopted in literature, were divided into criteria groups and classified. The definitions of green criteria were reviewed, and they were displayed by studies. This study contributes to the literature by proposing green criteria clustering with k-means algorithm. It is thought that the study reflects current trends since 50 articles between years 2014-2021 constitute the scope of the study. This clustering study is expected to be useful in determining the green criteria groups that should be used in future applications.

Issues, Controversies, Problems

The first Industrial Revolution enabled to use of steam power and made things easier for production. The second Industrial Revolution allowed for mass production through the division of labor and production lines, thus offering a capability to manufacture high amounts of goods at a low cost (Lins et al., 2018). The third Industrial Revolution integrated automation technologies into production and service industries. Nowadays, Industry 4.0, which is renowned for virtual platforms, use of big data, and cyber-physical systems (CPS), is widespread (Witkowski, 2017). In this process, most of the efforts focused on increasing production, and the environmental issue remained in the background. Waste generated in the process was released into the environment, polluting water reservoirs. Gases recklessly released caused the greenhouse effect and global warming. Global warming has increased the global temperature year in and year out, disrupting the natural balance and posing a threat to the habitat of creatures (Ehrlich, 2008).

Like all organizations, businesses have begun to include environmental concerns in their plans by studying environmental processes. In addition to management, production, and marketing concepts, supply chains have also been reorganized in a green-focused manner. Consequently, the concept of green supply chain has emerged (Srivastava, 2007).

Green Supply Chain

Before focus on green suppliers, it is essential to describe supply chains. A supply chain is a system of organizations where suppliers are involved as it stands for planning, design and inspection of any service, product and information flow from a manufacturer to a client (Srivastava, 2007). The concept of green supply chain is an environmentally friendly rendition of the aforementioned classic definition of a supply chain. Table 2 shows the transformation of supply chain operations into green supply chain operations. As displayed, procurement, production, distribution, logistics and packaging operations, which are part of supply chains, are replaced by green procurement, green production, green distribution, reverse logistics and green packaging, which are part of green supply chains (Büyüközkan & Vardaloğlu, 2008). Figure 1 displays green supply chain operations.

Table 2. Transformation of a supply chain to a green supply chain concept

Operation	Supply chain	Green supply chain	Change
1	Procurement	Green procurement	Priority attached to procurement of low resource consumption and environmentally friendly products, material and services.
2	Production	Green Production	Capability to not only manufacture high-quality, resilient and cost-effective items but do so without any environmental damage.
3	Distribution	Green Distribution	Establishment of a distribution network that generates little CO_2 and consumes environmentally friendly energy for the distribution of any service or product.
4	Logistics	Reverse Logistics	Establishment of a logistic infrastructure that covers re-use, recycling and recovery operations.
5	Packaging	Green Packaging	Use of recyclable materials for packaging.

Figure 1. Green supply chain operations (Büyüközkan and Vardaloğlu, 2008)

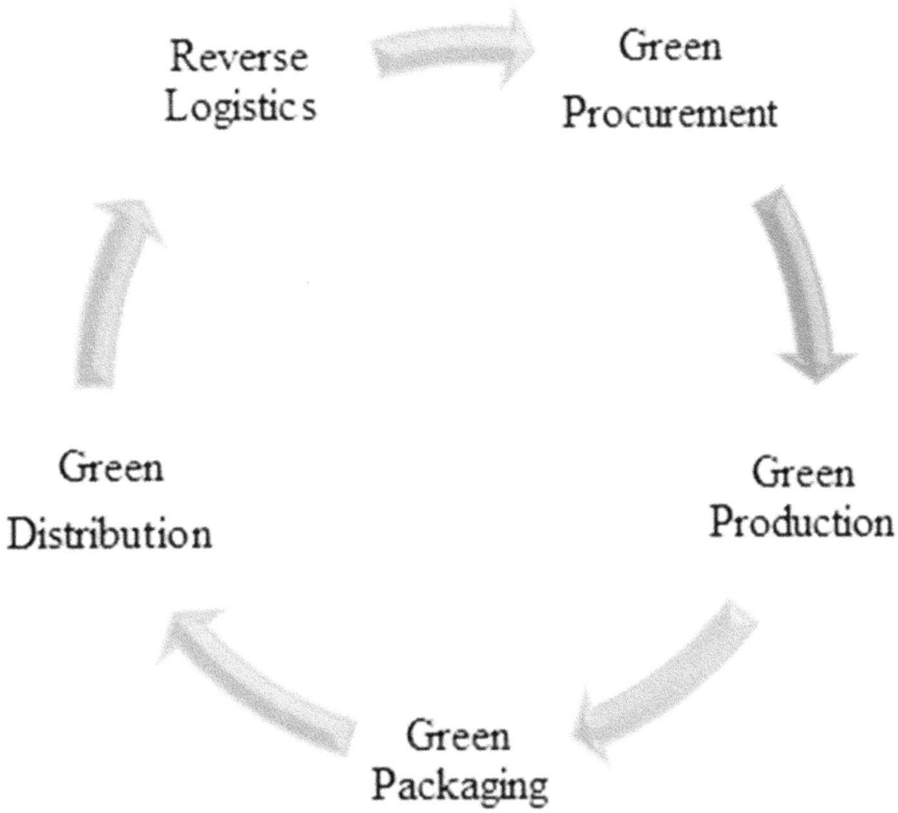

Green Supplier

A green supplier is a supplier that is part of a green supply chain. Green suppliers are supposed to be environmentally friendly, capable of managing their waste, aspire to reduce their carbon footprints, and respect to natural resources. A green supplier can be described in various ways depending on the industry involved, expectations of decision makers and perspectives of authors. The definitions that the authors used in their studies are displayed in Table 3.

The selection of suppliers brings about challenges for organizations (Govindan & Sivakumar, 2016). Involved in supply chains established by organizations to run their operations and allocated with a large number of financial resources, suppliers are expected to be successful when it comes to criteria such as quality, cost, and service. As attributes (criteria) usually contradict one another, it is essential to make an analytical assessment of criteria to be taken into account. This has entailed a solution to supplier selection problems through certain means. In the literature, numerous multi-criteria decision-making methods are widely used for green supplier selection. In addition, there are studies that adopted tools such as linear programming, artificial intelligence, optimization, and goal programming (Chen, 2011). Figure 2 displays the structure of green supplier selection problems.

Table 3. Some examples for the definition of a green supplier

Source	Definition of a green supplier
Lee et al. (2009)	A supplier that adapt to the environment and are expected to design efficient and green items and perform life cycle analyses.
Hu et al. (2015)	A supplier that are available in all supply chains to help saving cost and achieving environmental protection.
Wang Chen et al. (2016)	A supplier that comply with expectations and goals of a business to minimize its adverse environmental impact and maximize its economic performance.
Tsai et al. (2016)	A supplier that help businesses, once they are evaluated by proper criteria, improve their efficient use of resources, reduce their consumption of resources and create a green corporate image.
Pang et al. (2017)	A supplier with low-carbon ability selected by businesses to manage pressures imposed by governments and markets.
Lin et al. (2018)	A supplier that can provide environmentally friendly competencies.

Figure 2. The structure of green supplier selection problems (Konys, 2019)

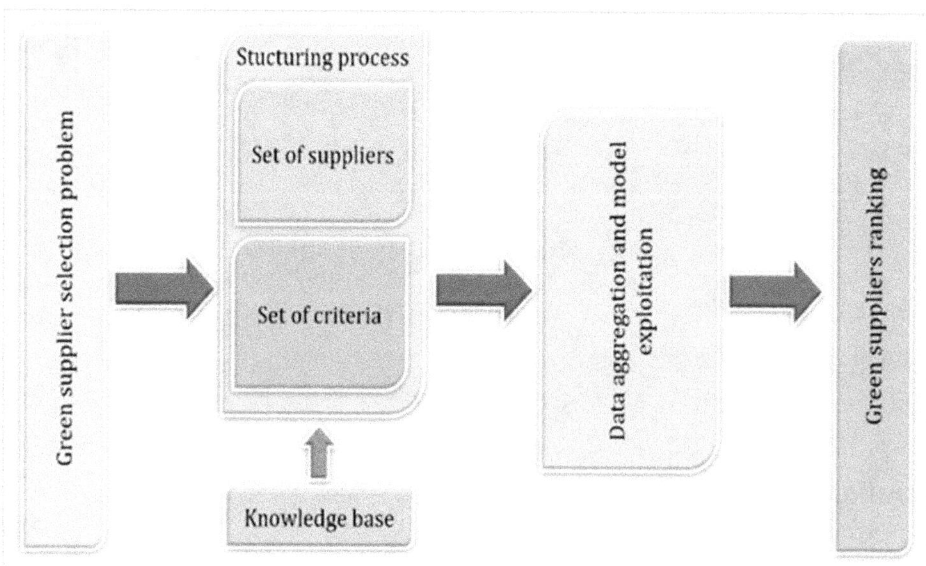

Green Criteria

A criterion is an attribute to be taken into consideration for the analysis of options. Criteria must be set as a part of the organizational setup for the selection of green suppliers. For conventional supplier selection, criteria such as cost, price, quality, service, and delivery are often adopted. In addition to the aforementioned criteria based on which suppliers are expected to deliver a good performance, there might be other criteria set in accordance with the requirements of a given industry or expectations of operators. By their nature, green criteria are taken into account for the selection of a green supplier. Although only green criteria can be used in the selection of green suppliers, it can also be used with dimensions that include economic and social criteria. Green criteria in the articles covering the scope of the study are shown in Table 4.

Table 4. Green criteria in the articles

No	Green criteria	Definition	References
1	Environmental management system and certification	Any certification system developed by organizations to find out and manage environmental effects of their operations in an effort to certify their environmental management skills. The most common is ISO14001 (Ahmadi et al., 2017).	Wang Chen et al. (2016), Guo et al. (2017), Banaeian et al. (2018), Abdullah et al. (2019)
2	Green image	Prioritization of the capability to manufacture green items in accordance with environmental standards and its perception in the eyes of consumers (Mousakhani et al., 2017).	Freeman and Chen (2015), Gupta et al. (2019), Ghorabaee et al. (2020)
3	Green packaging	Describes the use of environmentally friendly packaging and packaging materials that dissolve or degrade and do not adversely affect the nature (Uygun & Dede, 2016).	Kannan et al. (2015), Uygun and Dede (2016), Liang and Chong (2019)
4	Pollution control and prevention	Controlling and monitoring organizations so that the pollution generated by their operations does not do damage to the nature.	Fallahpour et al. (2016), Ulutaş et al. (2019), Gupta et al. (2019)
5	Green production systems	It refers to the effort to minimize the consumption of raw materials and energy for production.	Uygun and Dede (2016), Lo et al. (2018), Liou et al. (2019)
6	Green technology	Generating environmentally friendly outcomes by means of technology to be put to use for purposes such as protection of human health and environment, and promotion of sustainable economic development (Wu et al., 2019).	Ghorabaee et al. (2016), Tian et al. (2018), Chen et al. (2018)
7	Green purchasing	It is the observance of the environment in the purchasing and purchasing activities of reusable, recyclable and non-hazardous products and raw materials (Fallahpour et al., 2020).	Liou et al. (2019), Fallahpour et al. (2020)
8	Green product/ material	Any environmentally friendly product that can be monitored for its entire life cycle be free from any toxic material and made of recyclable elements.	Demir et al. (2018), Khorasani (2018) Kumari and Mishra (2020)
9	Management commitment to green supply chain	Having an environmentally friendly perspective and acting in line with a social responsibility within the body of an organization through the support and encouragement of high and medium level managers (Hashemi et al., 2015).	Hashemi et al. (2015), Haeri and Rezaei (2019), Krishankumar et al. (2020)
10	Energy/resource consumption	Any operational process requires minimizing energy consumption and gravitating toward renewable energy resources.	Cao et al. (2015), Bakeshlou et al. (2017), Hou and Xie (2019)
11	Green transportation	Use of less-polluting vehicles for transportation purposes and selection of alternative vehicles with less environmental damage.	Demir et al. (2018), Chatterjee et al. (2018), Ulutaş et al. (2019)
12	Reverse logistics	A green operation that stands for moving goods from their final destination (consumers) for the purpose of capturing value.	Cao et al. (2015), Chatterjee et al. (2018), Liang and Chong (2019)
13	Waste management	It is the reduction, measurement and monitoring of wastes by suppliers during their operations.	Yazdani (2014), Freeman and Chen (2015), Shojaei and bolvardizadeh (2020)
14	Hazardous/Toxic substances	Chemical and other materials that pose an environmental threat must be managed in all aspects such as storage, utilization, recycle and disposal.	Zhao and Guo (2014), Chatterjee et al. (2018), Gegovska et al. (2020)
15	Waste water	Any increase in the amount of waste water generated exacerbates environmental impact and causes an increase in ecological footprints. The flow of waste water into water reservoirs poses a risk for natural balance (Govindan et al., 2017).	Kannan et al. (2015), Govindan et al. (2017), Yazdani et al. (2019)

Table 4. Continued

No	Green criteria	Definition	References
16	Air emission	Taking under control, measuring, reducing and classifying any harmful quantities of air caused by chemicals and combustion operations that harm the ozone layer (Kuo et al., 2015).	Dobos and Vörösmarty (2019), Hou and Xie (2019), Yazdani et al. (2019)
17	Green market share	Criteria that project prospective changes in market share of businesses in consideration of their green reputation (Pati et al., 2016).	Kannan et al. (2015), Liang and Chong (2019)
18	Green design	Product design based on reuse, recovery and recycle of materials, use of fewer materials and less energy, prevention or reduction of the use of hazardous materials (Çalık, 2021).	Ecer (2020), Çalık (2021), Zhang et al. (2021)
19	Remanufacturing	Criteria to remanufacture items and recover their lost value through methods such as repairment of damaged components.	Kannan et al. (2015), Uygun and Dede (2016), Gegovska et al. (2020)
20	Reuse	The re-utilization of previously used items, and thus recovery of materials and energy in an effective fashion.	Yazdani (2014), Kannan et al. (2015), Dobos and Vörösmarty (2019)
21	Reducing activities	It demonstrates giving importance to mitigation efforts while conducting activities.	Uygun and Dede (2016)
22	Recycle	A common term that stands for processing used items to convert them into new ones.	Yu and Hou (2016), Hou and Xie (2019), Liang and Chong (2019)
23	Environmental training	The fact that individuals in an organization are equipped with environmental training in an effort to achieve green goals.	Qin et al. (2017), Gupta et al. (2019), Krishankumar et al. (2020)
24	Green R&D and innovation	This term is a token of research and development efforts exerted by a business for the development of environmentally friendly technologies, practices, procedures and methods, and continuous improvement of its current structure (Qin et al., 2017).	Kannan et al. (2015), Qin et al. (2017), Wu et al. (2019)
25	Eco-labeling	The usage of an environmentally friendly label by suppliers to deliver a high performance in line with green goals.	Liang and Chong (2019)
26	Green warehousing	A green criterion to save space and energy for all warehousing operations.	Kannan et al. (2015), Liou et al. (2019), Ulutaş et al. (2019)
27	Solid waste	Measurement and control of the amount and extent of solid waste generated.	Govindan et al. (2017), Liang and Chong (2019), Gegovska et al. (2020)
28	Environmental costs	Calculation of the cost of environmental damage caused by raw materials and goods	Cao et al. (2015), Wu et al. (2019), Ecer (2020)
29	Environmental policies	The efforts to adapt to applicable legislations and standards on environment.	Zhao and Guo (2014), Liou et al. (2016), Shojaei and bolvardizadeh (2020)
30	End-of-pipe practices	It shows investments in waste reduction and recycling management.	Ahmadi et al. (2017)
31	Green management and control	It describes the management and control activities carried out in order to have environmental targets, plans and policies.	Tsui and Wen (2014), Wan et al. (2017), Chatterjee et al. (2018)
32	Environmental competencies	Possession of sufficient ecological knowledge, socio-economic know-how, and skills for environmental conservation (Gitinavard et al., 2018).	Mousakhani et al. (2017), Gitinavard et al. (2018), Wan et al. (2020)
33	Pollution production	It refers to the amount of pollution that occurs in the measurements.	Ghorabaee et al. (2016), Haeri and Rezaei (2019), Wu et al. (2019)

Table 4. Continued

No	Green criteria	Definition	References
34	Environmental efficiency	Conducting environmentally efficient activities.	Kannan et al. (2015), Fallahpour et al. (2020)
35	Green degree level	It expresses the grade of environmental friendliness of the alternative.	Yu and Hou (2016)
36	Environmental performance	It is the efforts of the supplier to achieve its environmental goals.	Uygun and Dede (2016), Lo et al. (2018)
37	Environmental consciousness	It reflects the awareness of suppliers in the supply chain while implementing their operations.	Zhao and Guo (2014), Wan and Li (2015), Shojaei and bolvardizadeh (2020)
38	Green social responsibility	It is the social responsibility activities that suppliers show on environmental issues.	Liang and Chong (2019), Shojaei and bolvardizadeh (2020)
39	Green risk	It is to evaluate, measure, and prevent the dangers and risks that occur in the environment (Fallahpour et al., 2020).	Fallahpour et al. (2020)
40	Environmental collaboration	The capacity of collaborations on environmental assessment while carrying out the activities of the elements in the supply chain (Liou et al., 2019).	Liou et al. (2016), Liou et al. (2019), Fallahpour et al. (2020)
41	Green application	It is the environmental output of the supplier's practices.	Rouyendegh et al. (2020)

ANALYSIS

It is known that there is at least one green criterion in each of the green supplier selection practices that constitute the scope of the study. It was seen that only green criteria were used in some studies, and other criteria were used in addition to green criteria in others. In 42 of the 50 studies (84%), green criteria were used along with other criteria, while only green criteria were included in the remaining 8 studies (16%). As can be seen in Table 5, criteria other than green are generally traditional criteria such as cost, price,

Figure 3. Label cloud of criteria adopted in the studies

Table 5. The other most common criteria

Criteria	Reference
Cost	Mousakhani et al. (2017), Gitinavard et al. (2018), Chen et al. (2018), Gegovska et al. (2020)
Quality	Tsui and Wen (2014), Lo et al. (2018), Haeri and Rezaei (2019)
Price	Kannan et al. (2015), Banaeian et al. (2018), Gupta et al. (2019)
Delivery	Khorasani (2018), Hou and Xie (2019), Ghorabaee et al. (2020)
Technology	Wan and Li (2015), Wan et al. (2017), Bakeshlou et al. (2017)
Service	Ahmadi et al. (2017), Gegovska et al. (2020), Wan et al. (2020)

quality, service time, quality control. Figure 3 contains the tag cloud created according to the frequency of the criteria used in the studies. These criteria can be placed under the main criteria such as social, ethical, quality, cost, and it was noticed that green criteria and other criteria exist together without the main criteria. The intensive use of other criteria shows that the high rate of use continues in practices such as the recent selection of green suppliers.

To examine the green criteria, 28 green criteria groups were established. While determining groups, the most used green criteria are usually arranged as separate group titles. In cases where the environmental criteria were intertwined with each other, the descriptions of the authors who used the criteria were taken into consideration. If a definition of the criteria was not given in the study, it was tried to find out which criterion group it belongs to for the authors' point of view and the field of application. For each case, it was aimed to make an appropriate grouping study by making comparison with other studies. Each criterion was included in only one group in order to avoid a statistical error. The fact that the criteria names and definitions are included differently in most studies made it difficult to distribute the criteria. In order to overcome this difficulty, the criteria were examined one by one, even if they have different names, a common structure was created within the group by paying attention to the definitions and is shown in Table 6.

Two parameters named "frequency" and "appearance" were specified for evaluating these criteria groups. The "frequency" parameter refers to the total number of times each criterion group was seen in articles. "frequency" refers that the total number of groups in these articles. The other parameter, "appearance", is the total number of articles that include the first parameter. For example, there were criteria from 25 "eco-design" groups in the reviewed articles. Since there are criteria from more than one "eco-design" group in some studies, these criteria were included in the 19 articles.

SOLUTIONS AND RECOMMENDATIONS

The groups are classified by considering the two parameters mentioned above. Groups were intended to be divided into four classes, from the most frequently used to the least used. These groups are named "Very common", "Common", "Moderately common" and "Less common", respectively. The k-means clustering algorithm is used for this process. The k-Means clustering algorithm is an unsupervised algorithm that aims to divide n data into k clusters where each sample is closest to it (Likas et al., 2003). The clustering algorithm aims to minimize the sum of squares of various types (Wang et al., 2012). As

Table 6. Important criteria created for the content of the groups

Group no	Group	Criteria
G1	Environmental Management System	Environmental management system, ISO14001, green certification, environmental certification
G2	Green Image	Green image, green social responsibility, green enterprise image
G3	Green packaging and labeling	Green packaging, eco-packaging, eco-labeling
G4	Pollution Control	Pollution control, pollution prevention
G5	Green Manufacturing	Green manufacturing, cleaner production, green production
G6	Green Technology and R&D	Green technology, use of eco-friendly technologies, low-carbon technology
G7	Green Purchasing	Green purchasing, requirement of low-carbon purchasing
G8	Green Product/Material	Green product, green material, use of eco-friendly raw materials
G9	Management Commitment	Commitment of green supply chain management from managers, green supports by top managers of contractors companies
G10	Energy/Resource Consumption	Energy consumption, resource consumption, saving energy
G11	Green Distribution	Green transportation, reverse logistics, green warehousing
G12	Waste Management	Waste management
G13	Hazardous Substance	Hazardous substance, use of toxic/restricted substances, inventory of hazardous substances
G14	Amount of Waste	Waste water, waste production, solid waste
G15	Air Emission	Air emission, CO_2 emission, greenhouse gas emission
G16	Green Market Share	Ratio of green customers to total customers, green customers' market share
G17	Eco-Design	Green design, eco-design, ecological design
G18	Remanufacturing, Reuse and Recycle	Remanufacturing, recycle, reuse, recycling
G19	Staff Environmental Training	Green training, staff environmental training
G20	Environmental Costs	Environmental cost, total green product cost
G21	Environmental Policies	Environmental policies, Environmental permits and reporting, government regulations and legal compliance on the environment
G22	Green Management and Control	Green management, environmental management, environmental control, environmental planning
G23	Environmental Competencies	Environmental competencies, green competencies
G24	Pollution Production	Pollution production, pollution, Environmental pollution of production
G25	Green Risk Evaluation	Carbon risk evaluation
G26	Environmental Collaboration	Environmental collaboration, environmental collaboration and information sharing with firm
G27	Environmental Consciousness	Environmental consciousness, Environmental awareness
G28	Environmental Performance and Efficiency	Environmental performance, Environmental efficiency

Table 7. The frequency and appearance numbers of green criteria groups

Group no	Group	Frequency	Appearance
G1	Environmental Management System	31	20
G2	Green Image	14	14
G3	Green packaging and labeling	10	8
G4	Pollution Control	12	12
G5	Green Manufacturing	14	12
G6	Green Technology and R&D	17	14
G7	Green Purchasing	2	2
G8	Green Product/Material	13	12
G9	Management Commitment	7	7
G10	Energy/Resource Consumption	21	21
G11	Green Distribution	11	8
G12	Waste Management	6	6
G13	Hazardous Substance	9	7
G14	Amount of Waste	10	8
G15	Air Emission	9	9
G16	Green Market Share	3	2
G17	Eco-Design	25	19
G18	Remanufacturing, Reuse and Recycle	26	14
G19	Staff Environmental Training	8	8
G20	Environmental Costs	8	5
G21	Environmental Policies	9	7
G22	Green Management and Control	19	15
G23	Environmental Competencies	7	7
G24	Pollution Production	10	10
G25	Green Risk Evaluation	1	1
G26	Environmental Collaboration	3	3
G27	Environmental Consciousness	4	4
G28	Environmental Performance and Efficiency	4	4

an iterative algorithm, the closest one to the centroids is assigned with this method. The main function of k-means algorithm as follows (Eq. (1)):

$$J = \sum_{j=1}^{k}\sum_{i=1}^{n}\left\| x_i^{(j)} - c_j \right\|^2 \tag{1}$$

where $j=\{1,2,3,\ldots,k\}$ and $i=\{1,2,3,\ldots,n\}$. In Eq. (1), J represents main function, i represents number of data, c_j and x_i stand for centroid of *jth* cluster and data point, respectively. The application was made

Figure 4. The result of the algorithm taken from MATLAB

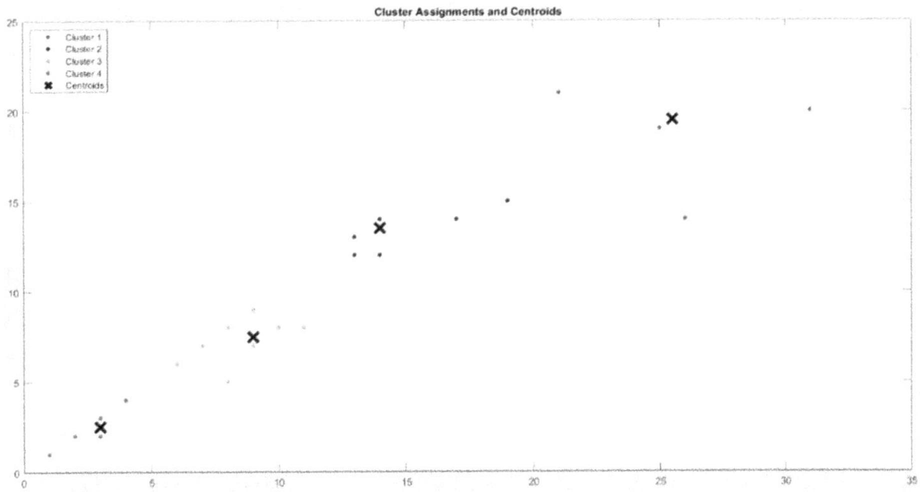

with MATLAB software. In the clustering study of the algorithm, three different centroids and four green criteria set groups were formed. The result of the algorithm taken from the software is shown in Figure 4.

As a result of the clustering in Figure 4, there are four green criteria groups in the first cluster, six in the second cluster, twelve in the third cluster, and six in the fourth cluster (Table 8).

Table 8. Clusters resulting from the k-means algorithm

The first cluster	The second cluster	The third cluster	The fourth cluster
Very common	**Common**	**Moderately common**	**Less common**
Environmental Management System	Green Image	Green packaging and labeling	Green Risk Evaluation
Energy/Resource Consumption	Pollution Control	Management Commitment	Environmental Collaboration
Eco-Design	Green Manufacturing	Green Distribution	Environmental Awareness
Remanufacturing, Reuse and Recycle	Green Technology and R&D	Waste Management	Environmental Performance and Efficiency
	Green Product/Material	Hazardous Substance	Green Market Share
	Green Management and Control	Amount of Waste	Green Purchasing
		Air Emission	
		Staff Environmental Training	
		Environmental Costs	
		Environmental Policies	
		Environmental Competencies	
		Pollution Production	

FUTURE RESEARCH DIRECTIONS

The number of studies on green supplier selection practices is increasing day by day. With this study, it was aimed to create a guide for green supplier selection practices. The number of articles investigated can be increased to achieve this goal. It may be possible to increase the accuracy of the results by including more articles in the study. Different databases can be scanned in addition to the Web of Science database. In addition, the work can be improved by increasing the number of parameters used in the k-means clustering algorithm. From another perspective, in addition to green criteria, criteria in economic and social dimensions can also be included in the analysis. By conducting surveys with businesses in the industry, the results obtained by looking at the use of green criteria can be compared. Some parameters such as methods and sectors in green supplier selection practices can also be added to the study to improve the study. Thus, a holistic view can be brought to green supplier selection practices.

CONCLUSION

Today, environmental issue occupies an important place in the agenda of people and organizations. Regardless of the sector, all businesses have responsibilities to restore the ecological balance. While establishing supply chains, it is necessary to organize the processes in line with the determined green targets (Akcan & Taş, 2019). At every moment of the implementation of the decisions taken, the environment should be cared for and the green should be taken into account. In this direction, the supply chains created should also be a "green supply chain". While establishing green supply chains, various criteria should be determined for the selection of suppliers and management of the chain, and the importance (weights) of these criteria should be determined and calculated in a deterministic manner.

In this study, the green criteria used in green supplier selection studies were analyzed. A taxonomy was intended to be made in which studies with similar names and/or definitions will be evaluated together. Since all of the green criteria aim to protect the environment and minimize damage, criteria that have close meanings with each other can be determined from time to time. In addition, it may be encountered with the situation of qualifying more than one green criterion in one criterion. In order to overcome this situation, the authors' definition of criteria, the markets where the applications are made, and the criteria setting structures were examined. To create a general framework, the green criteria are divided into 28 groups. These groups were analyzed with k-means clustering algorithm. MATLAB program was used for the application. According to this method, three centroids were formed and clusters were divided into four clusters according to their frequency of use. It was reached that first cluster are "environmental management system", "energy /resource consumption", "eco-design" and "remanufacturing, reuse and recycle". These groups can be referred to as very common green criteria. The second cluster is called common criteria included "green image", "pollution control", "green manufacturing", "Green technology and R&D", "green product/material", and "green management and control". Twelve and six groups took place in the third and fourth clusters, respectively.

The environmental management system group was determined as a green criteria set, including ISO14001, other environmental management systems and green certificates. It is deemed appropriate for businesses to request suppliers that have these, since they can be easily revealed, measured, and have certificates and systems issued by independent institutions and organizations. The fact that a large part of the green criteria still cannot be measured causes only 16% of the studies to be selected with completely green criteria. Price, cost, and service time have still an important place in applications as traditional criteria.

It is a complex task to determine the criteria for green supplier selection and green supply chain applications. Usually, the application parameters are determined by examining the past scientific literature or based on the experience of experts. Failure to create the most suitable structure for the desired purpose may cause the results to be inadequate or incorrect. In this study, green criteria in green supplier selection practices were examined. The results obtained reflect the green criteria trends in studies in the current literature. It is thought that the clusters obtained as a result of the algorithm can be used in determining the green criteria in the green supplier selection problems to be created.

ACKNOWLEDGMENT

This research received no specific grant from any funding agency in the public, commercial, or not-for-profit sectors.

REFERENCES

Abdullah, L., Chan, W., & Afshari, A. (2019). Application of PROMETHEE method for green supplier selection: A comparative result based on preference functions. *Journal of Industrial Engineering International*, *15*(2), 271–285. doi:10.100740092-018-0289-z

Ahmadi, H. B., Petrudi, S. H. H., & Wang, X. (2017). Integrating sustainability into supplier selection with analytical hierarchy process and improved grey relational analysis: A case of telecom industry. *International Journal of Advanced Manufacturing Technology*, *90*(9), 2413–2427. doi:10.100700170-016-9518-z

Akcan, S., & Taş, M. A. (2019). Green supplier evaluation with SWARA-TOPSIS integrated method to reduce ecological risk factors. *Environmental Monitoring and Assessment*, *191*(12), 736. doi:10.100710661-019-7884-3 PMID:31707512

Bakeshlou, E. A., Khamseh, A. A., Asl, M. A. G., Sadeghi, J., & Abbaszadeh, M. (2017). Evaluating a green supplier selection problem using a hybrid MODM algorithm. *Journal of Intelligent Manufacturing*, *28*(4), 913–927. doi:10.100710845-014-1028-y

Banaeian, N., Mobli, H., Fahimnia, B., Nielsen, I. E., & Omid, M. (2018). Green supplier selection using fuzzy group decision making methods: A case study from the agri-food industry. *Computers & Operations Research*, *89*, 337–347. doi:10.1016/j.cor.2016.02.015

Büyüközkan, G., & Vardaloğlu, Z. (2008). Yeşil tedarik zinciri yönetimi. *Lojistik Dergisi*, *8*, 66–73.

Çalık, A. (2021). A novel Pythagorean fuzzy AHP and fuzzy TOPSIS methodology for green supplier selection in the Industry 4.0 era. *Soft Computing*, *25*(3), 2253–2265. doi:10.100700500-020-05294-9

Cao, Q., Wu, J., & Liang, C. (2015). An intuitionsitic fuzzy judgement matrix and TOPSIS integrated multi-criteria decision making method for green supplier selection. *Journal of Intelligent & Fuzzy Systems*, *28*(1), 117–126. doi:10.3233/IFS-141281

Chatterjee, K., Pamucar, D., & Zavadskas, E. K. (2018). Evaluating the performance of suppliers based on using the R'AMATEL-MAIRCA method for green supply chain implementation in electronics industry. *Journal of Cleaner Production*, *184*, 101–129. doi:10.1016/j.jclepro.2018.02.186

Chen, J., Zeng, S., & Zhang, C. (2018). An OWA distance-based, single-valued neutrosophic linguistic topsis approach for green supplier evaluation and selection in low-carbon supply chains. *International Journal of Environmental Research and Public Health*, *15*(7), 1439. doi:10.3390/ijerph15071439 PMID:29986549

Chen, W. (2016). H. M., Chou, S. Y., Luu, Q. D., & Yu, T. H. K. (2016). A fuzzy MCDM approach for green supplier selection from the economic and environmental aspects. *Mathematical Problems in Engineering*, *2016*, 1–10. Advance online publication. doi:10.1155/2016/8097386

Chen, Y. J. (2011). Structured methodology for supplier selection and evaluation in a supply chain. *Information Sciences*, *181*(9), 1651–1670. doi:10.1016/j.ins.2010.07.026

Demir, L., Akpınar, M. E., Araz, C., & Ilgın, M. A. (2018). A green supplier evaluation system based on a new multi-criteria sorting method: VIKORSORT. *Expert Systems with Applications*, *114*, 479–487. doi:10.1016/j.eswa.2018.07.071

Dobos, I., & Vörösmarty, G. (2019). Inventory-related costs in green supplier selection problems with Data Envelopment Analysis (DEA). *International Journal of Production Economics*, *209*, 374–380. doi:10.1016/j.ijpe.2018.03.022

Ecer, F. (2020). Multi-criteria decision making for green supplier selection using interval type-2 fuzzy AHP: A case study of a home appliance manufacturer. *Operations Research*, •••, 1–35. doi:10.100712351-020-00552-y

Ehrlich, P. R. (2008). Key issues for attention from ecological economists. *Environment and Development Economics*, *13*(1), 1–20. doi:10.1017/S1355770X07004019

Espinilla, M., Halouani, N., & Chabchoub, H. (2015). Pure linguistic PROMETHEE I and II methods for heterogeneous MCGDM problems. *International Journal of Computational Intelligence Systems*, *8*(2), 250–264. doi:10.1080/18756891.2015.1001949

Fahimnia, B., Sarkis, J., & Davarzani, H. (2015). Green supply chain management: A review and bibliometric analysis. *International Journal of Production Economics*, *162*, 101–114. doi:10.1016/j.ijpe.2015.01.003

Fallahpour, A., Olugu, E. U., Musa, S. N., Khezrimotlagh, D., & Wong, K. Y. (2016). An integrated model for green supplier selection under fuzzy environment: Application of data envelopment analysis and genetic programming approach. *Neural Computing & Applications*, *27*(3), 707–725. doi:10.100700521-015-1890-3

Fallahpour, A., Wong, K. Y., Rajoo, S., & Mardani, A. (2020). An integrated fuzzy carbon management-based model for suppliers' performance evaluation and selection in green supply chain management. *International Journal of Fuzzy Systems*, *22*(2), 712–723. doi:10.100740815-019-00759-y

Freeman, J., & Chen, T. (2015). Green supplier selection using an AHP-Entropy-TOPSIS framework. *Supply Chain Management*, *20*(3), 327–340. doi:10.1108/SCM-04-2014-0142

Gegovska, T., Koker, R., & Cakar, T. (2020). Green Supplier Selection Using Fuzzy Multiple-Criteria Decision-Making Methods and Artificial Neural Networks. *Computational Intelligence and Neuroscience*, *8811834*, 1–26. Advance online publication. doi:10.1155/2020/8811834 PMID:33082775

Genovese, A., Koh, S. L., Bruno, G., & Bruno, P. (2010, October). Green supplier selection: A literature review and a critical perspective. In *2010 8th International Conference on Supply Chain Management and Information* (pp. 1-6). IEEE.

Ghadimi, P., Azadnia, A. H., Heavey, C., Dolgui, A., & Can, B. (2016). A review on the buyer–supplier dyad relationships in sustainable procurement context: Past, present and future. *International Journal of Production Research*, *54*(5), 1443–1462. doi:10.1080/00207543.2015.1079341

Ghorabaee, M. K., Zavadskas, E. K., Amiri, M., & Esmaeili, A. (2016). Multi-criteria evaluation of green suppliers using an extended WASPAS method with interval type-2 fuzzy sets. *Journal of Cleaner Production*, *137*, 213–229. doi:10.1016/j.jclepro.2016.07.031

Gitinavard, H., Ghaderi, H., & Pishvaee, M. S. (2018). Green supplier evaluation in manufacturing systems: A novel interval-valued hesitant fuzzy group outranking approach. *Soft Computing*, *22*(19), 6441–6460. doi:10.100700500-017-2697-1

Govindan, K., Kadziński, M., & Sivakumar, R. (2017). Application of a novel PROMETHEE-based method for construction of a group compromise ranking to prioritization of green suppliers in food supply chain. *Omega*, *71*, 129–145. doi:10.1016/j.omega.2016.10.004

Govindan, K., Rajendran, S., Sarkis, J., & Murugesan, P. (2015). Multi criteria decision making approaches for green supplier evaluation and selection: A literature review. *Journal of Cleaner Production*, *98*, 66–83. doi:10.1016/j.jclepro.2013.06.046

Govindan, K., & Sivakumar, R. (2016). Green supplier selection and order allocation in a low-carbon paper industry: Integrated multi-criteria heterogeneous decision-making and multi-objective linear programming approaches. *Annals of Operations Research*, *238*(1-2), 243–276. doi:10.100710479-015-2004-4

Guo, Z., Liu, H., Zhang, D., & Yang, J. (2017). Green supplier evaluation and selection in apparel manufacturing using a fuzzy multi-criteria decision-making approach. *Sustainability*, *9*(4), 650. doi:10.3390u9040650

Gupta, H., & Barua, M. K. (2017). Supplier selection among SMEs on the basis of their green innovation ability using BWM and fuzzy TOPSIS. *Journal of Cleaner Production*, *152*, 242–258. doi:10.1016/j.jclepro.2017.03.125

Gupta, S., Soni, U., & Kumar, G. (2019). Green supplier selection using multi-criterion decision making under fuzzy environment: A case study in automotive industry. *Computers & Industrial Engineering*, *136*, 663–680. doi:10.1016/j.cie.2019.07.038

Haeri, S. A. S., & Rezaei, J. (2019). A grey-based green supplier selection model for uncertain environments. *Journal of Cleaner Production*, *221*, 768–784. doi:10.1016/j.jclepro.2019.02.193

Hashemi, S. H., Karimi, A., & Tavana, M. (2015). An integrated green supplier selection approach with analytic network process and improved Grey relational analysis. *International Journal of Production Economics*, *159*, 178–191. doi:10.1016/j.ijpe.2014.09.027

Hou, Q., & Xie, L. (2019). Research on supplier evaluation in a green supply chain. *Discrete Dynamics in Nature and Society*, *2601301*, 1–14. Advance online publication. doi:10.1155/2019/2601301

Hu, Z., Rao, C., Zheng, Y., & Huang, D. (2015). Optimization decision of supplier selection in green procurement under the mode of low carbon economy. *International Journal of Computational Intelligence Systems*, *8*(3), 407–421. doi:10.1080/18756891.2015.1017375

Igarashi, M., de Boer, L., & Fet, A. M. (2013). What is required for greener supplier selection? A literature review and conceptual model development. *Journal of Purchasing and Supply Management*, *19*(4), 247–263. doi:10.1016/j.pursup.2013.06.001

Kannan, D., Govindan, K., & Rajendran, S. (2015). Fuzzy axiomatic design approach based green supplier selection: A case study from Singapore. *Journal of Cleaner Production*, *96*, 194–208. doi:10.1016/j.jclepro.2013.12.076

Keshavarz-Ghorabaee, M., Amiri, M., Hashemi-Tabatabaei, M., Zavadskas, E. K., & Kaklauskas, A. (2020). A New Decision-Making Approach Based on Fermatean Fuzzy Sets and WASPAS for Green Construction Supplier Evaluation. *Mathematics*, *8*(12), 2202. doi:10.3390/math8122202

Khorasani, S. T. (2018). Green supplier evaluation by using the integrated fuzzy AHP model and fuzzy copras. *Process Integration and Optimization for Sustainability*, *2*(1), 17–25. doi:10.100741660-017-0027-9

Konys, A. (2019). Green supplier selection criteria: From a literature review to a comprehensive knowledge base. *Sustainability*, *11*(15), 4208. doi:10.3390u11154208

Krishankumar, R., Gowtham, Y., Ahmed, I., Ravichandran, K. S., & Kar, S. (2020). Solving green supplier selection problem using q-rung orthopair fuzzy-based decision framework with unknown weight information. *Applied Soft Computing*, *94*, 106431. doi:10.1016/j.asoc.2020.106431

Kumar, P., Singh, R. K., & Vaish, A. (2017). Suppliers' green performance evaluation using fuzzy extended ELECTRE approach. *Clean Technologies and Environmental Policy*, *19*(3), 809–821. doi:10.100710098-016-1268-y

Kumari, R., & Mishra, A. R. (2020). Multi-criteria COPRAS method based on parametric measures for intuitionistic fuzzy sets: Application of green supplier selection. *Iranian Journal of Science and Technology. Transaction of Electrical Engineering*, *44*(4), 1645–1662. doi:10.100740998-020-00312-w

Lee, A. H., Kang, H. Y., Hsu, C. F., & Hung, H. C. (2009). A green supplier selection model for high-tech industry. *Expert Systems with Applications*, *36*(4), 7917–7927. doi:10.1016/j.eswa.2008.11.052

Liang, R., & Chong, H. Y. (2019). A hybrid group decision model for green supplier selection: A case study of megaprojects. *Engineering, Construction, and Architectural Management*, *26*(8), 1712–1734. doi:10.1108/ECAM-10-2018-0462

Likas, A., Vlassis, N., & Verbeek, J. J. (2003). The global k-means clustering algorithm. *Pattern Recognition*, *36*(2), 451–461. doi:10.1016/S0031-3203(02)00060-2

Lin, K. P., Hung, K. C., Lin, Y. T., & Hsieh, Y. H. (2018). Green suppliers performance evaluation in Belt and Road using fuzzy weighted average with social media information. *Sustainability*, *10*(1), 5. doi:10.3390u10010005

Lins, T., Oliveira, R. A. R., Correia, L. H., & Silva, J. S. (2018, November). Industry 4.0 Retrofitting. In *2018 VIII Brazilian Symposium on Computing Systems Engineering (SBESC)* (pp. 8-15). IEEE. 10.1109/SBESC.2018.00011

Liou, J. J., Chuang, Y. C., Zavadskas, E. K., & Tzeng, G. H. (2019). Data-driven hybrid multiple attribute decision-making model for green supplier evaluation and performance improvement. *Journal of Cleaner Production*, *241*, 118321. doi:10.1016/j.jclepro.2019.118321

Liou, J. J., Tamošaitienė, J., Zavadskas, E. K., & Tzeng, G. H. (2016). New hybrid COPRAS-G MADM Model for improving and selecting suppliers in green supply chain management. *International Journal of Production Research*, *54*(1), 114–134. doi:10.1080/00207543.2015.1010747

Lo, H. W., Liou, J. J., Wang, H. S., & Tsai, Y. S. (2018). An integrated model for solving problems in green supplier selection and order allocation. *Journal of Cleaner Production*, *190*, 339–352. doi:10.1016/j.jclepro.2018.04.105

Mousakhani, S., Nazari-Shirkouhi, S., & Bozorgi-Amiri, A. (2017). A novel interval type-2 fuzzy evaluation model based group decision analysis for green supplier selection problems: A case study of battery industry. *Journal of Cleaner Production*, *168*, 205–218. doi:10.1016/j.jclepro.2017.08.154

My Dung, T. T., Luan, N. M., & Quoc, L. H. (2016). The analytic approach applications in green supplier selection: A literature review. *Journal of Engineering and Applied Sciences (Asian Research Publishing Network)*, *11*(11), 6754–6762.

Pang, Q., Yang, T., Li, M., & Shen, Y. (2017). A fuzzy-grey multicriteria decision making approach for green supplier selection in low-carbon supply chain. *Mathematical Problems in Engineering*, *9653261*, 1–9. Advance online publication. doi:10.1155/2017/9653261

Qin, J., Liu, X., & Pedrycz, W. (2017). An extended TODIM multi-criteria group decision making method for green supplier selection in interval type-2 fuzzy environment. *European Journal of Operational Research*, *258*(2), 626–638. doi:10.1016/j.ejor.2016.09.059

Rouyendegh, B. D., Yildizbasi, A., & Üstünyer, P. (2020). Intuitionistic fuzzy TOPSIS method for green supplier selection problem. *Soft Computing*, *24*(3), 2215–2228. doi:10.100700500-019-04054-8

Sarkis, J., Zhu, Q., & Lai, K. H. (2011). An organizational theoretic review of green supply chain management literature. *International Journal of Production Economics*, *130*(1), 1–15. doi:10.1016/j.ijpe.2010.11.010

Shojaei, P., & bolvardizadeh, A. (2020). Rough MCDM model for green supplier selection in Iran: A case of university construction project. *Built Environment Project and Asset Management, 10*(3), 437–452. doi:10.1108/BEPAM-11-2019-0117

Soda, S., Sachdeva, A., & Garg, R. K. (2016). Literature review of multi-aspect research works carried out on the concept and implementation of GSCM. *International Journal of Industrial and Systems Engineering, 23*(2), 223–253. doi:10.1504/IJISE.2016.076404

Srivastava, S. K. (2007). Green supply-chain management: A state-of-the-art literature review. *International Journal of Management Reviews, 9*(1), 53–80. doi:10.1111/j.1468-2370.2007.00202.x

Tian, Z. P., Zhang, H. Y., Wang, J. Q., & Wang, T. L. (2018). Green supplier selection using improved TOPSIS and best-worst method under intuitionistic fuzzy environment. *Informatica (Vilnius), 29*(4), 773–800. doi:10.15388/Informatica.2018.192

Triantaphyllou, E. (2000). Multi-criteria decision making methods. In *Multi-criteria decision making methods: A comparative study* (pp. 5–21). Springer. doi:10.1007/978-1-4757-3157-6_2

Tsai, S. B., Wei, Y. M., Chen, K. Y., Xu, L., Du, P., & Lee, H. C. (2016). Evaluating green suppliers from a green environmental perspective. *Environment and Planning. B, Planning & Design, 43*(5), 941–959. doi:10.1177/0265813515600897

Tsui, C. W., & Wen, U. P. (2014). A hybrid multiple criteria group decision-making approach for green supplier selection in the TFT-LCD industry. *Mathematical Problems in Engineering, 709872*, 1–13. Advance online publication. doi:10.1155/2014/709872

Ulutaş, A., Topal, A., & Bakhat, R. (2019). An application of fuzzy integrated model in green supplier selection. *Mathematical Problems in Engineering, 2019*(4256359), 1–11. doi:10.1155/2019/4256359

Uygun, Ö., & Dede, A. (2016). Performance evaluation of green supply chain management using integrated fuzzy multi-criteria decision making techniques. *Computers & Industrial Engineering, 102*, 502–511. doi:10.1016/j.cie.2016.02.020

Villanueva-Ponce, R., Avelar-Sosa, L., Alvarado-Iniesta, A., & Cruz-Sánchez, V. G. (2015). The green supplier selection as a key element in a supply chain: A review of cases studies. *Dyna, 82*(194), 36–45. doi:10.15446/dyna.v82n194.54466

Wan, S. P., & Li, D. F. (2015). Fuzzy mathematical programming approach to heterogeneous multiattribute decision-making with interval-valued intuitionistic fuzzy truth degrees. *Information Sciences, 325*, 484–503. doi:10.1016/j.ins.2015.07.014

Wan, S. P., Lin, L. L., & Dong, J. Y. (2017). MAGDM based on triangular Atanassov's intuitionistic fuzzy information aggregation. *Neural Computing & Applications, 28*(9), 2687–2702. doi:10.100700521-016-2196-9

Wan, S. P., Zou, W. C., Zhong, L. G., & Dong, J. Y. (2019). Some new information measures for hesitant fuzzy PROMETHEE method and application to green supplier selection. *Soft Computing, 24*(12), 9179–9203. doi:10.100700500-019-04446-w

Wang, Q., Wang, C., Feng, Z. Y., & Ye, J. F. (2012). Review of K-means clustering algorithm. *Electronic Design Engineering, 20*(7), 21-24.

Witkowski, K. (2017). Internet of things, big data, industry 4.0–innovative solutions in logistics and supply chains management. *Procedia Engineering, 182*, 763–769. doi:10.1016/j.proeng.2017.03.197

Wu, Q., Zhou, L., Chen, Y., & Chen, H. (2019). An integrated approach to green supplier selection based on the interval type-2 fuzzy best-worst and extended VIKOR methods. *Information Sciences, 502*, 394–417. doi:10.1016/j.ins.2019.06.049

Yazdani, M. (2014). An integrated MCDM approach to green supplier selection. *International Journal of Industrial Engineering Computations, 5*(3), 443–458. doi:10.5267/j.ijiec.2014.3.003

Yazdani, M., Chatterjee, P., Pamucar, D., & Abad, M. D. (2019). A risk-based integrated decision-making model for green supplier selection: A case study of a construction company in Spain. *Kybernetes, 49*(4), 1229–1252. doi:10.1108/K-09-2018-0509

Yu, Q., & Hou, F. (2016). An approach for green supplier selection in the automobile manufacturing industry. *Kybernetes, 45*(4), 571–588. doi:10.1108/K-01-2015-0034

Zhang, S., Wei, G., Wang, R., Wu, J., Wei, C., Guo, Y., & Wei, Y. (2021). Improved CODAS Method Under Picture 2-Tuple Linguistic Environment and Its Application for a Green Supplier Selection. *Informatica (Vilnius), 32*(1), 195–216. doi:10.15388/20-INFOR414

Zhao, H., & Guo, S. (2014). Selecting green supplier of thermal power equipment by using a hybrid MCDM method for sustainability. *Sustainability, 6*(1), 217–235. doi:10.3390u6010217

ADDITIONAL READING

Arndt, H. (2008). *Supply chain management.* Wiesbaden: Gabler.

Beamon, B. M. (1999). Designing the green supply chain. *Logistics Information Management, 12*(4), 332–342. doi:10.1108/09576059910284159

Croxton, K. L., Garcia-Dastugue, S. J., Lambert, D. M., & Rogers, D. S. (2001). The supply chain management processes. *International Journal of Logistics Management, 12*(2), 13–36. doi:10.1108/09574090110806271

Davis, T. (1993). Effective supply chain management. *Sloan Management Review, 34*, 35–35.

Lambert, D. M., & Cooper, M. C. (2000). Issues in supply chain management. *Industrial Marketing Management, 29*(1), 65–83. doi:10.1016/S0019-8501(99)00113-3

Sellitto, M. A., Hermann, F. F., Blezs, A. E. Jr, & Barbosa-Póvoa, A. P. (2019). Describing and organizing green practices in the context of Green Supply Chain Management: Case studies. *Resources, Conservation and Recycling, 145*, 1–10. doi:10.1016/j.resconrec.2019.02.013

Seuring, S., & Müller, M. (2008). From a literature review to a conceptual framework for sustainable supply chain management. *Journal of Cleaner Production, 16*(15), 1699–1710. doi:10.1016/j.jclepro.2008.04.020

Werner, H. (2000). *Supply chain management*. Wiesbaden: Gabler.

Zhu, Q., Sarkis, J., & Lai, K. H. (2008). Green supply chain management implications for "closing the loop". *Transportation Research Part E, Logistics and Transportation Review*, *44*(1), 1–18. doi:10.1016/j.tre.2006.06.003

KEY TERMS AND DEFINITIONS

Environmental Management System: Certification system that measure the competencies of organizations and organizations in green management.

Green Criteria: Attributes or factors that evaluate environmental impact.

Green Distribution: Activities to minimize the negative impact on the environment in the movement of products and services from supplier to customer and vice versa.

Green Supplier: Suppliers that businesses cooperate to achieve environmental goals.

Green Supplier Selection: The process of enterprises to evaluate the suppliers on environmental issues.

Green Supply Chain: An environmentally friendly interpretation of the classic concept of supply chains.

Supply Chain: The movement of products and services and the chain or network of all the elements involved in this movement.

Chapter 13
Application of Fuzzy Topsis and Taguchi Methods for Optimization Problems With Disruptive Risk:
A Systematic Review

Tamer Aksoy
Ibn Haldun University, Turkey

Gencay Karakaya
Istanbul Commerce University, Turkey

Shahryar Ghorbani
Sakarya University, Turkey

ABSTRACT

Ranking and choosing research projects and analyzing experiments are usually difficult and complex responsibilities for professional research councils at universities and research centers. Its complexity stems from having more than one variable in each project, and the participation of many decision-makers in the ranking process and selection of research projects based on many variables. The fuzzy set theory provides the required flexibility to show the uncertainty about the lack of knowledge, and also it can manage the uncertainty in the real world that the values of criteria are not defined properly. For this purpose, in the environment where the criteria of research projects are vaguely defined, the ranking methods such as Taguchi, which can reduce the number of experiments and making process more efficient, can be used for quality design in designing and processing product. In this work, first of all, the authors review fuzzy TOPSIS technique and the Taguchi method as well; then they approach research efficiency and optimization of the level of effective parameters in an experiment.

DOI: 10.4018/978-1-7998-8900-7.ch013

INTRODUCTION

Modern organizations work in a dynamic and competitive environment in which critical factors for their survival and expansion, often result in their effort for efficiency and effectiveness. Despite the concepts of performance, being in harmony, well-organized organizations might consider the importance of their competitors and compare their situation with them. Dynamic job environment has caused the importance of considering the competitors, changes over time (Sergey & Kweku, 2013). One of the most important indexes for comparison is the efficiency rate that has a broad meaning in product literature.

The flexibility of the concept of efficiency and its ability to being common in people's daily life has allowed them to define this concept by their knowledge and attitude, which is different from other common definitions in other countries and majors (Chen & Richard, 2014). By using Taguchi and fuzzy TOPSIS models, efficiency in the organization and research can be evaluated and analyzed. In these methods, a set of goals and criteria are used to judge options, and by using a method of ranking, options are ranked according to the satisfaction of goals and criteria (Hajkowicz and et al., 2000). As a result, the option that can be ideal in every situation is more or less impossible, but choosing the most relevant result is relatively possible.

Fuzzy Mathematics

Fuzzy methodologies are efficient techniques for considering vagueness in the human decision. One of the branches of contemporary mathematics is fuzzy logic that determines complicated problems, and changes in the management environment can be more simply controlled by using this method.

Based on scientific principles, it has been proved that all things include an eternal principle which by this principle it is true or not. It may be possible to doubt 'trueness' or 'falseness' of something, but they do not have doubt about one thing, that each occurrence is either 'true' or 'false.' There are many examples like atoms vibrate or not, and grass is green and is not red or many other examples. In other words, the desired solution like the grass being green or red defines the trueness or falseness of a solution, should not be expanded to everything. The problem with science is expanding this matter to all occurrences. In logic and mathematics this principle was the norm, anything is either true or false, so in this way, relevant topics or mathematical expressions are either entirely right or completely false, black or white, one or zero, many examples can be presented (Wang & Elhag, 2006).

The problem with science has this kind of definition among all logical and mathematical matters. We will see that it should not be interpreted this way, but everything should be measured relatively and should be graded. Everything is 'relatively' right or wrong. In other words, real phenomena are not black or white but somehow 'gray.' Science shows gray realities with black and white tools and this was why we thought that reality is also black or white. Considering a person with age of more than 18 years as a mature one is not a true fact. The fact has some other particular concept. Between 15 to 16 years old, adults barely can be found and also amongst 23 or 24 years old there can be, though rare, that is not mature yet.

In the mathematics environment, there are many concepts, logics, and relations. Many of these relations and concepts are based on a two-value system meaning of true or false. For example, it is said: this system is stable (or unstable) because of some principle or relation or mathematical theory. Here only two values are considered, i.e. stable and unstable for a controllable system (it is covered in linear control course). In another example, software engineers say that particular software service is safe or unsafe and considering only two values: safe or unsafe, and value a system as safe or unsafe. Actually they use only two black and white colors; many examples exist in many engineering fields that cannot be covered here.

Using two value like black and white is not desirable and in a way an engineer that designs a system (such as circuit, building, mechanical system, hardware or software system) around two value mathematics is like a painter using only two colors like black and white (not even gray).

Fuzzy and Its History

If we take a closer look at systems and their operation and values, it will be a lot better and makes engineering science progress a lot faster. We can do this by setting aside the two value systems and replacing it with a system or mathematics with multiple values. This multiple value system was introduced by Prof. Lotfi Zadeh in 1965 in an article titled "Fuzzy sets" in the "information and control" journal.

He was a co-author in 1963 in the first book of "Linear systems theory" and this book is referred in the best universities around the world. Prof. Lotfi Zadeh is known by many engineering science activists, but at first his articles in 1965 were not paid much attention; of course, others had some theories in this area, but not using fuzzy as a term. Max Black, quantum philosopher in 1937 in an article titled 'vagueness' was published in the journal of science philosophy. Max Black used 'vagueness' because Charles Pearce and Bertrand Russell used this term to describe what we now call 'fuzzy.

Crisp Sets

Crisp sets are natural and normal sets that are introduced at the beginning of the classical sets' theory. Adding a crisp adjective can help us assume one of the basic concepts of fuzzy logic in our minds.

In crisp sets, membership function has only two value in its range (in mathematics, a range of a function equals to the sum of all outputs from that function). Yes or no (one or zero) are actually the two values that are in classical logic:

$$\text{ì}_A(x) = \begin{cases} 1, & if\ x \in A \\ 0, & if\ x \notin A \end{cases}.$$

Where, $\text{ì}_A(x)$ is the membership function of element x in the crisp set of A.

Fuzzy Sets

The range of membership function is converted from {0, 1} to a close range of [0, 1] for fuzzy sets (Gwo-Hshiung & Huang, 2011).

Fuzzy Topsis Technique

Fuzzy TOPSIS techniques were introduced by Chen in 1992. Based on fuzzy method teachings, human thoughts are with uncertainty, and this uncertainty is useful in decision making. So, as a result fuzzy decision-making methodis used and one of these methods is fuzzy TOPSIS technique for ranking op-

Table 1. Nominal terms and its relation to its values

	Nominal Terms				
	Very Low (VL)	**Low (L)**	**Medium (M)**	**High (H)**	**Very High (VH)**
Nominal Values	(0.0, 0.2)	(0.0, 0.2, 0.4)	(0.0, 0.4, 0.6)	(0.4, 0.6, 0.8)	(0.6, 0.8, 1)

tions. In this method, decision-making matrix elements or its associated weights or both of them are expressed as fuzzy terms or fuzzy numbers.

There are various methods for fuzzy TOPSIS. Moreover, all of them are derived by a small change in fuzzy TOPSIS method introduced by Chen. (Chen and et al., 1992). However, to give more explanation for the fuzzy TOPSIS introduced method by Chen will thoroughly explain in this review paper (Chen C. T., 2000).

In this research, fuzzy TOPSIS techniques introduced by Chen (2000), and to convert qualitative data and lingual terms to quantitative data, we use fuzzy numbers and spectrum introduced by Amiri (2010) that are shown in Table 1.

The first step in creating fuzzy decision-making matrix, assume there are m alternatives, n factors, and k decision-makers. Fuzzy multiple criteria decision-making problem can be expressed as the following matrix (Roszkowska E. 2011):

$$\tilde{D} = \begin{bmatrix} \tilde{x}_{11} & \tilde{x}_{12} & \cdots & \tilde{x}_{1j} & \cdots & \tilde{x}_{1n} \\ \tilde{x}_{21} & \tilde{x}_{22} & \cdots & \tilde{x}_{2j} & \cdots & \tilde{x}_{2n} \\ \vdots & \vdots & \ddots & \vdots & & \vdots \\ \tilde{x}_{i1} & \tilde{x}_{i1} & \cdots & \tilde{x}_{ij} & \cdots & \tilde{x}_{in} \\ \vdots & \vdots & \ddots & \vdots & & \vdots \\ \tilde{x}_{m1} & \tilde{x}_{m2} & \cdots & \tilde{x}_{mj} & \cdots & \tilde{x}_{mn} \end{bmatrix}$$

$$\tilde{W} = \begin{bmatrix} \tilde{W}_1 & \tilde{W}_2 & \cdots & \tilde{W}_j & \cdots & \tilde{W}_n \end{bmatrix}$$

In this matrix:

i = the number of inspecting elements (m)
j = the number of responders (n)
$\tilde{x}_{ij}$ is the opinion of j(th) person on i(th) element as the following fuzzy numbers:

$$\tilde{x}_{ij} = \left(a_{ij}, b_{ij}, c_{ij}\right).$$

$\tilde{w}_j$ is the importance of each person's opinion, as the following fuzzy numbers:

$$\widetilde{w}_j = \left(w_{j1}, w_{j2}, w_{j3}\right) \forall j \in n.$$

Second step

In this step, the fuzzy matrix of people's opinion should be converted to one of the scaleless fuzzy matrixes $\bar{R}$ like expressions below:

$$\bar{R} = \left[\overline{r_{ij}}\right]_{m \times n}$$
$$\overline{r_{ij}} = \left(\frac{a_{ij}}{cj}, \frac{b_{ij}}{cj}, \frac{c_{ij}}{cj}\right)$$

In this equation, cj^* for each person equals to:

$$cj^* = \frac{\max C_{ij}}{j}$$
$$\overline{r_{ij}} = \left(\frac{a_{ij}}{cj}, \frac{b_{ij}}{cj}, \frac{c_{ij}}{cj}\right) \quad (1)$$

In this equation $\overline{a_j}$ .value for each person is calculated by this equation:

$$cj^* = \frac{\max a_{ij}}{i}. \quad (2)$$

Third step

We create a fuzzy scale less weight matrix $\tilde{V}$ with $\widetilde{W_{ij}}$ as input vector:

$$\tilde{V} = \left[\tilde{v}_{ij}\right]_{m \times n}$$
$$i = 1, 2, \ldots, m \quad j = 1, 2, \ldots, n.$$
$$\tilde{v}_{ij} = \tilde{r}_{ij} \times \tilde{w}_j$$

Fourth step

The ideal for positive fuzzy A^+ and the ideal for negative fuzzy A^- are defined as the following terms:

$$A^+ = \left(\tilde{v}_{\bar{1}}, \tilde{v}_{\bar{2}}, \ldots, \tilde{v}_{\bar{m}}\right)$$
$$A^- = \left(\tilde{v}_{\bar{1}}, \tilde{v}_{\bar{2}}, \ldots, \tilde{v}_{\bar{n}}\right).$$

In this research, we use ideal positive fuzzy, and Chen (2000) ideal fuzzy.

Fifth step

If 'A' and 'B' are two fuzzy numbers as below, then their distance is derived by using equation 3:

$$\tilde{A} = (a_1, b_1, c_1)$$
$$\tilde{B} = (a_2, b_2, c_2)$$

$$D(A,B) = \sqrt{(a_1 - a_2)^2 + (b_1 - b_2)^2 + (c_1 - c_2)^2} \quad (3).$$

By taking into account the above statements about calculating the distance between two fuzzy numbers, we get the distance for each element from positive ideal and negative ideal:

$$d_i^* = \sum_{j=1}^{n} d(\tilde{v}_{ij} - \tilde{v}_j^*) \qquad i = 1, 2, \ldots m$$
$$d_{\bar{i}} = \sum_{j=1}^{n} d(\tilde{v}_{ij} - \tilde{v}_j^-) \qquad i = 1, 2, \ldots m$$

Sixth step

We calculate the relative nearness of i(th) element to the positive ideal like this:

$$CC_i = \frac{d_{\bar{i}}}{d_i^* + d_{\bar{i}}} \; i=1,2,\ldots m1.$$

Seventh step

CC_i is presented in descending order, and then we rank the existing options in the problem (Gwo-Hshiung& Huang. 2011).

Factors and Experimental Design

The technique of examining and identifying the multiple factors involved in an experiment under all conditions is known as experimental design (Roy, 1990). This technique is also included in the literature as a factorial design. The experimental design is a series of studies for the evaluation and design of experiments in order to maximize the information obtained from experiments outside the real process and production (Banks, 1984).

Experimental design techniques are not only a statistical approach, but they can also be used in all research and development activities, increase quality, reduce costs, improve the reliability of the results, support, and complement all other quality techniques. The advantages they bring in practice can be increased performance and quality, efficient use of resources, speeding up research and development activities, and values that determine the quality characteristics of the product and the process are less sensitive to uncontrollable or difficult and costly factors. If there is more than one factor in an experiment, factorial designs are used. This is a design where factors change together, influenced by each other, and these changes are presented separately (Hines and Montgomery, 1972).

Taguchi Method

In today's competitive world, quality is a fundamental matter that has come to attention since the 40s and 50s of the 20th century. Feigenbaum, the innovator of the word "Comprehensive quality management" in 1951, described it as: "quality means the ability of a product in bringing the desired goal with the least amount of cost of manufacturing." The main body related to quality science in the United Kingdom is expanded as experiments designer and the United States as statistical quality control. When Japan began its reconstruction after WWII, it faced a severe lack of resources, quality equipment, and experienced engineers and began to compete for producing quality products and continuing in quality increase in that state. Innovating a method to face competition problems was given to Dr. Genichi Taguchi, a telecom equipment development engineer in Electrical communication labs. He was given the Deming award, one of the most notable awards in quality management in 1962.

Dr. Taguchi described quality as the amount of loss forced upon the society by producing defective products, and on this basis; he named the method of improving quality as "Taguchi Method." The Taguchi method is entirely different from standard quality engineering methods. The methodology of Taguchi method emphasizes quality design in designing processes and products, but standard methods emphasize quality control in product manufacturing or after it. In reality Taguchi believed that the best way to increase quality is in designing the product itself.

Dr. Taguchi's theory (Genichi Taguchi) began in the early 50s as a means to optimize the telephone lines in a communication company. It evolved in the 80s and from then is used in major companies around the world.

Table 2. An example of companies using this theory

Dow Chemical	AT&T Bell Laboratories	Awrey Bakeries, Inc.
Ford Motor Company	Canon Business Machines	EASTMAN Kodak Company
Goodyear	General Electric Company	Kraft General Foods
McDonald's	IBM Corporation	Siemens Automotive
Motorola, Inc.	NASA Marshall Space Flight Center	Toyota Motor Corporate Services
Philips	Anheuser-Busch Company	Xerox Corporation
U.S. Army Headquarters	Chrysler Corporation	G.E. Aircraft Engines

One of the goals in designing this experiment is that by a known change in input variables, the output changes can be monitored. Many methods for designing this experiment exist. One of the first methods to be offered in this area was the Factorial method that acquired the number of experiments by an N=Lm equation. The main problem with this method was that a large number of variables and the experiments, which is not acceptable considering the time and costs. So methods of reducing the number of experiments were sought after. One of the reformations was the Taguchi method that we are going to explain here.

The Taguchi method is a standard method used to optimize the level of useful parameters in an experiment. In this method, using Orthogonal Arrays (OA), the number of experiments is reduced. These arrays with unique properties are chosen among the whole experiment in factorial method. It does not guarantee that the optimized answer is within the selected experiments, but by using calculations related to arrays experiment, we can determine the optimal conditions, and the answer in optimal conditions which at the end by confirmation test (repeating the test in optimal conditions and confirming the repeatable answer in these conditions) is validated. This method was originally a tool to improve product quality by means of using statistical and engineering concepts. As experimenting and using errors is often time-consuming and costly, the need to satisfy design goals with the least number of tests is an essential requirement. In experiment design methods, known changes are applied to input variables so that in this way the amount of changes in the output response is monitored. By using this method, we can systematically change the controllable input variables and monitor their effects on the output parameters (Cafer, Cengiz, & Da, 2003).

Orthogonal arrays are shown as (Xy) Ln, that L stands for Latin squares (arrays that are used in experiment design and have special properties), n is the number of experiments, x is the factor number and y is the least amount of factors that can be inspected with the arrays. For example, in (27) L8 the number of experiments is 9, and the maximum number of factors is 7 of the second-order can be examined with this array. Note that a maximum number of 7 factors can be examined, and fewer factors can be used too. Designing an experiment involves selecting the most appropriate orthogonal array, determining the factors with the appropriate columns and finally explaining the experiments with the environmental conditions.

Experiment design method steps by Taguchi method for their details and priorities are as follows:

1. Introducing the effective factors in the reaction
2. Number of required experiments
3. Results analysis
4. Optimal conditions evaluation

Table 3. Experiment design base table by Taguchi method

Number of Parameters	Experiment Design Data			
	2	3	4	5
2	L4	L9	L16	L25
3	L4	L9	L16	L25
4	L8	L9	L16	L5
5	L8	L18	L16	L25
6	L12	L18	L32	L25
7	L12	L18	L32	L50
8	L12	L27	L32	L50
9	L12	L27	L32	L50
10	L12	L27	L32	L50
11	L12	L27		L50
12	L16	L27		L50
13	L16	L27		

In the first phase, we determine the effective factors and consider multiple scenarios for each of them. By taking into account the number of useful parameters and their orders, the number of experiments is determined. This way the number of parameters is determined from the first horizontal row in table 1, and the order is determined by the vertical row in the right side, the intersection of these two rows determines the number of experiments.

After determining the number of experiments, we create a matrix; the rows of this matrix determine the experiment condition. To create these matrices, a couple of complex methods exist, but we can use various statistical software such as Minitab or Design expert. Designing experiments by the Taguchi method considers two sets of factors. The first set that is called controllable factors, having a defined order and are controllable. This set is used in experiment design and final product or process design. The second set called noise factors affect the reaction in a process but cannot be controlled economically. These factors are the main reason for variation. The goal of Taguchi experiment design is creating the best condition so that the least variables are affected by the noise factors (Yao & Sio, 2013).

The advantages of this method are as follows:

- Reduction in the costs and the number of experiments
- A means to evaluate discrete factors (material type, part color, etc.)
- Determining factors share
- A means to estimate results in an optimal condition
- A means to estimate results in desirable orders
- Determining error impact
- Determining considered colliding effects impact
- A means to acquire optimal conditions for multiple results
- A means to analyze factors of different orders
- Analyzing an infinite number of factors

In this method, we expect the following results from analyzing the answers

- Optimal conditions in which the desired quality is achieved
- The impact that each factor has on quality and operation, and which is the most effective factor?
- Estimating the results with optimal conditions (confirmation tests)

There are two methods to analyze the experiment

1. Standard method (ANOVA variance analysis)
2. Using a signal to noise ratio (S/N)

The amount of S/N defines the scarcity from a known value, or in other words, in what way our results changed due to the tests that were done.

However, how can we understand which value is better?

To acquire these results, there are 3 equations, and each one of them has a purpose in particular conditions. In the Taguchi method, a Loss function is used to calculate the deviated value from the expected value in the result, and this function has different conditions according to the problem conditions.

1. The smallest value is the best

$$SB = \frac{1}{n}\Sigma(y_i^2).$$

2. The biggest value is the best

$$LB = \frac{1}{n}\Sigma(\frac{1}{y_i^2}).$$

3. The nominal value is the best

$$NB = \frac{1}{n}\sum\nolimits_{k=0}^{n}\left(y_i + y_0\right)^{n2}.$$

Where n is the number of trials and y is the measured output (Gwo-Hshiung& Huang. 2011).

- The first equation is ideal where we are examining a property which is non-negative, its ideal value is zero where the result is expected to be near to zero

An example of this equation can be abrasion, contraction, and destruction

- The second equation is ideal where there is no predetermined condition, so a more significant value is more desirable

Like when we examine material resistance, longevity, and efficiency.

- The last equation is when we examine a specific property, and we do not want to deviate from the desired result.

In all of the above and as a whole, when we use the S/N value for analysis, tests are repeated multiple times, and finally, we acquire the optimal conditions for the experiment. However, these optimal conditions can be outside the tests that we have done. After calculating the loss function for each output, we calculate the Overall S/N ratio from the following formula:

$$SN = -10\ Log(L_i).$$

Finally, these conditions should be met, and tests should be done under these conditions so that we can see whether it gives us the desired output or not.

The general thinking about specification limits is the customer will satisfy as long as the differences remain within the required limits (Fig 1). If the difference exceeds the limits, then the client directly

Figure 1. Taguchi Loss function

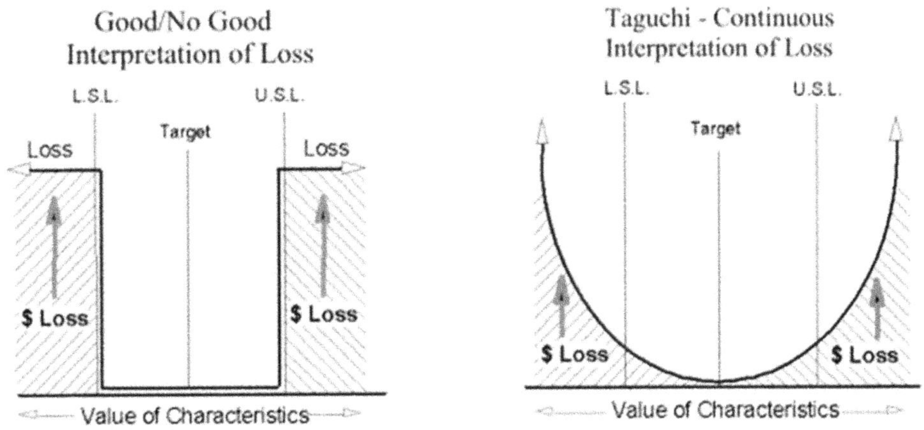

feels dissatisfied. The specification limits split fulfillment from dissatisfaction (Taguchi part formulas: (Jeyapaul and et al.2005)).

Variance Analysis

In order to find the relative importance of parameters in each output, we use a statistical method called ANOVA. In the ANOVA table, F test shows the effectiveness or ineffectiveness of tested parameters in the desired reliability level. Also, distribution percentage as a statistical parameter gives a better understanding of the impact of each parameter and also to the output. For parameters with high distribution percentage, small changes have a significant impact on the output.

The order of freedom for each input parameter is defined separately and equals to its orders minus one. Also the total order of freedom equals the number of tests minus one and the order of freedom for error equals the total order of freedom minus the sum of input variables.

Sum of total squared, errors, and parameters that are derived from the following formula (Gwo-Hshiung & Huang. 2011):

$$S_m = \text{Correction Factor} = \frac{\left(\Sigma \frac{S}{N_i}\right)^2}{N}$$

$$S_m = \text{Total sum of squares} = \left(\Sigma \frac{S}{N_i}\right)^2 - S_m$$

$$SS_A = \frac{A_1^2}{N_{A1}} + \frac{A_2^2}{N_{A2}} + \ldots + \frac{A_n^2}{N_{An}} - S_m$$

$$S_e = S_T - \Sigma SS_i$$

where N: number of signals to noises

Ai: the sum of signal to noise values for A factor in ith order
N_{Ai}: number of the signal to noise for A factor in ith order

Sum of total squares:

$$MS_i = \text{Mean sum of squares for factor i} = \frac{SS_i}{\text{DF of i}}.$$

Variance relativity (F-value):

$$F_i = \text{Variance ratio for factor i} = \frac{MS_i}{\text{MF of error}}.$$

Distribution percentage:

$$\tilde{n}(\%) = \frac{SS_i - (\text{DF of i})(\text{MF of error})}{S_T} \times 1002.$$

If we use variance analysis, we can understand that which factor has the most impact.
Taguchi method, like any other method has its limitations:

- We should consider the time in the experiment process.
- When designing systems, we should use this method from the start and if we use it in the middle of the experiment it is of no use
- After determining factors and their values, we use this method it is no longer useful.

Taguchi Experiment Design

Taguchi experiment design was developed as a process optimization technique in the 1950s (Roy, 1990). The method is based on statistical data analysis and simplifies the analysis and process optimization of complex systems. It can be applied in the process of modeling and modeling many other systems (Lochner and Matar, 1990). Taguchi offers new product development and complex problems, as well as cost-effective and time-consuming solutions to appropriate alternatives. It takes its technical strength from its discipline rather than technical experiments.

The Taguchi technique is implemented in four steps; To design the parameters by analyzing the quality characteristics that are important for the process and product by brainstorming, to design and implement the experiments, to analyze the results and to determine the optimum conditions and to evaluate the results by applying the optimum conditions. Brainstorming is a necessary and important step in practice. Brainstorming before an experimental design is a must for the Taguchi approach. However, Taguchi did not provide a guide or guide for brainstorming. The content and outputs of the brainstorming are mainly dependent on the nature of the process being studied and revealed by technical experience. The nature of the brainstorming may vary depending on the type of subject studied, its exact patterns not determined. Taguchi recommends the involvement of all relevant departments. Before starting the experiment, an approach will be developed in which the whole process will be discussed, how much will be done, how the results will be analyzed and which factors will be the most important. It is essential to consider and decide what, how and how many times the approach will be tested and when the results will be analyzed (Roy, 1990).

The Taguchi method aims to support high-quality process and product development in every aspect. This should include a minimum sensitivity against the production conditions and uncontrollable factors of the process or product, the provision of the required system, parameters and tolerances at the lowest cost, and the loss of loss of the product caused by the Taguchi loss function in the framework of a new quality cost understanding (Roy, 1990).

Taguchi Approach a Variability

Taguchi aims to increase the durability of the product and process design, system approach and design to the uncontrolled effects to improve the quality during the experimental design phase. The product and process design is a combination of accurate and appropriately tuned inputs, as well as the proper adjustment of mechanisms and process phases in the production process. The Taguchi technique is mainly based on improving quality by reducing the causes that cause variability (Roy, 1990).

Quality has not emerged to control the product or process; the aim is to design the quality to include it in the process to find the best way to improve beyond direct application to the product. Instead of improving quality through the inspection, screening, and control of processes, it should be targeted to bring the design to a level that will not be affected by changing, uncontrollable environmental factors in production processes. Quality improvement is achieved by minimizing deviation from the target. Taguchi claims that quality is directly related to deviation from the design parameters, not to compliance with fixed standards or specifications. The cost of quality should be measured as a function of deviation from the standard.

The product life cycle relates to the measurement of deviation from the parameters determined based on cost. These costs should include waste, recovery, inspection, warranty and service, recycling and product replacement costs. Cost is the factor that will guide the necessary parameters to be checked. Taguchi considered quality improvement efforts as a continuous effort. The first step is to bring the population distribution as close to the target value as possible. To achieve this, Taguchi design experiments use the orthogonal arrays (Roy, 1990).

Orthogonal Array, Effects and Interactions of Factors

The experiments are designed with strict rules, and the design of the experiment is fairly simplified with index systems that are rigid enough to leave no room for random situations and complex effects. An orthogonal index set is used to design the experiment. An orthogonal index can carry several alternatives, experimental alternatives. Commonly used orthogonal indexes are composed of 2, 3 and 4 alternative factors. Some directories may contain factors in the mixed alternative. In many cases, a standard orthogonal index is adapted to an assay containing factors in a mixed alternative. The experiment design process starts with selecting an appropriate orthogonal index, assigning the factors to the appropriate columns and determining the experimental conditions (Roy, 1990). To calculate the number of combinations of all effects, including main effects and interaction effects;

$$\sum_{a=1}^{k}\binom{k}{a}=\sum_{a=1}^{k}\frac{k!}{(k-a)!a!}=2^k-1.$$

The interactions between factors are established by taking all combinations into account, i.e., starting from binary interactions, all alternatives to the interaction effect that questions the interaction of the k factor, are determined;

$$\sum_{a=2}^{k}\binom{k}{a}=\sum_{a=2}^{k}\frac{k!}{(k-a)!a!}=2^k-k-1.$$

The Taguchi technique can be useful when applied in early stages in the design of product or process systems. On the other hand, Taguchi's approach does not involve self-modeling and optimization of the process; Analysis of the results of experiments is used to define the effects of factors, to plan additional experiments, and to improve performance (Simpson et al., 1997).

CONCLUSION

The present study marks our attempt to render a review of the previous literature on the concept of efficiency as related to various experiments, research programs, and projects. In the initial phase of the study, we analyzed fuzzy TOPSIS, a popular decision-making method, via reviewing fuzzy mathematics concepts. Next, for detecting the most efficient and error-free processes and tests, we explicate in detail Taguchi Experiment Design method. Our study shows that with a rise in the number of factors, the efficiency of the experimental design falls dramatically. This is, in part the result of an increase in interaction effects caused by the rise in the number of factors. Thus, the researcher should pay no attention to the interactions that pass a particular stage. In other words, as a result of time limit, the folding cost, and increased factors, the researcher will be unable to do a complete test of all interaction effects and necessary combinations when they reach a certain level (Uslu, 2007: 545-546).

On the other hand, within the framework of system integration and design optimization, we can see that technology-based road maps and investment planning methodologies, design optimization techniques, and design-targeted tools are used in combination with different disciplines (Olds, 2003). What these approaches imply is the assessment of the most ideal and practical methods possible to be used in various conditions. In other words, the final objective is to assess a variety of techniques with an eye to contextual conditions rather than relying on one particular method and to establish the best optimization method in the wake of the mental processes. The outcome of design optimization is a superior quality that will place manufacturers and process designers at an advantage enabling them to retain their fierce competition without much waste of time and money (Uslu, 2007: 549).

Simulation-based design optimizations, where different disciplines and approaches are used, are used to increase the efficiency of the calculations and to extend the application areas, such as the Taguchi orthogonal indexes, as well as various experimental design techniques, as well as the need for design optimization to bring together artificial neural networks and simulation applications. Complex engineering systems and applications make it necessary to formulate the utility function, environmental criteria and to test the game-theoretic approach simultaneously.

REFERENCES

Cafer, E. B., Cengiz, K., & Da, R. (2003). Fuzzy group decision making for selection among computer integrated manufacturing systems. *Computers in Industry*, *51*(1), 13–29. doi:10.1016/S0166-3615(03)00029-0

Chen, C. T. (2000). Extensions of the TOPSIS for group decision-making under fuzzy environment. *Fuzzy Sets and Systems*, *111*(1), 1–9. doi:10.1016/S0165-0114(97)00377-1

Chen, S. J., Hwang, C. L., Beckmann, M. J., & Krelle, W. (1992). *Fuzzy multiple attribute decision making: methods and applications*. Springer-Verlag. doi:10.1007/978-3-642-46768-4

Chen, T., & Richard, R. (2014). Forecasting the productivity of a virtual enterprise by agent-based fuzzy collaborative intelligence—With Facebook as an example. *Applied Soft Computing*, *24*, 511–521. doi:10.1016/j.asoc.2014.08.003

Gwo-Hshiung, T., & Jih-Jeng, H. (2011). *Multiple attribute decision making: methods and applications*. Academic Press.

Hajkowicz, S. A., Young, M., Wheeler, S., MacDonald, D. H., & Young, D. (2000). *Supportin decisions: understanding natural resource management assessment techniques. A Report to the Land and Water Resources Research and Development Corporation*. CSIRO.

Hines, W. W., & Montgomery, D. (1972). *Probability and Statistics in Engineering and Management Science* (3rd ed.). John Wiley and Sons.

Hong, C.-W. (2012). Using the Taguchi method for effective market segmentation. *Expert Systems with Applications*, *39*(5), 5451–5459. doi:10.1016/j.eswa.2011.11.040

Janssen, R. (1992). *Multiobjective Decision Support for Environmental Management Dordrecht*. Kluwer Academic Publishers. doi:10.1007/978-94-011-2807-0

Jeyapaul, R., Shahabudeen, P., & Krishnaiah, K. (2005). Quality management research by considering multi-response problems in the Taguchi method – a review. *International Journal of Advanced Manufacturing Technology*, *26*(11-12), 1331–1337. doi:10.100700170-004-2102-y

Kuo, Y., Yang, T., & Huang, G. W. (2008). The use of grey relational analysis in solving multiple attribute decision-making problems. *Computers & Industrial Engineering*, *55*(1), 80–93. doi:10.1016/j.cie.2007.12.002

Olds, J., Wei, Sh., Corin, S., Barry, D., Anthony, I., Bharat, S., Suresh, Ch., & Ziaul, H. (2003). *First Year Report*. The Institute For Future Space Transport A University Research Engineering And Technology Institute.

Rahul, D., & Pretesh, J. (2018). Application of Taguchi-Based Design of Experiments for Industrial Chemical Processes. *Intech Open*, 137-155.

Roszkowska, E. (2011). Multi-Criteria Decision- Making Models by Applying the TOPSIS Method to Crisp and Interval Data. Multiple Criteria Decision Making.

Roy, R. K. (1990). *A Primer on The Taguchi Method*. Van Nostrand Reinhold.

Sergey, S., & Kweku, M. (2013). Using Data Envelopment Analysis (DEA) for monitoring efficiency-based performance of productivity- driven organizations. Design and implementation of a decision support system. *Omega*, *41*(1), 131–142. doi:10.1016/j.omega.2011.02.010

Simpson, T. W., Peplinski, J., Koch, P. N., & Allen, J. K. (1997). On The Use Of Statistics In Design And The Implications For Deterministic Computer Experiments. *Proceedings of DETC'97, 1997 ASME Design EngineeringTechnical Conferences*.

Sorin, N., Simona, D., & Ioan, D. (2016). Fuzzy TOPSIS: A General View. *Information Technology and Quantitative Management*, 823-831.

Uslu, T. (2007). Faktöriyel Tasarım ve Süreç Optimizasyonu. *Dokuz Eylül Üniversitesi Sosyal Bilimler Enstitüsü Hakemli Dergisi*, *9*(1), 535–551.

Wang, Y.-M., & Elhag, T. (2006). Fuzzy TOPSIS method based on alpha level sets with an application to bridge risk assessment. *Expert Systems with Applications*, *31*(2), 309–319. doi:10.1016/j.eswa.2005.09.040

Wu, H. H. (2002). A comparative study of using grey relational analysis in multiple attribute decision making problems. *Quality Engineering*, *15*(2), 209–217. doi:10.1081/QEN-120015853

Yao, A., & Sio. (2013). Analysis and Design of a Taguchi–Grey Based Electricity Demand Predictor for Energy Management Systems. *Energy Conversion and Management*, *45*(7–8), 1205–1217.

Chapter 14
The Impact of Environmental Concern on Consumers' Attitude and Intention Toward Electric Vehicles:
The Role of Demographics

Niray Tunçel
Hacettepe University, Turkey

Esna Betül Buğday
Hacettepe University, Turkey

ABSTRACT

This research aims to analyze the impact of environmental concerns on Turkish consumers' attitudes toward and intention to purchase EVs. It is also aimed to investigate the role of demographical characteristics—gender, age, income, and education—on those impacts. Given the purpose, a descriptive study was conducted through an online survey with a sample of 334 consumers. The research findings indicate that environmental concern plays a significantly influential role in the attitude toward and intention to purchase EVs. The study also confirms that environmental concern's impact on the attitude and intention toward EVs differs in terms of gender, age, and income. Only women show a positive influence of environmental concern on the attitude toward EVs. There is no difference in the direct or indirect link between EV purchase intention and environmental concern for age groups. Environmental concern's indirect influence on the intention to buy EVs is the highest and significantly different for the consumers with 7500-9999 TL income.

DOI: 10.4018/978-1-7998-8900-7.ch014

INTRODUCTION

One of the major challenges facing our environment is the uncertainty of future access to fossil fuels and the significant volume of carbon dioxide (CO2) emissions. According to The World Resource Institute, road transportation accounts for 11.9% of global greenhouse emissions (Ritchie & Roser, 2021). Hence, the adoption of renewable energy vehicles instead of conventional ones may be considered a promising solution. One type of them, electric vehicles (EVs), produces zero direct emissions and helps protect natural sources and the environment (Shen et al., 2019, Tu & Yang, 2019).

Owing to the increasing environmental concerns of consumers, EVs are becoming an attractive alternative for transportation (Junquera et al., 2016). Various studies on EV adoption have assumed that EVs are eco-innovations with the ability to reduce the environmental problems of the transportation sector (Egbue & Long, 2012, Schuitema et al., 2013). According to the International Energy Agency (IEA), in 2020, over 3 million EVs were sold, with an increase of 43% compared to 2019 (IEA, 2020). EV sales rates in Turkey have also increased compared to 2019. The sales amount, which was 222 in 2019, increased to 844 in 2020 (TEHAD, 2021).

There has been an increasing amount of literature on the consumers' attitude and intention toward EVs (Asadi et al., 2021, Chu et al., 2018, Lai et al., 2015). However, the literature still lacks knowledge for the emerging markets. Since EVs are still a relatively new technology in those markets, few studies exist focusing on consumers' intention to adopt EVs. As an emerging market, Turkey, consumers' EV adoption is a brand-new subject. Besides, the relationship between environmental concern and EV acceptance has not been thoroughly investigated. Therefore, this paper aims to analyze the EC's impact on Turkish consumers' attitude and intention for EVs. We also aim to identify the role of demographics (i.e., gender, age, income, and education) on those effects.

This study has the following structure. The first section presents the theoretical background and hypotheses development. In the second part, the paper gives the methodology with the analysis and findings of the study that are followed by the conclusion and discussion section. In the last section of the study, the limitations and implications for future research are presented.

THEORETICAL BACKGROUND AND HYPOTHESIS DEVELOPMENT

Environmental concern (EC) refers to one's perception and knowledge about environmental issues, to what extent a person concerns for and is worried about threats to the environment, and supports efforts to solve them (Lin & Niu, 2018, Moyo, 2018). Environmentally concerned consumers are highly motivated and interested in purchasing environmentally friendly products (Moyo, 2018, Verma et al, 2019).

Previous studies have posited that EC has a direct positive effect on the attitude toward and intention of green consumption (Mainieri et al., 1997, Bamberg, 2003, Mcdonald et al., 2016, Mohd, 2016). McDonald et al. (2016) stated that people who are more concerned about the environment are more likely to respond to environmental issues and take actions in environmental protection. Also, some of the studies have shown that EC is a significant personal antecedent of consumers' preference for green consumption. For instance, it is confirmed that EC positively influences the attitude toward (Kirmani, 2016) and willingness to purchase (Mohd, 2006) green products.

The previous studies also demonstrated that there is a positive link between EC and EV purchase intention (Thananusak et al., 2017, Wu et al, 2019). Besides, there is some evidence for the impact of

EC on the attitude toward EVs. Lai et al. (2015) found that EC leads to a positive perception of EVs. Wu et al. (2019) showed that the more individuals are concerned about the environment, the more they perceive EVs as easy to use and environmentally valuable. Hence, we investigated the impact of EC on consumers' purchase intention and attitude toward EVs. Thus, the first two hypotheses of the study are built as below:

H1: EC has a positive impact on the intention to purchase EVs.
H2: EC has a positive impact on the attitude toward EVs.

Furthermore, the attitude is a powerful indicator of the intention (Fishbein & Azjen, 1975). The previous EV studies (Asadi et al., 2021, Degirmenci & Breitner, 2017, Yang et al., 2020) also confirmed a positive relationship between the attitude toward and intention to buy EVs. As aforementioned, EC positively influences consumers' attitude toward EVs (Lai et al., 2015, Wu et al., 2019) so the attitude can also act as a mediator on the link between EC and EV purchase intention. Thus, we proposed H3 and H4 as follow:

H3: Attitude toward EVs has a positive impact on the intention to purchase EVs.
H4: Attitude toward EVs mediates the link between EC and purchase intention of EVs.

Consumers with different demographics can vary in terms of their EC levels and so their inclinations for green consumption. Therefore, it is expected that demographical variables -gender, age, income and education- can alter the effect of EC on the attitude and intention toward EVs. As women and men go through the different socialization processes from their early childhood, they develop a distinct social expectations and values (Gilligan, 1982) which in turn differentiates their perception of environmental issues (Sreen et al., 2018). For instance, previous research has demonstrated that women are more environmentally concerned than men (e.g., (Goodale, 2021, Mueller & Mullenbach, 2018, Xiao & McCright, 2015). In addition to the empirical studies, the literature reviews (Liu et al., 2019) have also suggested that women are more environmentally concerned compared to men. Besides, the influence of EC on green consumption varies depending on gender. For instance, it is confirmed that EC's positive impact on green purchase behavior is stronger for women (Dagher et al., 2015). Another study (Nguyen et al., 2017) proved that women showed a stronger positive influence of environmental attitude on green purchase intention. Dimitropoulos (2014) found that potential adopters of full electric cars tend to be females, highly environmentally concerned and consider the environmental impact of car use. It is also confirmed that in general women put more emphasis on environmentally friendly mobility (Kawgan-Kagan, 2020). Thus, compared to men, women are more likely to be eco-friendly in their beliefs, attitudes, and choices (Lee & Holden, 1999) and help the environment (Zelezny et al., 2000). Whereas, men do not see a problem living a non-green life style (Tiller, 2014). Regarding this gender gap in sustainable consumption, with a series of experimental studies Brough et al. (2016) demonstrated that greenness and femininity are cognitively linked. Based on this reasoning, following three hypotheses of the study are built as below.

H5a: The strength/magnitude of EC's impact on the purchase intention of EVs is greater for women than men.
H5b: The strength/magnitude of EC's impact on the attitude toward EVs is greater for women than men.

H5c: The strength/magnitude of EC's indirect impact on the purchase intention of EVs through the attitude toward EVs is greater for women than men.

Prior research has also suggested that EC levels of consumers vary depending on age, so their attitude and intention toward purchasing green products differ accordingly. However, the literature presents conflicting findings about this issue. For instance, it is confirmed that as consumers age, they become less environmentally concerned (Gray et al., 2019, Marquart- Pyatt, 2012, Xiao & Dunlap, 2007). In contrast, it is also demonstrated that age positively influences EC (e.g., (Bouscasse et al., 2018, Hirsch, 2010, Urban & Ščasný, 2012)). Consumers' green product attitude and intention vary for different age groups as well. Lin and Syrgabayeva (2016) found that 18-24 age group has the highest willingness to pay more for renewable energy. Whereas, Newman and Fernandes (2016) confirmed that Gen X and Y consumers are less likely to engage in pro-environmental behavior than baby boomers. Consumers also differ in terms of EV adoption depending on their ages. For instance, early adopters of EVs are likely to be between 18 and 34 years of age (Tran et al., 2019). Other studies (Peters & Dütschke, 2014, Plötz et al., 2014) demonstrated that individuals who are highly interested in EVs tend to be middle-aged. Unlikely, Ghasri et al. (2019) found that Gen Y is more likely to adopt EVs compared to Gen X and Z. Sovacool et al. (2018) revealed that consumers below middle age are most inclined to purchase EVs.

In addition, age is a predictor of consumers' attitude toward EVs. Kim et al (2014) confirmed that the 36-50 age group has a more positive environmental attitude toward EVs than other age groups. However, another study (Chen et al., 2020) demonstrated that younger people have a more positive attitude for EV adoption. These studies show that although age alters consumers' EC, attitude and intention toward EVs, the literature presents mixed evidence (Coffman et al., 2017, Liao et al., 2017). Therefore, in this study, it is expected that age will moderate the EC's impact on the attitude toward EVs and EC's direct and indirect effects on the intention to buy EVs; however, due to the conflicting findings in the literature, the direction of the moderation effect could not be mentioned. Based on this, the hypotheses are built as follow:

H6a: Age alters the strength/magnitude of EC's impact on the purchase intention of EVs.
H6b: Age alters the strength/magnitude of EC's impact on the attitude toward EVs.
H6c: Age alters the strength/magnitude of EC's indirect impact on the purchase intention of EVs through the attitude toward EVs.

Consumers' EC level, attitude, and intention for green consumption may also vary depending on their income levels. For instance, Li and Chen (2018) demonstrated that a rise in both absolute and relative income increases consumers' EC levels. A similar research collected data from 40 countries also suggested a positive relationship between income and EC (Dorsch, 2014). However, another study investigated 50 nations confirmed a negative association between EC and individuals' income levels (Gelissen, 2007). In contrast to all these studies, Liu et al. (2014) did not find evidence for the effect of income on EC. The previous studies also found that consumers' intention to buy and attitude toward green consumption differ among income groups. For instance, Muhammad et al. (2015) confirmed that income positively relates to the willingness to pay for the organic food. Besides, income has a positive impact on energy efficiency (Li et al., 2019) and habitual green consumption (Zhang et al., 2019). Whereas some other studies found a negative relationship between income and green consumption. For instance, Zhao et al. (2018) demonstrated that income negatively affects the willingness to pay for carbon-labeled products and this negative impact is strongest for the lowest income level. Besides, Roberts (1996) proved a negative correlation between income and ecologically conscious consumption. Unlikely, some studies could not establish an association between income and green consumption (Elliot, 2003, Sun et al., 2019).

When it comes to the attitude toward green consumption, the previous research also presented conflicting findings; for instance, Zhao et al. (2014) found that income positively influences the attitude toward green consumption, however, Singhal and Malik (2018) confirmed that income does not make any difference in women's attitude toward green cosmetic products. Even though the effect of income on EC, attitude and intention toward green products is mixed, previous EV studies have shown that people with high incomes are likely to adopt EVs (Higueras-Castillo et al., 2020, Lee et al., 2019, Lin & Tan, 2017, Trommet et al., 2015, Vassileva & Campillo, 2017). EV adoption and EC are also positively associated as people with higher EC are more likely to purchase EVs (Wu et al., 2019, Krupa et al., 2014, Mohamed et al., 2016). Besides, it is also demonstrated that environmentally concerned consumers have a positive attitude toward EVs (Lai et al., 2015, Wu et al., 2019, Liu et al., 2015, Mohamed et al., 2016, He et al., 2018). Accordingly, it is expected that for the consumers with higher income levels, the impact of EC on the attitude toward EVs and its direct and indirect influence on EV purchase intention will be stronger. Thus, three hypotheses of the study are established as follow:

H7a: A rise in the income level increases the strength/magnitude of EC's impact on the purchase intention of EVs.
H7b: A rise in the income level increases the strength/magnitude of EC's impact on the attitude toward EVs.
H7c: A rise in the income level increases the strength/magnitude of EC's indirect impact on the purchase intention of EVs through the attitude toward EVs.

Previous research has confirmed that education is one of the factors influencing consumers' EC levels, attitude and intention for green consumption (Asunta, 2003, Harring & Jagers, 2017). Education can improve consumers' ability to establish a broad perspective on the importance of the natural environment and its interaction with humans. Ergen and Ergen (2011) stated that education is the most important factor to create consciousness about the environment. The strong association between education and EC is supported by different studies as well (Zelezny et al., 2000, Vicente-Molina et al., 2013). Some of the studies emphasized that level of education is considered to be linked to EC, environmental attitude and behavior (e.g., (Newell & Green, 1997, Roberts & Bacon, 1997)). The previous research agreed that education is positively correlated with these variables (Straughnan & Roberts, 1999, Daziano & Bolduc, 2013). For example, in their studies Liberty and HongJuan (2010) showed that individuals with higher levels of education tend to better understand environmental issues and thus become more concerned for environmental quality and more motivated to practice environmentally responsible behavior. Furthermore, Cerri et al. (2018) suggested that strong obstacles to engaging in pro-environmental behavior can be caused by the consumers' education level.

While most of the studies come up with a positive correlation between education and EC, environmental attitude and behavior, a few exceptions do exist. For instance, it is found that education is negatively correlated with environmental attitude (Samdahl & Robertson, 1989) and with consumers' willingness to seek environmentally friendly products (Jain & Kaur, 2006). Other studies (Chan, 1996, Diamantopoulos et al., 2003) could not find evidence for the impact of education on environmental attitude and behavior.

As regards to use environmentally friendly car, Dimitropoulos (2014) found that highly educated individuals are more concerned about the environmental impacts of car use than individuals who have achieved low education levels. Several studies found that a person with high level of education is likely to purchase EVs (Carley et al., 2013, Hackbarth & Madlener, 2013, Hidrue et al., 2011). In line with this, Kim et al. (2014) demonstrated that education level is significantly and positively influencing consumers' EV purchase intention. Morton et al. (2017) also found that early adopters of EVs are likely to hold a university-level education. On the contrary, Sierzchula et al. (2014) found that education is not significant in determining EV adoption. As can be seen, conflicting results are presented in the literature regarding education, EC and attitude and intention toward EVs. However, since several studies emphasized that there is strong positive link between education and green consumption (Vicente-Molina et al., 2013, Daziano & Bolduc, 2013, Cerri et al., 2018) and EV adoption is correlated with a high education level (Carley et al., 2013, Hackbarth & Madlener, 2013), it is expected that as the education level increases, the impact of EC on the attitude toward EVs and its direct and indirect influence on EV purchase intention will also increase. Accordingly, following hypotheses are built.

H8a: A rise in the education level increases the strength/magnitude of EC's impact on the purchase intention of EVs.
H8b: A rise in the education level increases the strength/magnitude of EC's impact on the attitude toward EVs.
H8c: A rise in the education level increases the strength/magnitude of EC's indirect impact on the purchase intention of EVs through the attitude toward EVs.

Based on the above hypotheses, we built the research model as in Figure 1.

Figure 1. Research model

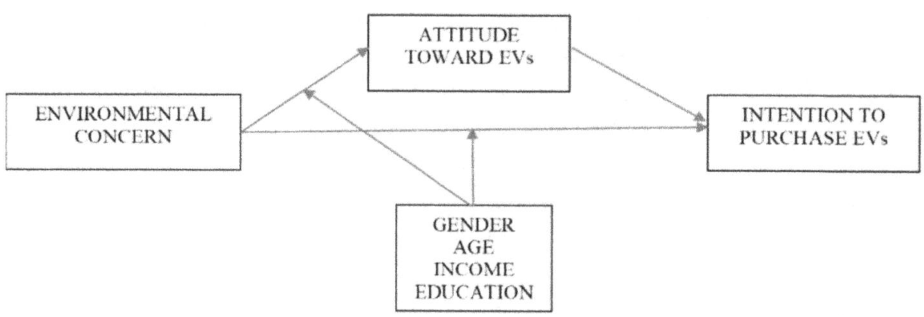

METHODOLOGY

Data Collection and Sampling

The data were collected from 334 consumers through an online survey and the questionnaire was designed on SurveyMonkey. Participants either received a survey link as a direct message or accessed to the link via different social media platforms (e.g., Linkedin, and Instagram). The online survey enabled

Table 1. Demographic characteristics of the participants

Gender	N	%	Education	N	%
Female	187	56.0	Secondary school and High school	56	16.8
Male	147	44.0	University	186	55.7
Total	334	100	Masters and PhD	92	27.5
			Total	334	100
Income (personal)			**Age**		
0-2499 TL	104	31.1	18-24	96	28.7
2500-4999 TL	71	21.3	25-31	86	25.7
5000-7499 TL	102	30.5	32-38	59	17.7
7500-9999TL	23	6.9	39-45	24	7.2
10000 TL<	34	10.2	46-52	29	8.7
Total	334	100	53<	40	12.0
			Total	334	100

researchers to reach the respondents from several cities of Turkey. Table 1 gives the demographics of the participants.

56% of the participants were women, and 44.0% were men, and most of the participants (83.2%) had university, master's or PhD degrees. Besides, 0-2499 TL consisted the 31.1%, 5000-7499 TL did 30.5% of the participants' income levels. Last, approximately half of the participants (54.4%) were between the ages of 18-31 (Table 1).

Measures

We adopted La Trobe and Acott's (2000) Modified NEP/DSP Environmental Attitudes Scale for EC, Degirmenci and Breitner's (2017) scale for the attitude toward and intention to purchase EVs. EC and intention to buy EVs were measured through a five-point Likert type scale (1: strongly disagree, 5: strongly agree). Besides, a five-point semantic differential scale (e.g., 1: bad, 5: good) measured the consumers' attitude toward EVs.

Analysis and Findings

First, we tested the measurement model through Confirmatory Factor Analysis (CFA) using AMOS 24. For the reliability of the scales, the values of Cronbach's Alpha (α) and Composite Reliability (CR) were calculated. Then, the Average Variance Extracted (AVE) values were checked for the convergent validity. Fornell-Larcker Criterion and Heterotrait-Monotrait Ratio of correlations (HTMT) method were applied for the discriminant validity. To calculate those values, we used the AMOS Plugin (Gaskin & Lim, 2016). Table 2 presents the results with the model fit indices.

The values of α and CR, which are greater than 0.7, demonstrate an appropriate level of reliability for the measures. Besides, the AVE values above 0.5 prove the convergent validity of the scales (Hair et al., 2014). Fornell Larcker Criterion confirms that the AVE value's square root for each measure

Table 2. CFA model-reliability and validity

						Fornell- Larcker			HTMT		
	α	CR	AVE	MSV*	MaxR(H)	EC	Attitude	Intention	EC	Attitude	Intention
EC	.844	.802	.519	.070	.863	***.721***					
Attitude	.913	.897	.690	.376	.914	.264	***.831***		.236		
Intention	.834	.866	.688	.376	.952	.239	.613	***.829***	.175	.666	
*Maximum Shared Variance χ2/df: 1.775, GFI: .928, AGFI: .905, CFI: .970, NFI: .935, RFI: .923, IFI: .970, TLI: .965, RMSEA: .048, SRMR: .049											

is higher than LVC values (Fornell & Larcker, 1981). In addition, HTMT signifies that all the values lower than 0.90 (Hair Jr. et al., 2017). Also, each scale's AVE value is greater than its MSV value. All of these indicate that the scales have decent discriminant validity. The model fit indices are within the ranges recommended in the literature (Hu & Bentler, 1999, Tabachnick & Fidell, 2007) representing a good model fit.

To test the hypotheses, we used Process Macro (Hayes, 2012). First, Model 4 was used to test H1, H2, H3, and H4. Then, Model 8 was selected for testing the hypotheses with moderating variables: for gender: H5a, H5b, H5c, for age: H6a, H6b, H6c, for income: H7a, H7b, H7c, and for education: H8a, H8b, H8c.

As depicted in Table 3, EC positively influences the intention to purchase ($\beta=.219$, $p<.05$) and attitude toward EVs ($\beta=.250$, $p<.05$), and the attitude positively affects the intention ($\beta=.749$, $p<.001$). Attitude also fully mediates the link between EC and intention; as attitude enters into the model, EC's direct impact on intention to buy EVs becomes insignificant. The confidence limits (BootLLCI: 021, BootULCI: .393) not including zero demonstrate the indirect effect's significance. Thus, H1, H2, H3, and H4 are confirmed.

The results for the moderated-mediation model are presented below for gender (Table 4), age (Table 5), income (Table 6), and education (Table 7).

Table 3. Regression model analysis (without moderators)

	Direct Effects			
	coeff	**se**	**t**	**p**
EC-->intention	.219	.094	2.324	.021
EC-->attitude	.250	.072	3.493	.001
	Mediated Model*			
	coeff	se	t	p
EC	.031	.079	.396	.692
attitude	.749	.059	12.636	.000
	Indirect Effect of EC on Intention			
	Effect	BootSE	BootLLCI	BootULCI
EC-->attitude-->intention	.187	.099	.021	.393
* Y: intention, X: EC, M: attitude, R^2=.336				

Table 4. Moderated-mediation model testing results for gender

Moderator: Gender	**coeff**	**se**	**t**	**p**		**effect**	**se**	**t**	**p**	**LLCI**	**ULCI**
DV: attitude											
EC*gender	-.500	.413	-3.491	.001**	female	.529	.105	5.031	.000*	.322	.736
					male	.030	.097	.305	.761	-.161	.220
DV: intention											
EC*gender	-.012	.159	-.078	.938	female	.067	.119	.654	.573	-.167	.302
					male	.055	.106	.518	.605	-.154	.263
EC--> attitude--> intention	**Index[1]**	**BootSE**	**Boot-LLCI**	**Boot-ULCI**		**effect**	**BootSE**	**Boot-LLCI**	**Boot-ULCI**		
gender	-.371	.175	-.681	-.017	female	.393	.152	.105	.671		
					male	.022	.083	-.125	.212		

[1] Index of moderated mediation
* p<.001 ** p<.01

As Table 4 exhibits, the interaction effect (EC*gender) on the intention is insignificant (β=-.012, p>.05). Therefore, contrary to the expectation, EC's impact on the intention to purchase EVs does not differ between genders, so H5a could not be supported. In contrast, EC significantly and positively influences the women's attitude toward EVs (β=.529, p<.001); however, this impact becomes insignificant for men's (β=.030, p>.05). Therefore, as it is expected, the strength/magnitude of EC's influence on the attitude toward EVs is higher for women compared to men, which confirms H5b. Besides, the index of moderated-mediation represents that the difference in conditional indirect effect is significant (BootLLCI: -.681, BootULCI: -.017); that is, the indirect effect of EC on the intention to buy EVs through the attitude varies depending on gender. Specifically, EC has a significant indirect impact on the women's intention to purchase EVs (BootLLCI: .105, BootULCI: .671), but an insignificant indirect impact for men's (BootLLCI: -.125, BootULCI: .212). This result indicates that the strength/magnitude of EC's effect on the intention to purchase EVs through the attitude is greater for women than men. Hence, H5c is also demonstrated.

When it comes to age (Table 5), for the direct and indirect effect of EC on the intention to buy EVs, none of the interaction effects (EC*age) is significant. Therefore, it seems that age does not make any difference in the EC's direct impact on the EV purchase intention and its indirect impact through the attitude toward EVs. Besides, the index of moderated mediation for each age group also confirms the indirect effects' resemblance. Thus, H6a and H6c could not be confirmed. In contrast, it is revealed that age makes a significant difference in the effect of EC on the attitude toward EVs (R^2 change=034, p<.05). Specifically, the first interaction effect (EC*w1) alters the relationship between the EC and the attitude (β=.563, p<.01). Explicitly, EC's impact on the attitude toward EVs is insignificant for the age group of 18-24, but it becomes significant for the one of 25-31. The other interaction effects are not statistically significant, so the EC's impact on the attitude toward EVs for all other age groups is similar to the one for the 18-24 age group. These results specify that the strength/magnitude of EC's influence on the attitude toward EVs changes (increases), only when the age range rises from 18-24 (effect=.109, p>.05) to 25-31 (effect=.673, p<.001). Thus, since age makes a significant positive difference in the effect of EC on the attitude toward EVs, H6b is supported.

Table 5. Moderated-mediation model testing results for age

Moderator: Age	coeff	se	t	p		effect	se	t	p	LLCI	ULCI
DV: attitude											
EC*w1	.563	.193	2.926	.004**	18-24	.109	.127	.862	.390	-.140	.359
EC*w2	.365	.238	1.536	.126	25-31	.673	.145	4.639	.000*	.387	.958
EC*w3	.136	.349	.390	.697	32-38	.474	.201	2.359	.019***	.079	.869
EC*w4	-.066	.206	.322	.747	39-45	.245	.325	.754	.451	-.395	.885
EC*w5	.023	.264	.087	.931	46-52	.043	.162	.265	.791	-.276	.362
EC*w: R^2 change=034, F=2.528, p=.029***					53<	.132	.232	.569	.569	-.324	.589
DV: intention											
EC*w1	.285	.219	1.303	.193	18-24	-.085	.142	-.601	.548	-.365	.194
EC*w2	.037	.267	.139	.890	25-31	.199	.168	1.189	.235	-.131	.529
EC*w3	.521	.391	1.332	.184	32-38	.048	.227	-.213	.832	-.495	.398
EC*w4	.199	.230	.864	.388	39-45	.436	.365	1.195	.233	-.282	1.153
EC*w5	.013	.296	.043	.966	46-52	.114	.182	.627	.531	-.243	.471
EC*w: R^2 change=007, F=.677, p=.641					53<	-.073	.260	-.279	.780	-.584	.439
EC--> attitude--> intention	**Index[1]**	**BootSE**	**Boot-LLCI**	**Boot-ULCI**		**effect**	**BootSE**	**Boot-LLCI**	**Boot-ULCI**		
w1	.414	.271	-.191	.793	18-24	.080	.062	-.022	.222		
w2	.268	.226	-.119	.762	25-31	.494	.265	-.089	.865		
w3	.100	.193	-.243	.478	32-38	.348	.218	-.007	.824		
w4	-.049	.120	-.221	.269	39-45	.180	.182	-.116	.550		
w5	.017	.353	-.599	.820	46-52	.032	.103	-.057	.341		
					53<	.097	.346	-.486	.897		

[1] Index of moderated mediation
* p<.001 ** p<.01 *** p<.05

Table 6 shows that the EC*income interaction effect makes no significant difference in the EC's direct impact on the EV purchase intention, so H7a could not be demonstrated. In contrast, the EC*income interaction effect alters the relationship between EC and attitude toward EVs (R2 change=081, p<.001). Specifically, EC*w2 (β=.349, p<.10) and EC*w3 (β=1.053, p<.001) interaction effects make a significant difference in the EC's influence on the attitude. That is, for the income group for 0-2499 TL, the impact of EC on the attitude toward EVs is not statistically significant (effect=.074, p>.05); however, EC positively influences the attitude toward EVs for the income groups of 5000-7499 TL (effect=.422, p<.01) and 7500-9999 TL (effect=1.109, p<.001). This positive impact is stronger for the latter income group. For the other income groups (2500-4999 TL and 10000 TL and more), EC's impact on the attitude toward EVs is similar to the one for the 0-2499 TL income group. Hence, it can be inferred that EC influences the attitude of the participants with higher income levels (i.e., 5000-7499 TL and 7500-9999 TL) more than it does the attitude of the participants with lower income levels (i.e., 0-2499 TL). However, since this case is only valid for the three income levels, H7b is partially supported. Last,

Table 6. Moderated-mediation model testing results for income

Moderator: Income	coeff	se	t	p		effect	se	t	p	LLCI	ULCI
DV: attitude											
EC*w1	-.053	.201	-.264	.792	0-2499 TL	.074	.134	.550	.582	-.190	.337
EC*w2	.349	.200	1.743	.082****	2500-4999 TL	.021	.150	.138	.891	-.274	.315
EC*w3	1.053	.230	4.507	.000*	5000-7499 TL	.422	.149	2.839	.005**	.130	.715
EC*w4	-.153	.215	-.711	.478	7500-9999 TL	1.109	.187	5.939	.000*	.742	1.476
EC*w: R^2 change=081, F=7.600, p=.000***					10000 TL<	-.079	.169	-.471	.638	-.411	.252
DV: intention											
EC*w1	-.221	.226	-.977	.329	0-2499 TL	.003	.151	.020	.984	-.293	.299
EC*w2	.262	.226	1.160	.247	2500-4999 TL	-.218	.168	-1.292	.197	-.549	.114
EC*w3	.145	.266	.545	.586	5000-7499 TL	.265	.169	1.566	.118	-.0.68	.599
EC*w4	-.030	.242	-.125	900	7500-9999 TL	.148	.221	.670	.503	-.287	.584
EC*w: R^2 change=009, F=1.107, p=.353					10000 TL<	-.027	.190	-.144	.886	-.401	.346
EC--> attitude--> intention	Index[1]	BootSE	Boot-LLCI	Boot-ULCI		effect	BootSE	Boot-LLCI	Boot-ULCI		
w1	-.039	.172	-.350	.332	0-2499 TL	.054	.087	-.114	.235		
w2	.254	.167	-.045	.597	2500-4999 TL	.015	.146	-.223	.355		
w3	.753	.273	.060	1.037	5000-7499 TL	.307	.145	.068	.643		
w4	-.111	.160	-.483	.199	7500-9999 TL	.807	.267	.151	1.076		
					10000TL <	-.058	.134	-.387	.224		

[1] Index of moderated mediation
* p<.001 ** p<.01 *** p<.05 **** p<.10

regarding EC's indirect impact on the intention, the conditional indirect effect is significantly different for the 7500-9999 TL income group (Index: .753, BootLLCI: .060, BootULCI: 1.037). It means that the strength/magnitude of EC's indirect influence on the purchase intention through the attitude toward EVs is more for the income level 7500-9999 TL (effect=.807, BootLLCI: .068, BootULCI: 643) than 0-2499 TL (effect=.054, BootLLCI: -.114, BootULCI: 235). Nevertheless, other income levels do not show any difference in this impact compared to the 0-2499 TL income level. Since there is an increase in the strength/magnitude of EC's indirect impact on the EV purchase intention only when the income level rises from 0-2499 TL to 7500-9999 TL, we found a partial support for the H7c.

Finally, none of the interaction effects is revealed to be significant for education (Table 7). Therefore, EC's impact on the attitude and its direct and indirect influence on the intention to buy EVs does not differ in terms of education. Hence, H8a, H8b, and H8c could not be supported.

Table 7. Moderated-mediation model testing results for education

Moderator: Education	**coeff**	**se**	**t**	**p**							
DV: attitude											
EC*w1	.175	.184	.947	.345							
EC*w2	.257	.236	1.091	.276							
EC*w: R^2 change=004, F=.668, p=.513											
DV: intention						**effect**	**se**	**t**	**p**	**LLCI**	**ULCI**
EC*w1	.031	.199	.156	.876	Secondary school and High school	.070	.172	.410	.682	-.267	.408
EC*w2	-.204	.254	-.801	.424	University	.101	.102	.996	.320	-.099	.302
EC*w: R^2 change=002, F=.616, p=.541					Masters and PhD	-.133	.189	-.707	.480	-.504	.238
EC--> attitude--> intention	**Index[1]**	**BootSE**	**Boot-LLCI**	**Boot-ULCI**		**effect**	**BootSE**	**Boot-LLCI**	**Boot-ULCI**		
w1	.130	.193	-.264	.495	Secondary school and Highschool	.092	.122	-.092	.384		
w2	.191	.206	-.221	.588	University	.222	.148	-.015	.531		
					Masters and PhD	.283	.165	-.004	.631		

[1] Index of moderated mediation

CONCLUSION AND DISCUSSION

The findings of this research show that EC positively affects the attitude toward and intention to purchase EVs. The results also confirm that the attitude toward EVs has a positive impact on the EV purchase intention and mediates the link between EC and this intention. When it comes to the moderation effects of the demographics, it is revealed that the gender, age, and income alter the EC's impact on the attitude toward EVs. Regarding gender, EC significantly and positively influences only the women's attitude toward EVs. In different age categories, EC's positive impact on the attitude toward EVs is the highest for the age group of 25-31. The rest of the age groups is similar in terms of this effect. Besides, it is revealed that compared to the consumers with 0-2499 TL income, ones with incomes of 5000-7499 TL and 7500-9999 TL show a stronger positive influence of EC on the attitude toward EVs, 7500-9999 TL group does the strongest one. Hence, it can be stated that an increase in income level leads to a more positive EC-attitude link. However, the strength/magnitude of this link does not differ considering education as EC's influence on the attitude toward EVs is insignificant for each education level.

Considering the EC's direct influence on the intention to purchase EVs, it is seen that none of the demographics alter this effect. For each gender, age, income and education group, this impact is similar. Therefore, regardless of their demographical characteristics, consumers are alike in terms of EC-purchase intention link. Last, it is demonstrated that EC's indirect effect on the intention to purchase EVs differs in terms of gender and income. Only women show a positive and significant impact of EC on the EV purchase intention through their attitude toward EVs. Besides, EC's positive indirect influence on the intention to buy EVs is higher for the consumers with 7500-9999 TL income and this influence is significantly different from the ones for other income groups.

The findings of this study support and contradict the ones in the current literature. For instance, it is unveiled that EC is an essential positive antecedent of the women's attitude toward and intention to purchase EVs, which supports the findings of several studies that have confirmed that women are more environmentally concerned (Goodale, 2021, Xiao & McCright, 2015) and show a stronger positive influence of environmental attitude on the green purchase intention (Nguyen et al., 2017) compared to men. In addition, Dimitropoulos also (2014) found that potential adopters of full electric cars are likely to be women and environmentally concerned. It was also found that 25-31 age group has the strongest positive link between EC and the attitude toward EVs. Our finding complies with the approach that younger people are likely to hold more positive attitude toward EV adoption (Sovacool et al., 2018, Chan et al, 2020) and to be environmentally concerned (Gray et al., 2019, Marquart-Pyatt, 2012). However, this finding contradicts the ones of Kim et al. (2014) as they found that 36-50 age group has a more favorable environmental attitude toward EVs than other age groups. We could not find evidence for the moderation effect of age on the EC's direct and indirect impact on EV purchase intention. This finding does not support the results of the studies, which demonstrated that age alters the purchase intention of EVs (Peters & Dütschke, 2014, Ghasri et al., 2019).

When it comes to the income level, it is found that 5000-7499 TL and 7500-9999 TL income group has the higher positive EC impact on the attitude toward EVs than the other income groups. It was also proved that EC's indirect effect of EV purchase intention through the attitude toward EVs is the highest for the 7500-9999 TL income group. Turkish Statistical Institute (TUIK) (2020) indicates that average personal monthly income is 2376 TL in Turkey, so our income groups can be considered high for Turkish population. These findings are congruent with the ones of previous research demonstrating that EV adoption positively relates to the high income (Higueras-Castillo et al., 2020, Lee et al., 2019) and EC levels

(Lai et al., 2015, He et al., 2018). As regards to education, we did not find evidence for the moderation effect of education on the influence of EC neither on the attitude toward nor on the purchase intention of EVs. This result is concordant with the previous studies (Chan, 1996) that could not support the impact on education on environmental attitude and intention. Besides, Sierzchula et al. (2014) confirmed that education is not a determinant of EV adoption. Thus, our finding is contradictory to the general point of view that EV adoption (Dimitropoulos, 2014, Morton et al., 2017) and EC (Harring & Jagers, 2017, Daziano & Bolduc, 2013) are positively correlated with education.

This study's novelty contributes to the existing literature in several respects theoretically. First, although EV has been a popular subject for developed countries, the present knowledge is still limited for the emerging markets. Thus, by focusing on Turkish consumers, this study enriches the current understanding of EV purchase intention's antecedents, particularly in an emerging market context. Second, even if EC has been studied as an indicator of willingness to purchase sustainable products, its role on consumers' preference for EVs needs further interest. Mainly, the current literature only supplies minimal information about the EC's impact on the attitude toward EVs. In this manner, the study at hand empirically contributes to a better understanding of this impact. Third, this study also provides a comprehensive approach providing empirical evidence for the relationships among EC, attitude toward and intention to purchase EVs by exploring them in terms of gender, age, income, and education. As the EV market is in its early phase for emerging economies, it is crucial to unveil the differences and similarities in consumers' insights from different demographics to present segment-specific implications. Last, the findings of the study support the Theory of Reasoned Action as consumers who are environmentally concern (i.e., individual factor) have a positive attitude toward buying EVs (i.e., attitude toward the behavior), which in turn, leads to a purchase intention (i.e., behavioral intention).

This study also has some practical implications for policy makers and the managers. The results show positive effects of EC on the attitude toward and intention to purchase EVs. Besides, the demographics mainly alter the relationship between EC and the attitude toward EVs. Therefore, EV companies and policy makers should take into account the demographical differences in this relationship. Hence, they should initially focus on women, 25-31 age and 5000-9999 TL income groups coming from all educational backgrounds for strengthening their positive attitude toward EVs through suitable environmental messages in Turkish market since they are most likely to be attracted by the promotional campaigns involving environmental messages. Hence, for these consumer segments, EVs should be specifically presented as eco-friendly products by highlighting its pro-environmental attributes such as producing zero direct emissions, reducing air pollution and enabling cleaner towns and cities. In this respect, EV companies, government and civil organizations should consider informing and educating these consumers about the benefits and advantages of EVs on environmental protection. Also, they should advertise more the environmental benefits of EVs through the integrated marketing communication channels that these consumers frequently use. For instance, for the age group 25-31, different social media platforms can be effective for promoting EVs as this age group is the primary audience of social media ads in Turkey. Considering women, YouTube may be a right channel to convey environmental benefits of EVs since 50% of Turkish women actively use this platform (We Are Social, 2021).

Besides, policy makers should cooperate with the environmental organizations and universities to conduct workshops on EVs' environmental advantages. Such events will be useful for consumers to clear up their doubts about EVs. Furthermore, since environmentally concerned Turkish consumers with high income levels (7500-9999) are more likely to purchase EVs, in order to convince them EV companies should arrange private and personalized promotional activities. For instance, inviting them to have a special EV driving experience and simultaneously informing them about how EVs protect the environment could help increase their willingness to purchase.

LIMITATIONS AND FUTURE RESEARCH

This study has some limitations. First, 83.2% of our sample consists of individuals with a university or higher education level. This is also a limitation for this study because only 21.6% of the population of Turkey has a university or higher education level. However, in the studies related to the attitude toward and purchase intention of EVs, it is seen that the majority of EV adopters have a high education level (Carley et al., 2013, Hackbarth & Madlener, 2013, Hidrue et al., 2011, Morton et al., 2017). In this sense, this does not constitute a problem in terms of representation of the sample considering EV adoption.

Second, the study tested the proposed model in Turkey. Similar studies can be carried out in different regions or countries with different samples. Third, EV market is in its early phase for emerging countries. As an emerging economy, the EV market in Turkey is not developed to include the actual purchase behavior in the model. Therefore, future studies are recommended to conduct with the aim of studying actual purchasing behavior. Fourth, this study investigated the effect of EC on the attitude toward and purchase intention of EVs. Further studies should account for both personal and social factors that affect consumers' pro-environmental behaviors such as perceived environmental benefits, environmental responsibility and price sensitivity etc. to widen the scope of the study. Such studies will contribute to consumers' awareness of EVs since consumers of emerging markets are unfamiliar and more skeptical about EVs (Rajper & Albrecht, 2020). Additionally, in order to popularize and raise interest for EVs, the comparative studies between EVs and fossil fuel cars should be conducted. Therefore, the differences in consumer perspectives among them could be revealed.

Further, the current EV knowledge presents limited studies for the moderating and mediating variables (Kumar & Alok, 2020). The current literature also gives mixed findings about the role of demographics on EV adoption and how they alter the relationships among personal factors, attitude toward and willingness to purchase EVs. Thus, the future research can focus on the demographic variables and examine their impact on the link between personal antecedents and EV adoption. In this manner, the role of demographics would be understood more clearly. Additionally, there are some other personal antecedents of EV purchase intention that still need further attention from the researchers such as psychological factors (Austmann, 2020) and personality traits (Qu et al., 2020). Last, the lack of cross-cultural research and EV studies for emerging markets are other gaps that the future studies should emphasize.

REFERENCES

Asadi, S., Nilashi, M., Samad, S., Abdullah, R., Mahmoud, M., Alkinani, M. H., & Yadegaridehkordi, E. (2021). Factors impacting consumers' intention toward adoption of electric vehicles in Malaysia. *Journal of Cleaner Production*, *282*(124474).

Asunta, T. (2003). *Knowledge of environmental issues. where pupils acquire information opinions, and how it affects their attitudes, and laboratory behavior*. University of Jyväskylä.

Austmann, L. M. (2020). Drivers of the electric vehicle market: A systematic literature review of empirical studies. *Finance Research Letters*, 1–10.

Bamberg, S. (2003). How does environmental concern influence specific environmentally related behaviors? A new answer to an old question. *Journal of Environmental Psychology*, *23*(1), 21–32. doi:10.1016/S0272-4944(02)00078-6

Bouscasse, H., Joly, I., & Bonnel, P. (2018). How does environmental concern influence mode choice habits? A mediation analysis. *Transportation Research Part D, Transport and Environment*, *59*, 205–222. doi:10.1016/j.trd.2018.01.007

Brough, A. R., Wilkie, J. E., Ma, J., Isaac, M. S., & Gal, D. (2016). Is eco-friendly unmanly? the green-feminine stereotype and its effect on sustainable consumption. *The Journal of Consumer Research*, *43*(4), 567–582. doi:10.1093/jcr/ucw044

Carley, S., Krause, R., Lane, B., & Graham, J. (2013). Intent to purchase a plug-in electric vehicle: A survey of early impressions in large US cities. *Transportation Research Part D, Transport and Environment*, *18*, 39–45. doi:10.1016/j.trd.2012.09.007

Cerri, J., Testa, F., & Rizzi, F. (2018). The more I care, the less I will listen to you: How information, environmental concern and ethical production influence consumers' attitudes and the purchasing of sustainable products. *Journal of Cleaner Production*, *175*, 343–353. doi:10.1016/j.jclepro.2017.12.054

Chan, T. S. (1996). Concerns for environmental issues and consumer purchase preferences: A two-country study. *Journal of International Consumer Marketing*, *9*(1), 43–55. doi:10.1300/J046v09n01_04

Chen, C. F., de Rubens, G. Z., Noel, L., Kester, J., & Sovacool, B. K. (2020). Assessing the socio-demographic, technical, economic and behavioral factors of Nordic electric vehicle adoption and the influence of vehicle-to-grid preferences. *Renewable & Sustainable Energy Reviews*, *121*, 109692. doi:10.1016/j.rser.2019.109692

Chu, W., Baumann, C., Hamin, H., & Hoadley, S. (2018). Adoption of environment-friendly cars: Direct vis-à-vis mediated effects of government incentives and consumers' environmental concern across global car markets. *Journal of Global Marketing*, *31*(4), 282–291. doi:10.1080/08911762.2018.1456597

Coffman, M., Bernstein, P., & Wee, S. (2017). Electric vehicles revisited: A review of factors that affect adoption. *Transport Reviews*, *37*(1), 79–93. doi:10.1080/01441647.2016.1217282

Dagher, G. K., Kassar, A. N., & Itani, O. (2015). The impact of environment concern and attitude on green purchasing behavior. *Contemporary Management Research*, *11*(2), 179–205. doi:10.7903/cmr.13625

Daziano, R. A., & Bolduc, D. (2013). Incorporating pro-environmental preferences towards green automobile technologies through a Bayesian hybrid choice model. *Transportmetrica A: Transport Science*, *9*(1), 74–106. doi:10.1080/18128602.2010.524173

Degirmenci, K., & Breitner, M. H. (2017). Consumer purchase intentions for electric vehicles: Is green more important than price and range? *Transportation Research Part D, Transport and Environment*, *51*, 250–260. doi:10.1016/j.trd.2017.01.001

Diamantopoulos, A., Schlegelmilchh, B. B., Sinkovics, R. R., & Bohlen, G. B. (2003). Can socio demographics still play a role in profiling green consumers? a review of the evidence and an empirical investigation. *Journal of Business Research*, *56*(6), 465–480. doi:10.1016/S0148-2963(01)00241-7

Dimitropoulos, A. (2014). The influence of environmental concerns on drivers' preferences for electric cars. In *Tinbergen Institute Discussion Papers 14-128/VIII*. Tinbergen Institute.

Dorsch, M. T. (2014). Economic development and determinants of environmental concern. *Social Science Quarterly*, *95*(4), 960–977. doi:10.1111squ.12071

Egbue, O., & Long, E. (2012). Barriers to widespread adoption of electric vehicles: An analysis of consumer attitudes and perceptions. *Energy Policy*, *48*, 717–729. doi:10.1016/j.enpol.2012.06.009

Elliott, R. (2013). The taste for green: The possibilities and dynamics of status differentiation through "green" consumption. *Poetics*, *41*(3), 294–322. doi:10.1016/j.poetic.2013.03.003

Ergen, B., & Ergen, Z. (2011). How does education affect environmental knowledge: A survey in urban and regional planning education? *US-China Education Review*, *7*, 924–931.

Fishbein, M., & Ajzen, I. (1975). *Belief, attitude, intention and behavior: An introduction to theory and research*. Addison-Wesley.

Fornell, C., & Larcker, D. F. (1981). Evaluating structural equation models with unobservable variables and measurement error. *JMR, Journal of Marketing Research*, *18*(1), 39–50. doi:10.1177/002224378101800104

Gaskin, J., & Lim, J. (2016). *Model fit measures*. AMOS Plugin. Gaskination's StatWiki.

Gelissen, J. (2007). Explaining popular support for environmental protection: A multilevel analysis of 50 nations. *Environment and Behavior*, *39*(3), 392–415. doi:10.1177/0013916506292014

Ghasri, M., Ardeshiri, A., & Rashidi, T. (2019). Perception towards electric vehicles and the impact on consumers' preference. *Transportation Research Part D, Transport and Environment*, *77*, 271–291. doi:10.1016/j.trd.2019.11.003

Gilligan, C. (1982). In a different voice: psychological theory and women's development. Harvard University Press.

Goodale, T. (2021). Multivariate analysis of the impact of gender and college major on student levels of environmental concern and knowledge. *International Electronic Journal of Environmental Education*, *11*(1), 1–12.

Gray, S. G., Raimi, K. T., Wilson, R., & Árvai, J. (2019). Will Millennials save the world? the effect of age and generational differences on environmental concern. *Journal of Environmental Management*, *242*, 394–402. doi:10.1016/j.jenvman.2019.04.071 PMID:31059952

Hackbarth, A., & Madlener, R. (2013). Consumer preferences for alternative fuel vehicles: A discrete choice analysis. *Transportation Research Part D, Transport and Environment*, *25*, 5–17. doi:10.1016/j.trd.2013.07.002

Hair, J. F., Black, W. C., Babin, B. J., & Anderson, R. E. (2014). *Multivariate data analysis: Pearson new international edition*. Pearson Education Limited.

Hair, J. F. Jr, Hult, G. T. M., Ringle, C. M., & Sarstedt, M. (2017). *A primer on partial least squares structural equation modeling (pls-sem)*. Sage.

Harring, N., & Jagers, S. C. (2017). Why do people accept environmental policies? the prospects of higher education and changes in norms, beliefs and policy preferences. *Environmental Education Research*, 1–16.

Hayes, A. F. (2012). *Process: A versatile computational tool for observed variable mediation, moderation, and conditional process modeling* [White paper]. http://www.afhayes.com/public/process2012.pdf

He, X., Zhan, W., & Hu, Y. (2018). Consumer purchase intention of electric vehicles in China: The roles of perception and personality. *Journal of Cleaner Production*, *204*, 1060–1069. doi:10.1016/j.jclepro.2018.08.260

Hidrue, M. K., Parsons, G. R., Kempton, W., & Gardner, M. P. (2011). Willingness to pay for electric vehicles and their attributes. *Resource and Energy Economics*, *33*(3), 686–705. doi:10.1016/j.reseneeco.2011.02.002

Higueras-Castillo, E., Molinillo, S., Coca-Stefaniak, J. A., & Liébana-Cabanillas, F. (2020). Potential early adopters of hybrid and electric vehicles in Spain—Towards a customer profile. *Sustainability*, *12*(11), 4345. doi:10.3390u12114345

Hirsh, J. B. (2010). Personality and environmental concern. *Journal of Environmental Psychology*, *30*(2), 245–248. doi:10.1016/j.jenvp.2010.01.004

Hu, L., & Bentler, P. M. (1999). Cutoff criteria for fit indexes in covariance structure analysis: Conventional criteria versus new alternatives. *Structural Equation Modeling*, *6*(1), 1–55. doi:10.1080/10705519909540118

Jain, S. K., & Kaur, G. (2006). Role of socio-demographics in segmenting and profiling green consumers. *Journal of International Consumer Marketing*, *18*(3), 107–146. doi:10.1300/J046v18n03_06

Junquera, B., Moreno, B., & Álvarez, R. (2016). Analyzing consumer attitudes towards electric vehicle purchasing intentions in Spain: Technological limitations and vehicle confidence. *Technological Forecasting and Social Change*, *109*, 6–14. doi:10.1016/j.techfore.2016.05.006

Kawgan-Kagan, I. (2020). Are women greener than men? a preference analysis of women and men from major German cities over sustainable urban mobility. *Transportation Research Interdisciplinary Perspectives*, 8.

Kim, J., Rasouli, S., & Timmermans, H. (2014). Expanding scope of hybrid choice models allowing for mixture of social influences and latent attitudes: Application to intended purchase of electric cars. *Transportation Research Part A, Policy and Practice*, *69*, 71–85. doi:10.1016/j.tra.2014.08.016

Kirmani, M. D., & Khan, M. N. (2016). Environmental concern to attitude towards green products: Evidences from India, *Serbian. Journal of Management*, *11*(2), 159–179.

Krupa, J. S., Rizzo, D. M., Eppstein, M. J., Lanute, D. B., Gaalema, D. E., Lakkaraju, K., & Warrender, C. E. (2014). Analysis of a consumer survey on plug-in hybrid electric vehicles. *Transportation Research Part A, Policy and Practice*, *64*, 14–31. doi:10.1016/j.tra.2014.02.019

Kumar, R. R., & Alok, K. (2020). Adoption of electric vehicle: A literature review and prospects for sustainability. *Journal of Cleaner Production*, *253*, 1–21. doi:10.1016/j.jclepro.2019.119911

La Trobe, H. L., & Acott, T. G. (2000). A modified NEP/DSP environmental attitudes scale. *The Journal of Environmental Education*, *32*(1), 12–20. doi:10.1080/00958960009598667

Lai, I., Liu, Y., Sun, X., Zhang, H., & Xu, W. (2015). Factors influencing the behavioural intention toward full electric vehicles: An empirical study in Macau. *Sustainability*, *7*(9), 12564–12585. doi:10.3390u70912564

Lee, J. A., & Holden, S. J. (1999). Understanding the determinants of environmentally conscious behavior. *Psychology and Marketing*, *16*(5), 373–392. doi:10.1002/(SICI)1520-6793(199908)16:5<373::AID-MAR1>3.0.CO;2-S

Lee, J. H., Hardman, S. J., & Tal, G. (2019). Who is buying electric vehicles in California? Characterising early adopter heterogeneity and forecasting market diffusion. *Energy Research & Social Science*, *55*, 218–226. doi:10.1016/j.erss.2019.05.011

Li, J., Zhang, J., Zhang, D., & Ji, Q. (2019). Does gender inequality affect household green consumption behaviour in China? *Energy Policy*, *135*, 111071. doi:10.1016/j.enpol.2019.111071

Li, W., & Chen, N. (2018). Absolute income, relative income and environmental concern: Evidence from different regions in China. *Journal of Cleaner Production*, *187*, 9–17. doi:10.1016/j.jclepro.2018.03.171

Liao, F., Molin, E., & van Wee, B. (2017). Consumer preferences for electric vehicles: A literature review. *Transport Reviews*, *37*(3), 252–275. doi:10.1080/01441647.2016.1230794

Liberty, M., & Hongjuan, W. (2010). Greening our future and environmental values: An investigation of awareness, attitudes and awareness of environmental issues in Zambia. *Environmental Values*, *19*(4), 485–516. doi:10.3197/096327110X531570

Lin, B., & Tan, R. (2017). Estimation of the environmental values of electric vehicles in Chinese cities. *Energy Policy*, *104*, 221–229. doi:10.1016/j.enpol.2017.01.037

Lin, C. Y., & Syrgabayeva, D. (2016). Mechanism of environmental concern on intention to pay more for renewable energy: Application to a developing country. *Asia Pacific Management Review*, *21*(3), 125–134. doi:10.1016/j.apmrv.2016.01.001

Lin, S. T., & Niu, H. J. (2018). Green consumption: Environmental knowledge, environmental consciousness, social norms, and purchasing behavior. *Business Strategy and the Environment*, *27*(8), 1679–1688. doi:10.1002/bse.2233

Liu, H., Sato, H., & Morikawa, T. (2015). Influences of environmental consciousness and attitudes to transportation on electric vehicle purchase intentions. *Asian Transport Studies*, *3*(4), 430–446.

Liu, X., Vedlitz, A., & Shi, L. (2014). Examining the determinants of public environmental concern: Evidence from national public surveys. *Environmental Science & Policy*, *39*, 77–94. doi:10.1016/j.envsci.2014.02.006

Mainieri, T., Barnett, E. G., Valdero, T. R., Unipan, J. B., & Oskamp, S. (1997). Green buying: The influence of environmental concern on consumer behavior. *The Journal of Social Psychology*, *137*(2), 189–204. doi:10.1080/00224549709595430

Marquart-Pyatt, S. T. (2012). Contextual influences on environmental concerns cross-nationally: A multilevel investigation. *Social Science Research*, *41*(5), 1085–1099. doi:10.1016/j.ssresearch.2012.04.003 PMID:23017919

McDonald, S., Oates, C. J., Young, C. W., & Hwang, K. (2006). Toward sustainable consumption: Researching voluntary simplifiers. *Psychology and Marketing*, *23*(6), 515–534. doi:10.1002/mar.20132

Mohamed, M., Higgins, C., Ferguson, M., & Kanaroglou, P. (2016). Identifying and characterizing potential electric vehicle adopters in Canada: A two-stage modelling approach. *Transport Policy*, *52*, 100–112. doi:10.1016/j.tranpol.2016.07.006

Mohd Suki, N. (2016). Consumer environmental concern and green product purchase in Malaysia: Structural effects of consumption values. *Journal of Cleaner Production*, *132*, 204–214. doi:10.1016/j.jclepro.2015.09.087

Morton, C., Anable, J., & Nelson, J. D. (2017). Consumer structure in the emerging market for electric vehicles: Identifying market segments using cluster analysis. *International Journal of Sustainable Transportation*, *11*(6), 443–459. doi:10.1080/15568318.2016.1266533

Moyo, N. (2018). Environmental concern and purchase intention of electric vehicles in the eastern part of China. *Archives of Business Research*, *6*(8), 297–309.

Mueller, J. T., & Mullenbach, L. E. (2018). Looking for a white male effect in generation Z: Race, gender, and political effects on environmental concern. *Society & Natural Resources*, *31*(8), 925–941. doi:10.1080/08941920.2018.1445331

Muhammad, S., Fathelrahman, E., & Ullah, R. U. T. (2015). Factors affecting consumers' willingness to pay for certified organic food products in United Arab Emirates. *Journal of Food Distribution Research*, *46*(1), 37–45.

Newell, S. J., & Green, C. L. (1997). Racial differences in consumer environmental concern. *The Journal of Consumer Affairs*, *31*(1), 53–69. doi:10.1111/j.1745-6606.1997.tb00826.x

Newman, T. P., & Fernandes, R. (2016). A re-assessment of factors associated with environmental concern and behavior using the 2010 General Social Survey. *Environmental Education Research*, *22*(2), 153–175. doi:10.1080/13504622.2014.999227

Nguyen, T. N., Lobo, A., & Greenland, S. (2017). The influence of cultural values on green purchase behaviour. *Marketing Intelligence & Planning*, *35*(3), 377–396. doi:10.1108/MIP-08-2016-0131

Peters, A., & Dütschke, E. (2014). How do consumers perceive electric vehicles? a comparison of German consumer groups. *Journal of Environmental Policy and Planning*, *16*(3), 359–377. doi:10.1080/1523908X.2013.879037

Plötz, P., Schneider, U., Globisch, J., & Dütschke, E. (2014). Who will buy electric vehicles? Identifying early adopters in Germany. *Transportation Research Part A, Policy and Practice*, *67*, 96–109. doi:10.1016/j.tra.2014.06.006

Qu, W., Sun, H., & Ge, Y. (2020). The effects of trait anxiety and the big five personality traits on self-driving car acceptance. *Transportation*, 1–17.

Rajper, S. Z., & Albrecht, J. (2020). Prospects of electric vehicles in the developing countries: A literature review. *Sustainability*, *12*(5), 1–19. doi:10.3390u12051906

Ritchie, H., & Roser, M. (2021). *Emissions by sector*. https://ourworldindata.org/emissions-by-sector

Roberts, J. A. (1996). Green consumers in the 1990s: Profile and implications for advertising. *Journal of Business Research*, *36*(3), 217–231. doi:10.1016/0148-2963(95)00150-6

Roberts, J. A., & Bacon, D. R. (1997). Exploring the subtle relationships between environmental concern and ecologically conscious consumer behavior. *Journal of Business Research*, *40*(1), 79–89. doi:10.1016/S0148-2963(96)00280-9

Samdahl, D. M., & Robertson, R. (1989). Social determinants of environmental concern: Specification and test of the model. *Environment and Behavior*, *21*(1), 57–81. doi:10.1177/0013916589211004

Schuitema, G., Anable, J., Skippon, S., & Kinnear, N. (2013). The role of instrumental, hedonic and symbolic attributes in the intention to adopt electric vehicles. *Transportation Research Part A, Policy and Practice*, *48*, 39–49. doi:10.1016/j.tra.2012.10.004

Shen, Z.-J. M., Feng, B., Mao, C., & Ran, L. (2019). Optimization models for electric vehicle service operations: A literature review. *Transportation Research Part B: Methodological*, *128*, 462–477. doi:10.1016/j.trb.2019.08.006

Sierzchula, W., Bakker, S., Maat, K., & Wee, B. (2014). The influence of financial incentives and other socio-economic factors on electric vehicle adoption. *Energy Policy*, *68*, 183–194. doi:10.1016/j.enpol.2014.01.043

Singhal, A., & Malik, G. (2018). The attitude and purchasing of female consumers towards green marketing related to cosmetic industry. *Journal of Science and Technology Policy Management*, 1-18.

Sovacool, B. K., Kester, J., Noel, L., & de Rubens, G. Z. (2018). The demographics of decarbonizing transport: The influence of gender, education, occupation, age, and household size on electric mobility preferences in the Nordic region. *Global Environmental Change*, *52*, 86–100. doi:10.1016/j.gloenvcha.2018.06.008

Sreen, N., Purbey, S., & Sadarangani, P. (2018). Impact of culture, behavior and gender on green purchase intention. *Journal of Retailing and Consumer Services*, *41*, 177–189. doi:10.1016/j.jretconser.2017.12.002

Straughan, R. D., & Roberts, J. A. (1999). Environmental segmentation alternatives: A look at green consumer behavior in the new millennium. *Journal of Consumer Marketing*, *16*(6), 558–575. doi:10.1108/07363769910297506

Sun, Y., Liu, N., & Zhao, M. (2019). Factors and mechanisms affecting green consumption in China: A multilevel analysis. *Journal of Cleaner Production*, *209*, 481–493. doi:10.1016/j.jclepro.2018.10.241

Tabachnick, B. G., & Fidell, L. S. (2007). *Using multivariate statistics*. Pearson Education.

TEHAD. (n.d.). http://tehad.org/2021/01/16/2020-yili-elektrikli-ve-hibrid-otomobil-satis-rakamlari-belli-oldu/

Thananusak, T., Rakthin, S., Tavewatanaphan, T., & Punnakitikashem, P. (2017). Factors affecting the intention to buy electric vehicles: Empirical evidence from Thailand. *International Journal of Electric and Hybrid Vehicles*, *9*(4), 361–381. doi:10.1504/IJEHV.2017.089875

Tiller, L. L. C. (2014). *Nationwide Survey Finds: Americans Believe Environmental Issues Pose Greater Long-Term Threat to Their Health and Well-Being Than Terrorism or Global Epidemics Like Ebola.* https://www.prnewswire.com/news-releases/nationwide-survey-finds-americans-believe-environmental-issues-pose-greater-long-term-threat-to-their-health-and-well-being-than-terrorism-or-global-epidemics-like-ebola-300014118.html

Tran, M., Banister, D., Bishop, J. D., & McCulloch, M. D. (2013). Simulating early adoption of alternative fuel vehicles for sustainability. *Technological Forecasting and Social Change*, *80*(5), 865–875. doi:10.1016/j.techfore.2012.09.009

Trommer, S., Jarass, J., & Kolarova, V. (2015). Early adopters of electric vehicles in Germany unveiled. *World Electric Vehicle Journal*, *7*(4), 722–732. doi:10.3390/wevj7040722

Tu, J. C., & Yang, C. (2019). Key factors influencing consumers' purchase of electric vehicles. *Sustainability*, *11*(14), 3863. doi:10.3390u11143863

TUIK. Gelir ve Yaşam Koşulları Araştırması, 2019. (2020). https://tuikweb.tuik.gov.tr/PreHaberBultenleri.do?id=33820

Urban, J., & Ščasný, M. (2012). Exploring domestic energy-saving: The role of environmental concern and background variables. *Energy Policy*, *47*, 69–80. doi:10.1016/j.enpol.2012.04.018

Vassileva, I., & Campillo, J. (2017). Adoption barriers for electric vehicles: Experiences from early adopters in Sweden. *Energy*, *120*, 632–641. doi:10.1016/j.energy.2016.11.119

Verma, V. K., Chandra, B., & Kumar, S. (2019). Values and ascribed responsibility to predict consumers' attitude and concern toward green hotel visit intention. *Journal of Business Research*, *96*, 206–216. doi:10.1016/j.jbusres.2018.11.021

Vicente-Molina, M. A., Fernandez-Sainz, A., & Izagirre-Olaizola, J. (2013). Environmental knowledge and other variables affecting pro-environmental behavior: Comparison of university students from emerging and advanced countries. *Journal of Cleaner Production*, *61*, 130–138. doi:10.1016/j.jclepro.2013.05.015

We are Social. Digital 2021: Turkey. (n.d.). https://datareportal.com/reports/digital-2021-turkey

WE Outlook International Energy Agency (2020). https://www.iea.org/reports/world-energy-outlook-2020

Wu, J., Liao, H., Wang, J. W., & Chen, T. (2019). The role of environmental concern in the public acceptance of autonomous electric vehicles: A survey from China. *Transportation Research Part F: Traffic Psychology and Behaviour*, *60*, 37–46. doi:10.1016/j.trf.2018.09.029

Xiao, C., & Dunlap, R. E. (2007). Validating a comprehensive model of environmental concern cross-nationally: A US-Canadian comparison. *Social Science Quarterly*, *88*(2), 471–493. doi:10.1111/j.1540-6237.2007.00467.x

Xiao, C., & McCright, A. M. (2015). Gender differences in environmental concern: Revisiting the institutional trust hypothesis in the USA. *Environment and Behavior*, *47*(1), 17–37. doi:10.1177/0013916513491571

Yang, C., Tu, J. C., & Jiang, Q. (2020). The influential factors of consumers' sustainable consumption: A case on electric vehicles in China. *Sustainability*, *12*(8), 3496. doi:10.3390u12083496

Zelezny, L. C., Chua, P., & Aldrich, C. (2000). Elaborating on gender differences in environmentalism-statistical data included. *The Journal of Social Issues*, *56*(3), 443–445. doi:10.1111/0022-4537.00177

Zhang, M., Guo, S., Bai, C., & Wang, W. (2019). Study on the impact of haze pollution on residents' green consumption behavior: The case of Shandong Province. *Journal of Cleaner Production*, *219*, 11–19. doi:10.1016/j.jclepro.2019.02.077

Zhao, H. H., Gao, Q., Wu, Y. P., Wang, Y., & Zhu, X. D. (2014). What affects green consumer behavior in China? a case study from Qingdao. *Journal of Cleaner Production*, *63*, 143–151. doi:10.1016/j.jclepro.2013.05.021

Zhao, R., Geng, Y., Liu, Y., Tao, X., & Xue, B. (2018). Consumers' perception, purchase intention, and willingness to pay for carbon-labeled products: A case study of Chengdu in China. *Journal of Cleaner Production*, *171*, 1664–1671. doi:10.1016/j.jclepro.2017.10.143

KEY TERMS

Electric Vehicles: Electric vehicles (EVs) have one or more electric motors, using energy stored in rechargeable batteries for propulsion. EVs consume no-petroleum-based fuel and have no tailpipe emissions.

Environmental Concern: Environmental concern (EC) refers to one's perception and knowledge about environmental issues, to what extent a person concerns about the environmental consequences of their own or others' behavior and supports efforts to prevent threats to the environment.

Chapter 15
Adoption of Wearable Technology Devices:
A Cross-Cultural Study

Bengi Meriç Benderlioğlu
Bogazici University, Turkey

U. Zeynep Ata
https://orcid.org/0000-0002-5493-0276
Bogazici University, Turkey

ABSTRACT

With rapid change in technology worldwide, innovative products such as wearable technology devices tend to have an uprising trend. Consumers, however, are not necessarily adaptive in their nature and their perception is shaped by many factors. The aim of this research is to investigate the consumer acceptance of wearable technology devices, specifically smartwatches. The study extends the widely used technology acceptance model with the introduction of new variables. For the purpose of the study, survey data was collected from German and Turkish university students. The overall results provide validation to previous literature while introducing new factors for consumer acceptance of technology products, wearable technology devices, and smartwatches. Importance of this research comes from the innovative and promising nature of the wearable technology devices concept, the lack of work on smartwatches in literature, as well as the cross-cultural nature of the study. The study also has managerial implications for technology companies who chase after growth in their businesses.

INTRODUCTION

Disruptive innovation creates new value, eventually changing existing networks and establishing new ones. Technology continuously modifies our lives in subtle ways yet disruptive technologies such as artificial intelligence, 3D printing and advanced virtual reality have really made major impacts. Many digital technologies such as smartphones, social media, big data, predictive analytics and cloud are radi-

DOI: 10.4018/978-1-7998-8900-7.ch015

cally different from preceding IT-based technologies and products such as wearable technology devices tend to have an uprising trend that comes with wider variety nowadays. Wearable technology is any type of technology that is incorporated in electronics that can be worn on the body, either as an accessory or as part of materials used in clothing. One of the major features of wearable technology is its ability to connect to the Internet, enabling data to be exchanged between a network and the device.

With internet usage rates increasing rapidly since the beginning of the 90s, the usage capacity of information increased exponentially and technology spread rapidly throughout the world. This reflected on the amount of space and scope allocated to technology products in our day-to-day lives. Along with this, the number of wearable technology devices produced by companies has been increasing in recent years and these products are seen as the new era of growth for technology (Stinson, 2013). There are various forms in which wearable technology devices can be worn (Ko, et al., 2005) and despite being a relatively new product in its early diffusion stage in the global market, smartwatches are categorized as one of the most popular wearable technology devices in today's world (Chuah, et al., 2016).

The aim of this study is to understand the effects of specific factors on consumers' acceptance of wearable technology devices. The study also investigates the difference between the acceptance patterns of cultures, taking Turkey and Germany as sample populations. The model of this study is an extended version of the technology acceptance model of Davis (1989). External variables are added to the original model, while also aiming to test the validity of the original model.

WEARABLE TECHNOLOGY

The term wearable technology and its practices are relatively new in today's world, thus it is not possible to find an established description in literature. As the topic is an emerging one not only in consumer behavior industry but also in the technology industry, a variety of understanding of the topic has been presented in recent years. Although it is not likely to describe the meaning and scope of wearable technology in a standardized way, some terms have many close meanings, including 'wearable electronics', 'wearable devices' and 'wearable computers'. Although the topic has recently started to be a trending topic in both academic research and managerial application, the history of wearable technology dates back years ago starting with the head-mounted displays developed for pilots in the 1960s (European Commission, 2015).

According to Dunne (2004), "wearable technology is a term used to describe many different forms of body-mounted technology, including wearable computers, smart clothing, and functional clothing". Ko, et al. (2005) describe wearable technology devices as electronic devices that people continually wear as unhindered as clothes providing intelligence support that increases memory, intellect, communication, physical senses and creativity. European Commission's (2015) report on the internet of things and wearable technology states the following:

Wearable technology is a type of technology that is incorporated in electronics that can be worn on the body, either as an accessory or as part of materials used in clothing. One of the major features of wearable technology is its ability to connect to the Internet, enabling data to be exchanged between a network and the device.

Furthermore, another understanding and exposition of the concept of wearable technology devices is their usage as a fashion item allowing customers to reflect their style and characteristics into their daily life. Being available in a variety of customized opportunities such as different colors and sizes, a fringe benefit of wearable technology devices is their ability to be used as accessories. This trendy approach combines 'fashion' and 'technology' and often regarded as 'fashnology'. Brem, Ro, & Rauschnabel (2006) regard wearable technology as the new form of fashion accessories for the users. Supporting the theory of Dion, Berscheid, & Walster (1972) which suggests that the consumers have a tendency to pick objects that are perceived as aesthetically pleasing, Bajarin (2014) concludes that, "while people buy watches to tell time . . . the number one criteria in choosing a watch for most people is how it will look. It's a fashion statement, not a technology one".

There are various forms that wearable technology devices can be worn such as a wristwatch, a badge, a ring, jewelry, shoes, clothing or eyeglasses (Ko, et al., 2005). The current existence of a wide range of wearable technology and smart devices is constantly in a rising trend thanks to the extended use of these products and developments in the technology industry. Being already available in many forms, it is not hard to guess that smart products will be available for their users in many different forms and preferences in the coming future.

Today, wearable technology devices are available for many different purposes for numerous customer groups all around the world. Although the literature differs between the respective industries in which the smart devices are being used, in general wearable technology devices can be categorized into five main product groups based on their usage methods, which are smartwatches, smart wristbands, smart glasses (including all types of head-mounted displays), smart clothes and smart accessories.

Despite being a relatively new product in its early diffusion stage in the global market, smartwatches are categorized as one of the most popular wearable technology devices in today's world (Chuah, et al., 2016). As technology evolved to become an inevitable and crucial part of our lives, the advantages we have taken from the practical and life-easing features of them have been increasing respectively. Gaining popularity and rapidly saturating the wearable technology devices market, smartwatches are one of the devices people use in their daily life in order to gain a wide range of advantages offered by the products.

The benefits of smartwatches are numerous depending on the aimed area of usage and initial expectation from the product. Users of smart wearable devices, smartwatch owners in particular, might have different motivations leading them to use or consider to use these devices including; being in more control of their actual vital indicators through health activity tracking, ability to understand the extent to which the user is participating in sports activities, and increasing efficiency in work and personal life through having continuous availability and experiencing constant mobility. Connectedness, with minimum possibility of discontinuance, is a valuable asset that smartwatches offer to the owner. Belk (2013) has stated that the meaning of everyday objects changes according to who and what connects them and to where they are connected to. According to Verhoef, et al. (2016), "the ability of smart objects to connect not only with consumers and each other but also with other virtual and physical devices on the Internet will speed up the pace of change in their meaning".

Smartwatches are considered among the most important developments in the information technology industry having plentiful functions in addition to showing time (Chuah, et al., 2016). Smartwatches that are in contact with a phone allow users to track their activities through the sensors transmitting information in contrast to classical watches transmitting mainly time information only (Akbulut & Akan, 2015). Smart wristbands, on the other hand, are usually light-weighted and used for tracking and monitoring daily physical activities. These activities include but are not limited to sleeping patterns, the number of calories spent, number of steps taken, water consumption and pulse tracking. There is a variety of sensors embedded inside the smart wristbands, leading the device to have numerous diverse functions (Nanda, 2017). In addition to the activity tracking features of the smart wristbands, when connected to a mobile phone and Bluetooth, these smart devices also act as notification devices transmitting messages from various applications.

Smart glasses are like mini-computers with high-definition images in the outside world. By processing and capturing its user's physical environment and augmenting with virtual elements, smart glasses are considered as new wearable augmented reality devices (Rauschnabel, Brem, & Ro, 2015). Having internal and external sensors that can collect data from computers, smartphones and other electronic devices, smart glasses enable their users to have a wireless connection to GPS, Wi-Fi and Bluetooth, leading users to be able to connect to the Internet and watch videos or gather information. Some versions of smart glasses and head-mounted displays include face recognition software, built-in cameras, GPS and other applications. The aim behind the usage of the smart glasses is rather different from head-mounted displays since they are mostly used in order to reach more information regarding the environment in which the user is, instead of isolating the user from the outer world.

Smart clothes, or interactive or digital clothing, is defined as a "garment-integrated device which augments the functionality of clothing or which imparts information-processing functionality to a garment" (Dunne et al., 2005). According to Hwang (2014), "science has combined with fashion where the property of clothes and various information technology (IT) functions coexist together in this new conceptual wear". Ranging a high variety of products, smart clothes can be in many different forms such as pants, underwear, socks, suits, hats, shoes and so on.

Smart accessories, although not as commonly used as other wearable technology devices, mostly function as to be synchronized with other mobile devices. Since smart accessories are designed relatively smaller to be worn on the body, their relative connectivity and ability to operate various applications are limited. However, at the same time, they allow the users to take advantage of the accessories' usage as electronic identity, signature and application locker. Smart accessories such as smart rings and smart necklaces can be featured with GPS modules that allow the devices to be used as location-direction detection for individuals with special needs.

Wearable technology devices are in an increasing trend all around the world, currently capturing a favorable future potential in the market for technology products. As of the beginning of 2016, 15% of overall U.K. population was characterized as smart wearable technology product users allowing a strong rise of the share of these devices on online sales stores (Mills, et al., 2016). Providing numerous advantages for consumers, wearable technology devices can be stated as one of the most personal accessories available in many shapes and forms. The new concept of wearable technology devices shows that the consumer can benefit from a range of wearables on their bodies no matter which part of the body is the area of interest for taking the maximum advantage on the usage of the device (Robson, Kietzmann, & Pitt, 2016).

TECHNOLOGY ACCEPTANCE MODELS

As a result of the advanced and emerging technologies' continuous development in a fast-paced and constantly changing digital world, numerous models have been proposed to better understand consumer adoption of new technologies and their implications (Chuah, et al., 2016). Diffusion and acceptance

Figure 1. The original technology acceptance model (Davis, 1989)

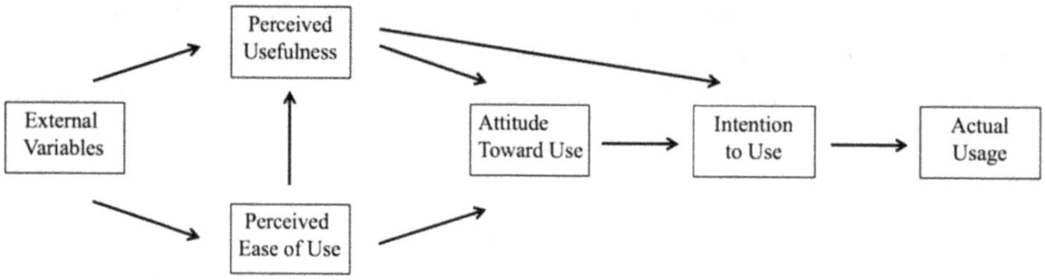

of innovation and technology are particularly important for social scientists in order to understand the perception of the consumer and serve their needs better. Understanding and analyzing the cultural and social changes while trying to better explain the tools of change is an area of interest for researchers interested in social science.

There is a vast amount of research in literature which tries to meaningfully understand and make sense of user's adoption of technology such as innovation diffusion model of Rogers (1962), theory of reasoned action (TRA) (Ajzen & Fishbein, 1980), technology acceptance model of Davis (1989), extended version of the technology acceptance model (Venkatesh & Davis, 2000; Venkatesh & Bala, 2008), theory of planned behavior (Ajzen, 1991), unified theory of acceptance and use of technology (Venkatesh, et al., 2003) and extended version of the unified theory of acceptance and use of technology (Venkatesh, Thong, & Xu, 2012). According to literature, the Technology Acceptance Model (TAM) of Davis (1989) is believed to be one of the most validated models investigating and explaining the intention to adopt the technology. Being an evolved version of Fishbein and Ajzen's (1975) theory of reasoned action, the TAM model introduces new belief variables, perceived usefulness and perceived ease of use. The original model consists of five elements: external variables, perceived usefulness, perceived ease of use, attitude and intention. The model is widely used in order to understand the technology acceptance and adoption patterns of consumers. According to Fishbein and Ajzen's (1975) theory of reasoned action, beliefs are one of the influencing sources of attitude, which then shape the intention. Following, behaviors are generated through intention (Ma & Liu, 2004). As the technology acceptance model takes its roots from the theory of reasoned actions and is further developed, today the model is applicable in not only consumer behavior area but also in social psychology. Davis, in 1989, introduced two new dimensions to the model being perceived usefulness and perceived ease of use. Figure 1 presents the original TAM model of Davis (1989).

With the new constructs of the model - perceived usefulness and perceived ease of use - consumers' beliefs on technology are measured, resulting in explaining and predicting attitude towards technology and therefore predicting acceptance of technology (Ma & Liu, 2004). As a result of the analysis conducted on the model, Davis (1989) concluded that both perceived usefulness and perceived ease of use have a relationship with actual use, with perceived usefulness being significantly more correlated. To conclude, by combining an already existing model of TRA with two new constructs, the original TAM model was developed and tested many times by researchers trying to understand the patterns of consumer acceptance of the technology. It was found that when perceived usefulness and perceived ease of use - variables that are successfully adapted to the model - are enhanced, they positively influence attitude and intention (Kim & Shin, 2014).

Figure 2. Research model

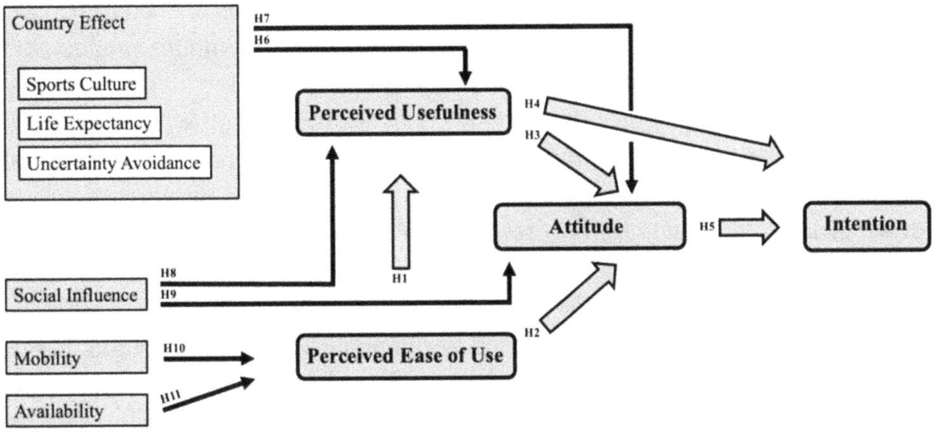

RESEARCH MODEL AND METHODOLOGY

The structure of this study is based on the technology acceptance model of Davis, with the aim of confirming the findings of previous research while contributing to the literature with its cross-cultural nature and specific focus on wearable technology devices and smartwatches in particular. The model in Figure 2 represents the basis of this study and is constructed by the addition of new variables that are believed to have a relationship with consumer acceptance of wearable technology devices to the basic Technology Acceptance Model of Davis (1989).

Perceived ease of use of a product represents the degree to which a consumer perceives a product easy to operate or service easy to receive. Davis (1989) defines perceived ease of use as “the degree to which a person believes that using a particular system would be free of efforts”. If a specific service which includes technological aspects is perceived as easy to use and take advantage of, users have an enhanced belief of that technology or service is more useful and they are more likely to have a positive attitude towards that technology product or service (Kim & Shin, 2014). As perceived ease of use is proved to be effective in technology acceptance, our study aims to explore if such a relationship is established in acceptance to wearable technology devices in a cross-cultural comparison perspective.

According to Davis (1989, 1993), perceived usefulness is one of the main psychological determinants of attitude towards technology and the intention to use technology. Perceived usefulness is defined by Davis (1989) as “the extent to which a person believes that using particular technology will enhance his/her job performance”. As stated by Kim and Shin (2014), “when users tend to believe that the technology is useful, they form favorable attitudes towards it”. In the basis of the study of Davis, together with the importance of opportunity cost of using any service, perceived usefulness is believed and proved to have effects on the attitude towards a technology product and the intention to use that product.

Attitude is defined as the positive or negative tendency the consumers gain against an object over time by learning. The object mentioned here could be a human, product, service, location, event, advertising and similar elements (Schiffman & Kanuk, 1978). Attitude, as stated in the definition, is not inherent in consumers upon birth, rather they tend to learn it later. For example, the idea that a brand or a product has better quality than the others is not an innate thought of the consumer (Solomon, 2009). According to Fishbein and Ajzen (1975), attitude towards use is defined as “an individual’s positive or negative feelings (evaluative affect) about performing the target behavior". It is necessary to draw attention to the various characteristics of attitude. Firstly, attitudes are formed by consumers through the process of learning, so attitude is learned tendencies. If a consumer is not satisfied with the product, the consumer may gain a negative attitude towards that product. Attitudes refer to trends over time and they often stay unchanged and continuous because they develop slowly over time (Hoyer & MacInnis, 2007). Therefore, when consumers have a negative attitude towards an object, it is hard for marketers to change this negative attitude. Consumers make purchasing decisions based on their attitudes; a consumer who is satisfied with the quality of a brand may choose the brand again for their next purchase or recommend it to others (Hanna & Wozniak, 2017).

Intention is defined as "the strength of one's intention to perform a specified behavior" by Fishbein and Ajzen (1975). Multi-attribute models are often used to anticipate behavioral intention. In these models, the focus is on the technology related beliefs of users (Fishbein & Ajzen, 1975). According to the TAM model of Davis (1989), intention is shaped by attitude and factors affecting attitudes such as perceived usefulness and perceived ease of use and these as a result influence actual use of a technology product. Additionally, TAM proposes that technologies are perceived as more useful when they are easier to use, and that usefulness also directly influences usage intention. In addition, as per the theory of the reasoned actions suggests, the subjective standard determines the intended meaning of the behavior (Ajzen & Fishbein, 1980) and actions are taken with more motivation by the individuals when it is believed that the others expect them to perform these actions (Venkatesh & Davis, 2000).

Following the discussions above, the following hypotheses are proposed:

Hypothesis 1: Perceived ease of use will have a positive effect on perceived usefulness of wearable technology devices.

Hypothesis 2: Perceived ease of use will have a positive effect on attitude towards wearable technology devices.

Hypothesis 3: Perceived usefulness will have a positive effect on attitude towards wearable technology devices.

Hypothesis 4: Perceived usefulness will have a positive effect on the intention to use wearable technology devices.

Hypothesis 5: Attitude will have a positive effect on intention to use wearable technology devices.

The next set of hypotheses reflects that this study also aims to focus on the differences between cultures/countries and the effects of these differences on acceptance to wearable technology devices. Following the review of the literature, three main aspects are identified that potentially will lead to differences between cultures/countries: sports culture representing the involvement in sports activities, life expectancy rates showing the long term expectations and uncertainty avoidance indicating extent to which individuals embrace ambiguity. First, sports culture in this study refers to the extent to which different cultures' attitude towards participation in sports or physical activities. Second, if improvements in technology affect life expectancy, it is worth to further investigate this effect the other way around; a potential alteration on the levels of consumer belief in terms of increased usefulness of wearable technology products, when it is believed that there will be a longer life span ahead. Third, the level of uncertainty avoidance, which is a function of the degree to which a social group feels threatened by ambiguous, uncertain, unknown situations (Mooij & Hofstede, 2002). Societies with higher uncertainty avoidance scores tend to minimize the unknown, as they feel more comfortable with clarity and certainty.

Hence, the following hypotheses are proposed:

Hypothesis 6: Country difference has an effect on the perceived usefulness of wearable technology devices.
Hypothesis 7: Country difference has an effect on attitude towards wearable technology devices.

Finally, smartwatches are as much a fashion item as a technological device. Thus it can be expected that fashion-wise social influence will have an effect on the consumer decision-making process and technology-wise the mobility and availability provided by using a smartwatch should be considered.

Social influence, as defined by Turner (1991), "is the processes whereby people directly or indirectly influence the thoughts, feelings and actions of others". In order to feel socially accepted, people tend to focus on and implement other people's opinion in their daily lives. Whether being a compliment or a suggestion, or just sharing of a positive or undesired experience, people affect each other's purchase, consumption and disposal decisions. Having said that purchase decisions are shaped around the environment and trusted people in the eyes of the consumers, peer groups are not the only way to be exposed to social influence on decision-making. Online consumer reviews are also considered as a kind of consumer-created review that provide the indirect experience of products and services to potential buyers influencing their purchase decisions (Park, Lee, & Han, 2014).

Furthermore, mobility and availability of product or service are more important than ever in today's fast-paced world with numerous information reaching a single consumer on a day-to-day basis and the ability of the consumer to be responsive constantly. As the mobility and availability of a service or a product is now a crucial concept, the applications and implementations of increasing the mobility and availability of offered services and products are very broad mainly in the areas of entertainment, knowledge and health (Kirstein, et al., 2005). The implications of mobility and availability features of the wearable technology devices in the health sector include monitoring health indicators such as tracking pulse, water consumption and blood pressure rates in addition to general monitoring and checking of physical sports activities. Due to the shifting focus on technical concerns to consumer-oriented marketing strategies, it has been a fundamental issue for researchers to reach higher mobility (Ariyatum & Holland, 2003). It is further argued by Kim and Sundar (2015) that "effects of mobile communications factors mainly result from the mobility levels of the device giving a feel of availability . . . mobility sense tends to be reinforced in wearable communication having firm effects on perceived hedonic quality of smart wearable technology devices including smartwatches".

Along these lines, the following hypotheses are proposed:

Hypothesis 8: Social influence will have a positive effect on the perceived usefulness of wearable technology devices.
Hypothesis 9: Social influence will have a positive effect on attitude towards wearable technology devices.
Hypothesis 10: Mobility will have a positive effect on perceived ease of use of wearable technology devices.
Hypothesis 11: Availability will have a positive effect on perceived ease of use of wearable technology devices.

Methodology and Study Setting

As the aim of the study is to investigate and explore factors affecting the consumer acceptance of wearable technology devices, a questionnaire is prepared in order to deeper understand these variables and their effects on a cross-cultural level between German people and Turkish people. As the nature of the study required us to do so, quantitative research methods are applied.

Data collection was through a face-to-face survey that was administered in 2019 at two universities, one in Turkey and the other in Germany. An introduction statement was presented at the beginning of the survey summarizing the aim of the study, confidentiality matters and the study concept. In terms of scale development for the survey, most of the questions are adopted from literature and used after slight adaptations to the context of this study and sample groups. Following the collection of the data, responses are transcribed into data analysis software and made ready for the statistical analysis process. The model and proposed hypotheses are tested through statistical analysis techniques. Statistical analysis methods applied in this research range from descriptive statistics to chi-square tests, regression analysis and independent samples t-tests.

ANALYSIS AND FINDINGS

The overall sample size for the study is 221 individuals, among which 101 are Turkish students (46%) and 120 German students (54%). 12 Turkish and 17 German participants are smartwatch owners. The average age among Germans is found to be 26 with Turkish average being 22. 58% of Turkish and 53% of German respondents are female. In terms of area of study, almost all participants are from social science studies such as business administration, economics and law in Turkey, whereas in Germany participants are more heterogeneous coming from the following areas: social science, communications, media and arts, natural science and engineering, information technologies and medicine.

First contingency analysis was conducted to see if the use of wearable technology devices and social media usage are country dependent or not. Chi-Square analysis findings below show that while use of wearable technology devices was independent of country, social media usage was found to be more frequent and popular in Turkey compared to Germany. Sports culture, the degree to participate in sports activities, in particular, is found to be independent between countries following the same patterns. However, when further investigated it is found that on a weekly basis, German people are more involved in sports activities.

Table 1. Chi-square tests

Country * Use of Wearable Technology Devices		Country * Social Media Usage		Country * Sports Frequency		Country * Weekly Sports Involvement	
Chi-Square	Significance (2-Sided)	Chi-Square	Significance (2-Sided)	Chi-Square	Significance (2-Sided)	Chi-Square	Significance (2-Sided)
0.369	0.543	20.843	0.000	9.309	0.054	11.638	0.009

Hypothesis Testing

In order to test the reliability and trustworthiness of the instruments used in this study, Cronbach's alpha reliability test has been applied to the data collected from the surveys. Table 2 presents the results which show that intention (α=0.946), attitude (α=0.895), and perceived usefulness (α=0.761) are found to have the highest internal reliability. These are followed by social influence (α=0.630) and availability (α=0.615). The items, perceived ease of use (α=0.538) and mobility (α=0.505) have lower reliability than expected. This relatively poorer internal reliability results of perceived ease of use and mobility might be resulting from the number of items used to measure the variable or sample size.

Table 2. Reliability analysis of the scales

Name of the Variable	Number of Items	Cronbach's Alpha
Perceived Usefulness	3	0.761
Social Influence	2	0.630
Perceived Ease of Use	2	0.538
Mobility	2	0.505
Availability	2	0.615
Attitude	3	0.895
Intention	2	0.946

Moving forwards, the first five hypothesis are tested to see the significance of the positive effects of variables on each other. Below, Table 3 and Table 4 summarize the findings and show that the hypothesized positive effects are all found to be significant. The analysis results show that the hypotheses have statistical significance both overall and at the specific country level. We can summarize the results as:

- Perceived ease of use has a positive effect on perceived usefulness of and attitude towards wearable technology devices (H1 and H2)
- Perceived usefulness has a positive effect on attitude and intention towards wearable technology devices (H3 and H4)
- Attitude has a positive effect on intention to use wearable technology devices (H5)

Moving on to the next set of hypotheses, when the effect of country difference was investigated through independent samples tests, while there seems to be no effect of the country on the perceived usefulness of these devices (H6), it was found that the country difference had an effect on attitude towards wearable technology devices (H7). The results are gathered in Table 5.

With an aim to confirm the positive effect of social influence on perceived usefulness of wearable technology devices (H8), simple linear regression analysis was conducted. The results are summarized in Table 6. Model summary shows that 12.3% of the variation on the perceived usefulness of wearable technology devices can be attributed to social influence. This hypothesis is statistically supported. Similarly, results in Table 7 show that social influence is found to be positively affecting the attitude towards

Table 3. Regression model summary

Hypotheses	R Square	F	Sig.	Turkey (n = 101)		Germany (n = 120)	
				R Square	Sig.	R Square	Sig.
H1: Perceived ease of use and Perceived usefulness	0.105	25.781	0.000	0.122	0.000	0.090	0.001
H2: Perceived ease of use and Attitude	0.151	38.859	0.000	0.153	0.000	0.104	0.000
H3: Perceived usefulness and Attitude	0.381	134.524	0.000	0.315	0.000	0.460	0.000
H4: Perceived usefulness and Intention	0.250	72.929	0.000	0.220	0.000	0.277	0.000
H5: Attitude and Intention	0.547	264.939	0.000	0.520	0.000	0.518	0.000

Table 4. Regression coefficients

Hypotheses	Constant	Beta	Std. Error
H1: Perceived ease of use and Perceived usefulness	0.105	25.781	0.000
H2: Perceived ease of use and Attitude	0.151	38.859	0.000
H3: Perceived usefulness and Attitude	0.381	134.524	0.000
H4: Perceived usefulness and Intention	0.250	72.929	0.000
H5: Attitude and Intention	0.547	264.939	0.000

Table 5. Group statistics and independent samples tests

		Group statistics	Independent samples t-test for equality of means		
		Mean	t	Std. Error Difference	Sig (2-tailed)
H6: Perceived Usefulness	Turkey (n = 101)	3.3663	0.826	0.11731	0.410
	Germany (n = 120)	3.2694			
H7: Attitude	Turkey (n = 101)	3.5710	4.819	0.12310	0.000
	Germany (n = 120)	2.9778			

Table 6. Regression results – social influence and perceived usefulness

Model	Unstandardized Coefficients	Sig.	Model Summary
(Constant)	2.633	0.000	R Square: 0.123
Social Influence	0.320	0.000	F = 30.841

Table 7. Regression results – social influence and attitude

Model	Unstandardized Coefficients	Sig.	Model Summary
(Constant)	2.183	0.000	R Square: 0.240
Social Influence	0.502	0.000	F = 69.115

Table 8. Regression results – mobility and availability on perceived ease of use

Model	Unstandardized Coefficients		Standardized Coefficients	t	Sig.
	B	Std. Error	Beta		
(Constant)	1.886	0.213		8.849	0.000
Availability	0.108	0.056	0.131	1.934	0.054
Mobility	0.362	0.059	0.413	6.098	0.000

Adjusted R Square: 0.233 – F: 34.130 – Significance: 0.000
Dependent Variable: Perceived Ease of Use

wearable technology devices (H9). Model summary shows that almost one quarter of variations on the attitude towards wearable technology devices results from social influence.

When we examine the effect of mobility and availability on perceived ease of use, these two variables together are found to be explaining almost one fourth of the variations on the perceived ease of use of wearable technology devices with an R Square of .240. However, mobility has a significant effect on the perceived ease of use of wearable technology devices (H10), whereas availability is not found to have a positive effect on perceived ease of use (H11). On the other hand, when only measured in Turkey, availability was found to be significantly affecting perceived ease of use. Below findings in Table 8 represent the analysis.

CONCLUSION

First contribution of the study is that technology acceptance model of Davis (1989) is proved to be valid and applicable for wearable technology products. This study has shown that perceived ease of use has an effect on perceived usefulness, the more a consumer perceives a smartwatch as easy to use the more likely that they will also perceive this product as useful. Moreover, when a product is found to be easy to use, consumers also build a positive attitude towards it. Validating the results of Davis (1989), the current study has affirmed that perceived usefulness of smartwatches effects both attitude towards them and the intention to own and use a smartwatch. The strongest and the most significant relationship is found to be between attitude and intention, again validating the results on the original TAM of Davis (1989). The respondents of this study had a tendency to fully associate intention with attitude, when positive attitudes are formed towards smartwatches the intention to buy one increases respectively.

In this cross-cultural study, we have also investigated the country effect on perceived usefulness and attitude. Country differentials were sports culture, life expectancy and uncertainty avoidance which differ between the two countries that comprise our sample, Germany and Turkey. According to the findings, while perceived usefulness did not differ in Germany in Turkey, attitude on the other hand is found to be shaped by the country and thus culture. When means are compared and the results of the independent samples t-tests are taken into consideration, it is found that Turkish people hold a more positive attitude towards smartwatches than the Germans.

Social influence received from the environment is found to have a positive relationship with perceived usefulness of smartwatches and attitude towards them. This relationship is stronger between social influence and attitude compared to social influence and perceived usefulness. The study shows that when people receive positive influence about a smartwatch from their environment, there is higher likelihood of them to shape a more positive attitude towards smartwatches, which than leads to positive intention to use and purchase one. More interestingly, our study has revealed that Turkish people are more likely to be affected by social influence and they form a more positive attitude towards smartwatches. Furthermore, the social media usage in both cultures were also measured, suspecting a potential positive correlation with smartwatch usage and social media usage. Results have proven our suspects correct, as there is a more positive attitude towards smartwatches in Turkey with much higher social media usage. The more consumers are exposed to positive social influence on smartwatch usage and the more they are active on social media, the stronger positive attitudes are formed.

Last but not the least, as mobility is a very popular and essential consideration in our age, our study also investigated the relationship between mobility provided by smartwatches and their perceived ease of use. Our study results show that while mobility has a significant positive relation with the perceived usefulness of smartwatches, whereas availability has no relationship. However, when the results were analyzed at the country level, it was observed that Turkish respondents positively associated smartwatches' perceived usefulness and their availability. Overall, according to the statistical findings of this study, mobility is one of the features shaping the perception towards usefulness of smartwatches. Consumers who think a smartwatch has good mobility, also think that it is useful.

This study contributes to a limited body and extent of research, as academic studies on smartwatches are still rare. In addition to once more validating the well-known and widely used TAM model, this study particularly introduces sports culture, life expectancy and uncertainty avoidance all under country effect, measuring their potential relationship with acceptance of smartwatches. Availability and mobility variables are also used as a part of the extended model, through which new independent factors are added, revealing more findings on consumer acceptance patterns for technology products and smartwatches in particular.

The study also has a contribution towards the industry as the insights and findings lead us to make further comments on the overall effect of consumer behavior on adaption and purchasing decision making, especially for a disruptive technology product. To some extent, managers of technology companies who chase after growth in their businesses in terms of acquiring bigger market share and creating further demand can use this study as an exploratory tool that allows them with insights and findings on consumer adaptation to technology devices and intention to own one. Furthermore, by getting closer to exploring the factors behind wearable technology acceptance and the effect of culture, companies may be able to understand the consumers better and serve them accordingly, with customized features and characteristics per culture and target group. Despite the fact that the focus of this research is on smartwatches, managerial implications are believed to be transferrable to other wearable technology devices in general.

When interpreting this study, it should be kept in might that the study has limitations and room for future research. As a limitation, we can present the differences between the sample groups in Turkey and Germany. Further, the ability to show this study as a pure comparison between Turkish and German culture is limited due to all samples being students from selected universities, which result in the decreased representation of the overall culture. In potential research, the scale of this study can be enlarged and made more representative by having more homogenous samples. In addition, there exists potential to explore more about consumer acceptance of other innovative products and services, and the research instrument can be tailored and extended to fit other industries.

REFERENCES

Ajzen, I. (1991). The theory of planned behavior. *Organizational Behavior and Human Decision Processes*, *50*(2), 179–211. doi:10.1016/0749-5978(91)90020-T

Ajzen, I., & Fishbein, M. (1980). *Understanding attitudes and predicting social behavior*. Prentice Hall.

Akbulut, F., & Akan, A. (2015). Smart wearable patient tracking systems. In *Medical Technologies National Conference*. Bodrum: IEEE.

Ariyatum, B., & Holland, R. (2003). A strategic approach to new product development in smart clothing. *Proceedings of the 6th Asian Design Conference*, 70.

Bajarin, T. (2014). *3 Things Smartwatches Need to Be Ready for Prime Time*. https://time.com/6545/3-things-smartwatches-need-to-be-ready-for-prime-time/

Belk, R. (2013). Extended self in a digital world. *The Journal of Consumer Research*, *40*(3), 477–500. doi:10.1086/671052

Brem, A., Ro, Y., & Rauschnabel, P. (2006). Augmented reality smart glasses: An investigation of technology acceptance drivers. *International Journal of Technology Marketing*, *11*(2), 123–148.

Chuah, S.-W., Rauschnabel, P., Krey, N., Nguyen, B., Ramayah, T., & Lade, S. (2016). Wearable technologies: The role of usefulness and visibility in smartwatch adoption. *Computers in Human Behavior*, *65*, 276–284. doi:10.1016/j.chb.2016.07.047

Davis, F. (1989). Perceived usefulness, perceived ease of use, and user acceptance of information technology. *Management Information Systems Quarterly*, *13*(3), 319–340. doi:10.2307/249008

Davis, F. (1993). User acceptance of information technology: System characteristics, user perceptions and behavioral impacts. *International Journal of Man-Machine Studies*, *38*(3), 475–487. doi:10.1006/imms.1993.1022

Dion, K., Berscheid, E., & Walster, E. (1972). What is beautiful is good. *Journal of Personality and Social Psychology*, *24*(3), 285–290. doi:10.1037/h0033731 PMID:4655540

Dunne, L. (2004). *The Design of Wearable Technology: Addressing The Human-Device Interface through Functional Apparel Design* (Unpublished master thesis). https://ecommons.cornell.edu/handle/1813/150

Dunne, L., Ashdown, S., & Smyth, B. (2005). Expanding garment functionality through embedded electronic technology. *Journal of Textile and Apparel Technology and Management*, *4*(3), 1–11.

European Commission. (2015). *Internet of Things: Wearable technology*. European Commission.

Fishbein, M., & Ajzen, I. (1975). *Belief, attitude, intention, and behavior: An introduction to theory and research*. Addison-Wesley Pub. Co.

Hanna, N., & Wozniak, R. (2017). *Consumer behavior an applied approach*. Kendall Hunt Pub Co.

Hoyer, W., & MacInnis, D. (2007). Consumer behavior (4th ed.). Boston: Houghton Mifflin Co.

Hwang, C. (2014). *Consumers ' acceptance of wearable technology: Examining solar-powered clothing* (Unpublished master thesis). https://lib.dr.iastate.edu/cgi/viewcontent.cgi?article=4957&context=etd

Kim, K., & Shin, D.-H. (2014). An acceptance model for smart watches. *Internet Research, 25*(4), 527–541. doi:10.1108/IntR-05-2014-0126

Kim, K., & Sundar, S. (2015). Mobile persuasion: Can screen size and presentation mode make a difference to trust? *Human Communication Research, 42*(1), 45–70. doi:10.1111/hcre.12064

Kirstein, T., Cottet, D., Grzyb, J., & Tröster, G. (2005). Wearable computing systems – electronic textiles. In T. Xiaoming (Ed.), *Wearable electronics and photonics*. Woodhead Publishing.

Ko, F., El-Aufy, A., Lam, H., & Macdiarmid, A. (2005). Electrostatically generated nanofibres for wearable electronics. In X. Tao (Ed.), *Wearable Electronics and Photonics*. Woodhead Publishing.

Ma, Q., & Liu, L. (2004). The technology acceptance model: A meta-analysis of empirical findings. *Journal of Organizational and End User Computing, 16*(1), 59–72. doi:10.4018/joeuc.2004010104

Mills, A., Watson, R., Pitt, L., & Kietzmann, J. (2016). Wearing safe: Physical and informational security in the age of the wearable device. *Business Horizons, 59*(6), 615–622. doi:10.1016/j.bushor.2016.08.003

Mooij, M., & Hofstede, G. (2002). Convergence and divergence in consumer behavior: Implications for international retailing. *Journal of Retailing, 78*(1), 61–69. doi:10.1016/S0022-4359(01)00067-7

Nanda, S. (2017). A study on the customer awareness of wearable technology. *International Journal of Multidisciplinary and Current Research, 5*, 293–298.

Park, D.-H., Lee, J., & Han, I. (2014). The effect of on-line consumer reviews on consumer purchasing intention: The moderating role of involvement. *International Journal of Electronic Commerce, 11*(4), 125–148. doi:10.2753/JEC1086-4415110405

Rauschnabel, P., Brem, A., & Ro, Y. (2015). *Augmented Reality Smart Glasses: Definition, Conceptual Insights, and Managerial Importance*. Retrieved from Semantic Scholar: https://pdfs.semanticscholar.org/b2d8/bbbf1c654f8bb157ad83ffdc9543598c6724.pdf

Robson, K., Kietzmann, J., & Pitt, L. (2016). APC forum: Extending business values through wearables. *MIS Quarterly Executive, 15*(2), 169–179.

Rogers, E. (1962). *Diffusion of innovations*. The Free Press.

Schiffman, L., & Kanuk, L. (1978). *Consumer Behavior*. Prentice Hill.

Solomon, M. (2009). Consumer Behavior: Buying, Having, and Being (8th ed.). Prentice Hall.

Stinson, B. (2013). *Google Glass and its rival smart glasses: everything you need to know*. https://www.stuff.tv/features/google-glass-and-its-rival-smart-glasses-everything-you-need-know

Venkatesh, V., & Bala, H. (2008). Technology acceptance model 3 and a research agenda on interventions. *Journal of the Decision Science Institute*, *39*(2), 273–315. doi:10.1111/j.1540-5915.2008.00192.x

Venkatesh, V., & Davis, F. (2000). A theoretical extension of the technology acceptance model: Four longitudinal field studies. *Management Science*, *46*(2), 169–332. doi:10.1287/mnsc.46.2.186.11926

Venkatesh, V., Morris, M., Davis, G., & Davis, F. (2003). User acceptance of information technology: Toward a unified view. *Management Information Systems Quarterly*, *27*(3), 425–478. doi:10.2307/30036540

Venkatesh, V., Thong, J., & Xu, X. (2012). Consumer acceptance and use of information technology: Extending the unified theory of acceptance and use of technology. *Management Information Systems Quarterly*, *36*(1), 157–178. doi:10.2307/41410412

Verhoef, P., Stephen, A., Kannan, P., Luo, X., Abhishek, V., Andrews, M., Bart, Y., Datta, H., Fong, N., Hoffman, D. L., Hu, M. M., Novak, T., Rand, W., & Zhang, Y. (2017). Consumer connectivity in a complex, technology-enables, and mobile-oriented world with smart products. *Journal of Interactive Marketing*, *40*, 1–72. doi:10.1016/j.intmar.2017.06.001

KEY TERMS AND DEFINITIONS

Attitude: Positive or negative tendency of a consumer against a product or service.

Availability: Ability to operate when needed, connectedness.

Intention: Strength of purpose to perform a specified behavior.

Mobility: Ease of movement.

Perceived Ease of Use: The degree to which a consumer perceives a technology easy to use, product easy to operate or service easy to receive.

Perceived Usefulness: The extent to which a person believes that using a particular technology or tool will contribute to achieving their goal.

Smartwatch: Wearable technology device in the form of a watch that is in contact with a smart phone and allows users to track their activities through sensors transmitting information.

Wearable Technology: Any type of technology that is incorporated in electronics that can be worn on the body, either as an accessory or as part of materials used in clothing.

Compilation of References

Abdel-Basset, M., Manogaran, G., & Mohamed, M. (2018). Internet of Things (IoT) and its impact on supply chain: A framework for building smart, secure and efficient systems. *Future Generation Computer Systems*, *86*, 614–628. doi:10.1016/j.future.2018.04.051

Abdullah, M. S., & Keshminder, J. S. (2020). What drives green sukuk? A leader's perspective. *Journal of Sustainable Finance & Investment*, 1-21.

Abdullah, N., & Nayan, M. A. (2020). *Green Sukuk: Financing The Future To Sustainable Environment*. Academic Press.

Abdullah, L., Chan, W., & Afshari, A. (2019). Application of PROMETHEE method for green supplier selection: A comparative result based on preference functions. *Journal of Industrial Engineering International*, *15*(2), 271–285. doi:10.100740092-018-0289-z

Abishek, A., Bharathwaj, M., & Bhagyalaksgmi, L. (2016). Agriculture marketing using web and mobile based technologies. *Proceedings of the 2016 IEEE Technological Innovations in ICT for Agriculture and Rural Development (TIAR)*, 41-44. 10.1109/TIAR.2016.7801211

Abubakar, L., & Handayani, T. (2020, May). Green Sukuk: Sustainable Financing Instruments for Infrastructure Development in Indonesia. In *1st Borobudur International Symposium on Humanities,Economics and Social Sciences (BIS-HESS 2019)* (pp. 983-987). Atlantis Press.

Acs, Z. J., & Audretsch, D. B. (2003). Innovation and Technological Change. In Z. J. Acs & D. B. Audretsch (Eds.), *Handbook of Entrepreneurship Research* (pp. 55–79). Kluwer Academic Publishers. doi:10.1007/b105789

Adegbite, O. O., & Machethe, C. L. (2020). Bridging the financial inclusion gender gap in smallholder agriculture in Nigeria: An untapped potential for sustainable development. *World Development*, *127*, 104755.

Adolph, B., Allen, M., Beyuo, E., Banuoku, D., Barrett, S., Bourgou, T., Bwanausi, N., Dakyaga, F., Derbile, E. K., Gubbels, P., Hié, B., Kachamba, C., Naazie, G. K., Niber, E. B., Nyirengo, I., Tampulu, S. F., & Zongo, A. F. (2020). Supporting smallholders' decision making: Managing trade-offs and synergies for sustainable agricultural intensification. *International Journal of Agricultural Sustainability*, 1–18. doi:10.1080/14735903.2020.1786947

Ágnes, U. T. (2017). The Examination of the Importance of the Thermál Panzió in Mórahalom. *EBEEC Conference Proceedings The Economies of Balkan and Eastern Europe Countries in the Changed World*, 226. 10.18502/kss.v1i2.659

Ahi, P., & Searcy, C. (2015). Assessing sustainability in the supply chain: A triple bottom line approach. *Applied Mathematical Modelling*, *39*(10-11), 2882–2896. doi:10.1016/j.apm.2014.10.055

Ahmad, F., Zayed, N. M., & Harun, M. (2013). Factors behind the adoption of green banking by Bangladeshi commercial banks. *ASA University Review*, *7*(2), 241–255.

Ahmadi, H. B., Petrudi, S. H. H., & Wang, X. (2017). Integrating sustainability into supplier selection with analytical hierarchy process and improved grey relational analysis: A case of telecom industry. *International Journal of Advanced Manufacturing Technology*, *90*(9), 2413–2427. doi:10.100700170-016-9518-z

Aiginger, K. (2006). Competitiveness: From a dangerous obsession to a welfare creating ability with positive externalities. *Journal of Industry, Competition and Trade*, *6*(2), 161–177. doi:10.100710842-006-9475-6

AIM2Flourish. (2021). *BIOFASE: The solution from Avocado Seed*. Retrieved from https://aim2flourish.com/innovations/biofase-the-solution-from-avocado-seed

Ajzen, I. (1991). The theory of planned behavior. *Organizational Behavior and Human Decision Processes*, *50*(2), 1–63. doi:10.1016/0749-5978(91)90020-T

Ajzen, I., & Fishbein, M. (1980). *Understanding attitudes and predicting social behavior*. Prentice Hall.

Akbulut, F., & Akan, A. (2015). Smart wearable patient tracking systems. In *Medical Technologies National Conference*. Bodrum: IEEE.

Akcan, S., & Taş, M. A. (2019). Green supplier evaluation with SWARA-TOPSIS integrated method to reduce ecological risk factors. *Environmental Monitoring and Assessment*, *191*(12), 736. doi:10.100710661-019-7884-3 PMID:31707512

Al Amri, M. M., & Almaiah, M. A. (2020). The Use of Mobile Gamification Technology for Sustainability Learning in Saudi Higher Education. *International Journal of Advanced Trends in Computer Science and Engineering*, *9*(5).

Alam, N., Duygun, M., & Ariss, R. T. (2016). Green Sukuk: An Innovation in Islamic Capital Markets. In *Energy and Finance* (pp. 167–185). Springer. doi:10.1007/978-3-319-32268-1_10

Alber, N., & Dabour, M. (2020). The Dynamic Relationship between FinTech and Social Distancing under COVID-19 Pandemic: Digital Payments Evidence. *International Journal of Economics and Finance*, *12*(11), 109–109.

Aldás-Manzano, J., Lassala-Navarré, C., Ruiz-Mafé, C., & Sanz-Blas, S. (2009). The role of consumer innovativeness and perceived risk in online banking usage. *International Journal of Bank Marketing*, *27*(1), 53–75. doi:10.1108/02652320910928245

Alexander, N., & McKenna, A. (1998). Rural tourism in the heart of England. *International Journal of Contemporary Hospitality Management*, *10*(5), 203–207. doi:10.1108/09596119810227820

Alinvi, F. (2009). *Customers' Expectations of Banks Becoming Cashless-How could Banks meet customer's expectations when changing from cash services to deeper customer relationship?* Retrieved in July 20, 2016 from www.divaportal.org/smash/get/diva2:220556/FULLTEXT01.pdf

Allen, C., Metternicht, G., & Wiedmann, T. (2016). National pathways to the Sustainable Development Goals (SDGs): A comparative review of scenario modelling tools. *Environmental Science & Policy*, *66*, 199–207.

Allen, C., Metternicht, G., & Wiedmann, T. (2018). Initial progress in implementing the Sustainable Development Goals (SDGs): A review of evidence from countries. *Sustainability Science*, *13*(5), 1453–1467.

Almeida, F., Oliveira, J., & Cruz, J. (2011). Open Standards and Open Source: Enabling Interoperability. *International Journal of Software Engineering and Its Applications*, *2*(1), 1–11.

Anggraini, Y. (2018). Peran green sukuk dalam memperkokoh posisi Indonesia di pasar keuangan syariah global. *El-Barka: Journal of Islamic Economics and Business*, *1*(2), 251–268. doi:10.21154/elbarka.v1i2.1453

Ankit, S. (2011). Factors Influencing Online Banking Customer Satisfaction and Their Importance in Improving Overall Retention Levels: An Indian Banking Perspective. *Information and Knowledge Management*, *1*(1), 45–54.

Annosi, M. C., Brunetta, F., Bimbo, F., & Kostoula, M. (2021). Digitalization within food supply chains to prevent food waste. Drivers, barriers and collaboration practices. *Industrial Marketing Management, 83*, 208–220. doi:10.1016/j.indmarman.2021.01.005

Antonucci, T. C., Ajrouch, K. J., & Manalei, J. A. (2017). Social Relations and Technology: Continuity, Context, and Change. *Innovation in Aging, 1*(3), 1–9. doi:10.1093/geroni/igx029 PMID:29795794

Anugrahaeni, P. (2017). Analysis of Indonesian Sovereign Green Bond and Green Sukuk Initiatives. *Kajian Ekonomi dan Keuangan, 1*(1), 1-24.

Arable. (2021). *How we turn data into insights*. Retrieved from https://arable.com/

Aragon-Sanchez, A., Baixauli-Soler, S., & Carrasco-Hernandez, A. J. (2017). A missing link: The behavioral mediators between resources and entrepreneurial intentions. *International Journal of Entrepreneurial Behaviour & Research, 23*(5), 752–768. Advance online publication. doi:10.1108/IJEBR-06-2016-0172

Ariyatum, B., & Holland, R. (2003). A strategic approach to new product development in smart clothing. *Proceedings of the 6th Asian Design Conference*, 70.

Armitage, C. J., & Conner, M. (2001). Efficacy of the theory of planned behavior: A meta-analytic review. *British Journal of Social Psychology, 40*(4), 471–499. doi:10.1348/014466601164939 PMID:11795063

Armstrong, H., & Taylor, J. (2006). *Regional Economics and Policy*. Blackwell Publishing.

Arumugam.D.,& Chirute.T.(2018). Factors determining the adoption of green banking amongst commercial banks in Malaysia. *Electronic Journal of Business & Management,* 50 - 62.

Arvanitis, K. G., & Symeonaki, E. G. (2020). Agriculture 4.0: The Role of Innovative Smart Technologies Towards Sustainable Farm Management. *The Open Agriculture Journal, 14*(1), 130–135. doi:10.2174/1874331502014010130

Asadi, S., Nilashi, M., Samad, S., Abdullah, R., Mahmoud, M., Alkinani, M. H., & Yadegaridehkordi, E. (2021). Factors impacting consumers' intention toward adoption of electric vehicles in Malaysia. *Journal of Cleaner Production, 282*(124474).

Aschemann-Witzel, J., De Hooge, I., Amani, P., Bech-Larsen, T., & Oostindjer, M. (2015). Consumer-related food waste: Causes and potential for action. *Sustainability, 7*(6), 6457–6477. doi:10.3390u7066457

Asunta, T. (2003). *Knowledge of environmental issues. where pupils acquire information opinions, and how it affects their attitudes, and laboratory behavior*. University of Jyväskylä.

Atkinson, A.B., & Stiglitz, J.E. (2015). *Lectures on Public Economics Updated edition*. Economics Books, Princeton University Press.

Audate, P., Fernandez, M., Cloutier, G., & Lebel, A. (2018). Impacts of Urban Agriculture on the Determinants of Health: Scoping Review Protocol. *JMIR Research Protocols, 7*(3), e89. doi:10.2196/resprot.9427 PMID:29588270

Austmann, L. M. (2020). Drivers of the electric vehicle market: A systematic literature review of empirical studies. *Finance Research Letters*, 1–10.

Azhgaliyeva, D., Kapoor, A., & Liu, Y. (2020). Green bonds for financing renewable energy and energy efficiency in South-East Asia: A review of policies. *Journal of Sustainable Finance & Investment, 10*(2), 113–140. doi:10.1080/20430795.2019.1704160

Babu, M. M., Dey, B. L., Rahman, M., Roy, S. K., Alwi, S. F. S., & Kamal, M. M. (2020). Value co-creation through social innovation: A study of sustainable strategic alliance in telecommunication and financial services sectors in Bangladesh. *Industrial Marketing Management*, *89*, 13–27.

Bacco, M., Barsocchi, P., Ferro, E., Gotta, A., & Ruggeri, M. (2019). The digitisation of agriculture: a survey of research activities on smart farming. *Array*, 3-4.

Badia-Melis, R., Mishra, P., & Ruiz-García, L. (2015). Food traceability: New trends and recent advances. A review. *Food Control*, *57*, 393–401. doi:10.1016/j.foodcont.2015.05.005

Bag, S., Wood, L. C., Xu, L., Dhamija, P., & Kayikci, Y. (2020). Big data analytics as an operational excellence approach to enhance sustainable supply chain performance. *Resources, Conservation and Recycling*, *153*, 104559. doi:10.1016/j.resconrec.2019.104559

Bai, C., Dallasega, P., Orzes, G., & Sarkis, J. (2020). Industry 4.0 technologies assessment: A sustainability perspective. *International Journal of Production Economics*, *229*, 107776.

Bajarin, T. (2014). *3 Things Smartwatches Need to Be Ready for Prime Time.* https://time.com/6545/3-things-smartwatches-need-to-be-ready-for-prime-time/

Bakeshlou, E. A., Khamseh, A. A., Asl, M. A. G., Sadeghi, J., & Abbaszadeh, M. (2017). Evaluating a green supplier selection problem using a hybrid MODM algorithm. *Journal of Intelligent Manufacturing*, *28*(4), 913–927. doi:10.100710845-014-1028-y

Bamberg, S. (2003). How does environmental concern influence specific environmentally related behaviors? A new answer to an old question. *Journal of Environmental Psychology*, *23*(1), 21–32. doi:10.1016/S0272-4944(02)00078-6

Banaeian, N., Mobli, H., Fahimnia, B., Nielsen, I. E., & Omid, M. (2018). Green supplier selection using fuzzy group decision making methods: A case study from the agri-food industry. *Computers & Operations Research*, *89*, 337–347. doi:10.1016/j.cor.2016.02.015

Barney, J. (1991). Firm Resources and Sustained Competitive Advantage. *Journal of Management*, *17*(1), 99–120. doi:10.1177/014920639101700108

Barney, J. (2001). Is the resource-based "view" a useful perspective for strategic management research? *Academy of Management Review*, *26*(1).

Barreto, L., Amaral, A., & Pereria, T. (2017). Industry 4.0 implications in logistics: an overview. *Proceedings of the Manufacturing Engineering Society International Conference MESIC 2017.*)10.1016/j.promfg.2017.09.045

Barrows, G., Sexton, S., & Zilberman, D. (2014). Agricultural Biotechnology: The Promise and Prospects of Genetically Modified Crops. *The Journal of Economic Perspectives*, *28*(1), 99–120. doi:10.1257/jep.28.1.99

Battaglia, M., Gragnani, P., & Annesi, N. (2020). Moving Businesses toward Sustainable Development Goals (SDGs): Evidence from an Italian "Benefit-For-Nature" Corporation. *Entrepreneurship Research Journal*, *20190305*(4). Advance online publication. doi:10.1515/erj-2019-0305

Bayar, Y. (2017). Individual Pension Funds and Capital Market Development in Turkey. *Review of Economic and Business Studies*, *9*(2), 95–109. doi:10.1515/rebs-2016-0036

Beckhard, R. (1975). Strategies for Large System Change. *Sloan Management Review*, *16*(2).

Behsudi, A. (2020). Wish you were here - tourism-dependent economies are among those harmed the most by the pandemic. *Finance & Development*. https://www.imf.org/external/pubs/ft/fandd/2020/12/pdf/impact-of-the-pandemic-on-tourism-behsudi.pdf

Beier, G., Niehoff, S., & Xue, B. (2018). More sustainability in industry through industrial internet of things? *Applied Sciences (Basel, Switzerland)*, *8*(2), 219.

Bekun, F. V., Alola, A. A., & Sarkodie, S. A. (2019). Toward a sustainable environment: Nexus between CO2 emissions, resource rent, renewable and nonrenewable energy in 16-EU countries. *The Science of the Total Environment*, *657*, 1023–1029. doi:10.1016/j.scitotenv.2018.12.104 PMID:30677870

Belakeri, P., Prasad, C., Bajantri, S., Mahantesh, M., Maruthi, S., & Rudresh, G. (2017). Trends of Mobile Applications in Farming. *International Journal of Current Microbiology and Applied Sciences*, *6*(7), 2499–2512. doi:10.20546/ijcmas.2017.607.295

Belij, M. S. (2016). *Transformation of Serbian spas into modern centers of spa and wellness tourism: dissertation*. Faculty of Geography, University of Belgrade.

Belk, R. (2013). Extended self in a digital world. *The Journal of Consumer Research*, *40*(3), 477–500. doi:10.1086/671052

Bihari, M. (2021). *Using Health Information Technology to information management*. https://bs.approby.com/koriscenje-zdravstvene-informacione-tehnologije-za-upravljanje-vasim-informacijama/

Bihari, S. C., & Pandey, B. (2015). Green banking in India. *Journal of Economics and International Finance*, *7*(1), 1–17. doi:10.5897/JEIF2014.0599

Biswas, N. (2011). Sustainable Green Banking Approach: The Need of the Hour. *Business Spectrum*, *1*, 32–38.

Blakstad, S., & Allen, R. (2018). Green Fintech. In *FinTech Revolution* (pp. 183–199). Palgrave Macmillan.

Blomsma, F., Pieroni, M., Kravchenko, M., Pigosso, D. C., Hildenbrand, J., Kristinsdottir, A. R., ... Li, J. (2019). Developing a circular strategies framework for manufacturing companies to support circular economy-oriented innovation. *Journal of Cleaner Production*, *241*, 118271. doi:10.1016/j.jclepro.2019.118271

Boardman, A. E., Greenberg, D. H., Vining, A. R., & Weimer, D. L. (2018). *Cost-Benefit Analysis. Concepts and Practice* (5th ed.). Cambridge University Press. doi:10.1017/9781108235594

Bodeker, G., & Cohen, M. (2008). Understanding the Global Spa Industry. Elsevier.

Boks, C. (2006). The soft side of ecodesign. *Journal of Cleaner Production*, *14*(15-16), 1346–1356. doi:10.1016/j.jclepro.2005.11.015

Bolos, L. A., Lagerkvist, C. J., & Nayga, R. M. (2019). Consumer choice and food waste. *Choices*, *34*(1), 1–7.

Bondarenko, T. G., Zhdanova, O. A., & Klimova, N. (2019). Multi-Criteria Mechanism for Selecting Projects by Fintech Accelerators. In *International Science and Technology Conference" FarEastCon"(ISCFEC 2019)* (pp. 6-10). Atlantis Press.

Boratyńska, K. (2019). Impact of digital transformation on value creation in Fintech services: An innovative approach. *Journal of Promotion Management*, *25*(5), 631–639.

Borhan, N. A., & Ahmad, N. (2018). Identifying the determinants of Malaysian corporate Sukuk rating. *International Journal of Islamic and Middle Eastern Finance and Management*.

Bossle, M. B., de Barcellos, M. D., Vieira, L. M., & Sauvée, L. (2016). The drivers for adoption of eco-innovation. *Journal of Cleaner Production*, *113*, 861–872. doi:10.1016/j.jclepro.2015.11.033

Bouma, J. J., Jeucken, M., & Klinkers, L. (2001). *Sustainable Banking: The Greening of Finance*. Greenleaf Publishing Limited.

Bouscasse, H., Joly, I., & Bonnel, P. (2018). How does environmental concern influence mode choice habits? A mediation analysis. *Transportation Research Part D, Transport and Environment*, *59*, 205–222. doi:10.1016/j.trd.2018.01.007

Boutti, R., & El Mosaid, F. (2015). Sukuk or The Green Sukuk: A Promising Opportunity For The Financing of Renewable Energies. *Revue de Recherches en Economie et en Management Africain*, *2*(2).

Braczyk, H.-J., Cooke, P., & Heidenreich, M. (1998). *Regional Innovation Systems*. UCL Press.

Braun, A. T., Colangelo, E., & Steckel, T. (2018). Farming in the Era of Industrie 4.0. *Procedia CIRP*, *72*, 979–984. doi:10.1016/j.procir.2018.03.176

Brem, A., Ro, Y., & Rauschnabel, P. (2006). Augmented reality smart glasses: An investigation of technology acceptance drivers. *International Journal of Technology Marketing*, *11*(2), 123–148.

Bronisz, U., Heijman, W., & Miszczuk, A. (2008). Regional competitiveness in Poland: Creating an index. *Jahrbuch für Regionalwissenschaft*, *28*(2), 133–143. doi:10.100710037-008-0026-y

Brooksbank, D. J., & Pickernell, D. G. (1999). Regional competitiveness indicators: A reassessment of method. *Local Economy*, *13*(4), 310–326. doi:10.1080/02690949908726459

Brophey, G., Baregheh, A., & Hemsworth, D. (2013). Innovation Process, Decision-Making, Perceived Risks And Metrics: A Dynamics Test. *International Journal of Innovation Management*, *17*(3), 1–22. doi:10.1142/S1363919613400148

Brough, A. R., Wilkie, J. E., Ma, J., Isaac, M. S., & Gal, D. (2016). Is eco-friendly unmanly? the green-feminine stereotype and its effect on sustainable consumption. *The Journal of Consumer Research*, *43*(4), 567–582. doi:10.1093/jcr/ucw044

Burgueño Salas, E. (2020). *Volumen de producción mensual de fertilizantes fosfatados en México de enero de 2019 a julio de 2020 (en toneladas métricas)*. Statista. Retrieve on the 2 of July 2021 from https://es.statista.com/estadisticas/594554/volumen-de-produccion-de-fertilizantes-superfosfatados-en-mexico/

Burton, R. M. (2013). An Interpretative Synthesis in Three Themes. *Journal of Organization Design*, *1*(2), 42–44. doi:10.7146/jod.7363

Bušelić, M. (2007). Knowledge and competitiveness. Pula, Croatia: University "Juraj Dobrila" in Pula, Department of Economics and Tourism "Dr. Mijo Mirkovic". (in Croatian)

Büyüközkan, G., & Vardaloğlu, Z. (2008). Yeşil tedarik zinciri yönetimi. *Lojistik Dergisi*, *8*, 66–73.

Cady, S. H., Jacobs, R. J., Koller, R., & Spalding, J. (2014). The Change Formula: Myth, Legend, or Lore? *OD Practitioner*, *46*(3).

Cafer, E. B., Cengiz, K., & Da, R. (2003). Fuzzy group decision making for selection among computer integrated manufacturing systems. *Computers in Industry*, *51*(1), 13–29. doi:10.1016/S0166-3615(03)00029-0

Caiado, R. G., Filho, W. L., Quelhas, O. L. G., de Mattos, D. L., & Avila, L. V. (2018). A literature-based review on potentials and constraints in the implementation of the sustainable development goals. *Journal of Cleaner Production*, *198*, 10.

Çalık, A. (2021). A novel Pythagorean fuzzy AHP and fuzzy TOPSIS methodology for green supplier selection in the Industry 4.0 era. *Soft Computing*, *25*(3), 2253–2265. doi:10.100700500-020-05294-9

Candraningrat, I., Abundanti, N., Mujiati, N., & Erlangga, R. (2021). The role of financial technology on development of MSMEs. *Accounting*, *7*(1), 225–230.

Cantú, P. (2008). Desarrollo sustentable: conceptos y reflexiones. Universidad Autónoma de Nuevo León. Colección: Tendencias Científicas.

Cao, Q., Wu, J., & Liang, C. (2015). An intuitionsitic fuzzy judgement matrix and TOPSIS integrated multi-criteria decision making method for green supplier selection. *Journal of Intelligent & Fuzzy Systems*, *28*(1), 117–126. doi:10.3233/IFS-141281

Carley, S., Krause, R., Lane, B., & Graham, J. (2013). Intent to purchase a plug-in electric vehicle: A survey of early impressions in large US cities. *Transportation Research Part D, Transport and Environment*, *18*, 39–45. doi:10.1016/j.trd.2012.09.007

Carree, M. A., & Thurik, A. R. (2010). The impact of entrepreneurship on economic growth. In J. Zoltan & D. Audretsch (Eds.), *Handbook of entrepreneurship research.* Springer. doi:10.1007/978-1-4419-1191-9_20

Carrillo-Hermosilla, J., del Río, P., & Könnölä, T. (2010). Diversity of eco-innovations: Reflections from selected case studies. *Journal of Cleaner Production*, *18*(10), 1073–1083. doi:10.1016/j.jclepro.2010.02.014

Castro, A. (1998). *Hacia el desarrollo sostenible y la ecoeficiencia: integración de las normas ISO 9000, ISO 14000 e ISO 18000 diseño de un sistema de gestión ambiental certificable* (Tesis de maestría inédita). Obtenido de Facultad Regional de Buenos Aires: http://posgrado.frba.utn.edu.ar/investigacion/tesis/MIA-1998-Castro.pdf

Cecere, G., Corrocher, N., & Mancusi, M. L. (2018). Financial constraints and public funding of eco-innovation: Empirical evidence from European SMEs. *Small Business Economics*, •••, 1–18.

Central Bank of Turkey. (2020). *Financial Stability Report.* Author.

Cerri, J., Testa, F., & Rizzi, F. (2018). The more I care, the less I will listen to you: How information, environmental concern and ethical production influence consumers' attitudes and the purchasing of sustainable products. *Journal of Cleaner Production*, *175*, 343–353. doi:10.1016/j.jclepro.2017.12.054

Chan, T. S. (1996). Concerns for environmental issues and consumer purchase preferences: A two-country study. *Journal of International Consumer Marketing*, *9*(1), 43–55. doi:10.1300/J046v09n01_04

Chatnani, N. N. (2018). Receivables management and supply chain finance for MSMEs: Analysis of TREDS. *Academy of Strategic Management Journal*, *17*(3), 1–8.

Chatterjee, K., Pamucar, D., & Zavadskas, E. K. (2018). Evaluating the performance of suppliers based on using the R'AMATEL-MAIRCA method for green supply chain implementation in electronics industry. *Journal of Cleaner Production*, *184*, 101–129. doi:10.1016/j.jclepro.2018.02.186

Chen, C. F., de Rubens, G. Z., Noel, L., Kester, J., & Sovacool, B. K. (2020). Assessing the socio-demographic, technical, economic and behavioral factors of Nordic electric vehicle adoption and the influence of vehicle-to-grid preferences. *Renewable & Sustainable Energy Reviews*, *121*, 109692. doi:10.1016/j.rser.2019.109692

Chen, C. T. (2000). Extensions of the TOPSIS for group decision-making under fuzzy environment. *Fuzzy Sets and Systems*, *111*(1), 1–9. doi:10.1016/S0165-0114(97)00377-1

Cheng, F., Lin, M., Yüksel, S., Dincer, H., & Kalkavan, H. (2020). A hybrid hesitant 2-tuple IVSF decision making approach to analyze PERT-based critical paths of new service development process for renewable energy investment projects. *IEEE Access.*

Chen, J., Zeng, S., & Zhang, C. (2018). An OWA distance-based, single-valued neutrosophic linguistic topsis approach for green supplier evaluation and selection in low-carbon supply chains. *International Journal of Environmental Research and Public Health*, *15*(7), 1439. doi:10.3390/ijerph15071439 PMID:29986549

Chen, S. J., Hwang, C. L., Beckmann, M. J., & Krelle, W. (1992). *Fuzzy multiple attribute decision making: methods and applications.* Springer-Verlag. doi:10.1007/978-3-642-46768-4

Chen, T., & Richard, R. (2014). Forecasting the productivity of a virtual enterprise by agent-based fuzzy collaborative intelligence—With Facebook as an example. *Applied Soft Computing*, *24*, 511–521. doi:10.1016/j.asoc.2014.08.003

Chen, W. (2016). H. M., Chou, S. Y., Luu, Q. D., & Yu, T. H. K. (2016). A fuzzy MCDM approach for green supplier selection from the economic and environmental aspects. *Mathematical Problems in Engineering*, *2016*, 1–10. Advance online publication. doi:10.1155/2016/8097386

Chen, Y. J. (2011). Structured methodology for supplier selection and evaluation in a supply chain. *Information Sciences*, *181*(9), 1651–1670. doi:10.1016/j.ins.2010.07.026

Chhipa, H. (2017). Nanofertilizers and nanopesticides for agriculture. *Environmental Chemistry Letters*, *15*(1), 15–22. doi:10.100710311-016-0600-4

Chik, M. N. (2012). *Sukuk: Shariah guidelines for Islamic bonds.* Centre for Banking Studies, Central Bank of Sri Lanka.

Chuah, S.-W., Rauschnabel, P., Krey, N., Nguyen, B., Ramayah, T., & Lade, S. (2016). Wearable technologies: The role of usefulness and visibility in smartwatch adoption. *Computers in Human Behavior*, *65*, 276–284. doi:10.1016/j.chb.2016.07.047

Chumbler, M., Negro, S., & Bechler, L. (2016). *Urban Agriculture: Policy, Law, Strategy, and Implementation.* American Bar Association.

Chu, W., Baumann, C., Hamin, H., & Hoadley, S. (2018). Adoption of environment-friendly cars: Direct vis-à-vis mediated effects of government incentives and consumers' environmental concern across global car markets. *Journal of Global Marketing*, *31*(4), 282–291. doi:10.1080/08911762.2018.1456597

Cicculllo, F., Caglianoa, R., Bartezzaghia, G., & Perego, A. (2021). Implementing the circular economy paradigm in the agri-food supply chain: The role of food waste prevention technologies. *Resources, Conservation and Recycling*, *164*, 105114. doi:10.1016/j.resconrec.2020.105114

Coffman, M., Bernstein, P., & Wee, S. (2017). Electric vehicles revisited: A review of factors that affect adoption. *Transport Reviews*, *37*(1), 79–93. doi:10.1080/01441647.2016.1217282

Cohen, S. (2017). *The Sustainable City.* Columbia University Press. doi:10.7312/cohe18204

Collart, A. J., & Interis, M. G. (2018). Consumer imperfect information in the market for expired and nearly expired foods and implications for reducing food waste. *Sustainability*, *10*(11), 3835. doi:10.3390u10113835

Collste, D., Pedercini, M., & Cornell, S. E. (2017). Policy coherence to achieve the SDGs: Using integrated simulation models to assess effective policies. *Sustainability Science*, *12*(6), 921–931. PMID:30147764

Commission of the European Communities. (1999). *Regional Policy and Cohesion Sixth Periodic Report on the Social and Economic Situation and Development of the Regions of the European Union.* Commission of the European Communities.

Cong, L. W., & He, Z. (2019). Blockchain disruption and smart contracts. *Review of Financial Studies*, *32*(5), 1754–1797.

Consejo Mundial Empresarial para el Desarrollo Sostenible. (1991). Ecoeficiencia y sustentabilidad. *Ciencia UANL, 35.*

Constantin, D. L., & Banica, G. C. (2007, August). Human Resources and Regional Competitiveness. *The Case of Romania.* Paper presented at the meeting of the European Regional Science Association, Paris, France.

Cosh, A., Cumming, D., & Hughes, A. (2009). Outside enterpreneurial capital. *Economic Journal (London), 119*(540), 1494–1533. doi:10.1111/j.1468-0297.2009.02270.x

Costantini, V., Crespi, F., Martini, C., & Pennacchio, L. (2015). Demand-pull and technology-push public support for eco-innovation: The case of the biofuels sector. *Research Policy, 44*(3), 577–595. doi:10.1016/j.respol.2014.12.011

Costopoulou, C., Ntaliani, M., & Karetsos, S. (2016). Studying Mobile Apps for Agriculture. *IOSR Journal of Mobile Computing and Application, 3*(6), 44–49.

COVID-19 Crisis in Serbia. (2020). *OECD.* https://www.oecd.org/south-east-europe/COVID-19-Crisis-in-Serbia.pdf

COVID-19 Crisis in the Western Balkans: economic impact, policy responses and short-term sustainable solutions. (2020). *OECD.* https://www.oecd.org/south-east-europe/COVID-19-Crisis-Response-Western-Balkans.pdf

Croatian Chamber of Commerce. (2019). *Foreign trade of counties in 2019 - provisional data.* Croatian Chamber of Commerce. (in Croatian)

Croatian Chamber of Commerce. (2019). *Foreign trade of counties with foreign countries in 2019 - provisional data.* Croatian Chamber of Commerce. (in Croatian)

Csapó, J., & Marton, G. (2017). The Role and Importance of Spa and Wellness Tourism in Hungary'sTourism Industry. *Czech Journal of Tourism, 6*(1), 55–68. doi:10.1515/cjot-2017–0003

Cupic, M., Tummala, R., & Suknovic, M. (2001). *Decision Making: Formal Approach.* SZR Zuhra.

Currie, B. A., French, A. D., & Ülkü, M. A. (2020). Sustainability, Big Data, and Consumer Behavior: A Supply Chain Framework. In I. Rahimi, A. H. Gandomi, S. J. Fong, & M. A. Ülkü (Eds.), *Big Data Analytics in Supply Chain Management* (pp. 109–132). CRC Press. doi:10.1201/9780367816384-7

Dagher, G. K., Kassar, A. N., & Itani, O. (2015). The impact of environment concern and attitude on green purchasing behavior. *Contemporary Management Research, 11*(2), 179–205. doi:10.7903/cmr.13625

Dailey, L. C., & Ülkü, M. A. (2018). Retailers beware: On denied product returns and consumer behavior. *Journal of Business Research, 86*, 202–209. doi:10.1016/j.jbusres.2018.01.064

Dangelico, R. M., Pujari, D., & Pontrandolfo, P. (2017). Green product innovation in manufacturing firms: A sustainability-oriented dynamic capability perspective. *Business Strategy and the Environment, 26*(4), 490–506. doi:10.1002/bse.1932

Dannemiller, K. D., & Jacobs, R. W. (1992, December). Changing the Way Organizations Change: A Revolution of Common Sense. *The Journal of Applied Behavioral Science, 28*(4), 480–498. doi:10.1177/0021886392284003

Dasić, B., Dević, Z., Denić, N., Zlatković, D., Ilic, I. D., Cao, Y., Jermsittiparsert, K., & Le, H. V. (2020). Human development index in a context of human development: Review on the western Balkans countries. *Brain and Behavior, 10*(9), 17–55. doi:10.1002/brb3.1755 PMID:33460259

Das, S. R. (2019). The future of fintech. *Financial Management, 48*(4), 981–1007.

Davies, E. T., & Gilbert, D. C. (1992). A case study of the development of farm tourism in Wales. *Tourism Management, 13*(1), 56–63.

Davis, F. (1989). Perceived usefulness, perceived ease of use, and user acceptance of information technology. *Management Information Systems Quarterly*, *13*(3), 319–340. doi:10.2307/249008

Davis, F. (1993). User acceptance of information technology: System characteristics, user perceptions and behavioral impacts. *International Journal of Man-Machine Studies*, *38*(3), 475–487. doi:10.1006/imms.1993.1022

Daziano, R. A., & Bolduc, D. (2013). Incorporating pro-environmental preferences towards green automobile technologies through a Bayesian hybrid choice model. *Transportmetrica A: Transport Science*, *9*(1), 74–106. doi:10.1080/18128602.2010.524173

De Clercq, M., Vats, A., & Biel, A. (2018). Agriculture 4.0: The Future of Farming Technology. *World Government Summit.* Retrieved from https://www.oliverwyman.com/our-expertise/insights/2018/feb/agriculture-4-0--the-future-of-farming-technology.html

De Clercq, M., Vats, A., & Biel, A. (2018). *Agriculture 4.0: The future of farming technology*. World Government Summit in Collaboration with Oliver Wyman.

De Jong, K. (2016). *Evolutionary Computation: A Unified Approach*. Bradford Books.

De la Vega, J. C. S., Azorín, J. D. B., Segura, A. C. F., & Yago, M. E. (2019). A new measure of regional competitiveness. *Applied Economic Analysis*, *27*(80), 108–126. doi:10.1108/AEA-07-2019-0010

De los Rios, I. C., & Charnley, F. J. (2017). Skills and capabilities for a sustainable and circular economy: The changing role of design. *Journal of Cleaner Production*, *160*, 109–122. doi:10.1016/j.jclepro.2016.10.130

De Marchi, V. (2012). Environmental innovation and R&D cooperation: Empirical evidence from Spanish manufacturing firms. *Research Policy*, *41*(3), 614–623. doi:10.1016/j.respol.2011.10.002

De Rus, G. (2008). *Análisis Coste-beneficio. Evaluación económica de políticas y proyectos de inversión* (3rd ed.). Ariel Economía.

Deepa, P., & Karpagam, C. R. (2018). A study on customer's awareness on green banking in selected public and private sector banks with reference to Tirupur. *International Journal of Advanced Research and Development*, *3*(1), 58–63.

Degirmenci, K., & Breitner, M. H. (2017). Consumer purchase intentions for electric vehicles: Is green more important than price and range? *Transportation Research Part D, Transport and Environment*, *51*, 250–260. doi:10.1016/j.trd.2017.01.001

Deka, G. (2015). Green Banking Practices: A Study on environmental strategies of banks with special reference to State bank of India. *Indian Journal of Commerce and Management Studies*, *6*(3), 11–19.

Deka, G. (2018). Customers' perception on green banking practices of state bank of India: A SWOC Analysis. *International Journal of Management Studies*, *5*(1), 20–42. doi:10.18843/ijms/v5iS1/04

Del Vecchio, P., Mele, G., Ndou, V., & Secundo, G. (2018a). Open innovation and social big data for sustainability: Evidence from the tourism industry. *Sustainability*, *10*(9), 3215.

Del Vecchio, P., Secundo, G., & Passiante, G. (2018b). Analyzing big data through the lens of customer knowledge management: Evidence from a set of regional tourism experiences. *Kybernetes*, *47*(7), 1348–1362.

Demirel, P., & Danisman, G. O. (2019). Eco-innovation and firm growth in the circular economy: Evidence from European small-and medium-sized enterprises. *Business Strategy and the Environment*, *28*(8), 1608–1618. doi:10.1002/bse.2336

Demirel, P., & Kesidou, E. (2019). Sustainability-oriented capabilities for eco-innovation: Meeting the regulatory, technology, and market demands. *Business Strategy and the Environment*, *28*(5), 847–857. doi:10.1002/bse.2286

Demir, L., Akpınar, M. E., Araz, C., & Ilgın, M. A. (2018). A green supplier evaluation system based on a new multi-criteria sorting method: VIKORSORT. *Expert Systems with Applications*, *114*, 479–487. doi:10.1016/j.eswa.2018.07.071

Deng, X., Huang, Z., & Cheng, X. (2019). FinTech and sustainable development: Evidence from China based on P2P data. *Sustainability*, *11*(22), 6434.

Deutz, P., McGuire, M., & Neighbour, G. (2013). Eco-design practice in the context of a structured design process: An interdisciplinary empirical study of UK manufacturers. *Journal of Cleaner Production*, *39*, 117–128. doi:10.1016/j.jclepro.2012.08.035

Devkota, N., & Phuyal, R. K. (2017). An analysis of Nepalese youth understanding level on climate change. *Asian Journal of Economic Modelling*, *5*(3), 342–353. doi:10.18488/journal.8.2017.53.342.353

Diamantopoulos, A., Schlegelmilchh, B. B., Sinkovics, R. R., & Bohlen, G. B. (2003). Can socio demographics still play a role in profiling green consumers? a review of the evidence and an empirical investigation. *Journal of Business Research*, *56*(6), 465–480. doi:10.1016/S0148-2963(01)00241-7

Diekmann, L., Gray, L., & Baker, G. (2020). Growing 'good food': Urban gardens, culturally acceptable produce and food security. *Renewable Agriculture and Food Systems*, *35*(2), 169–181. doi:10.1017/S1742170518000388

Digitalizacija u zdravstvu [Digitalization in healthcare]. (2019). PC Press.https://pcpress.rs/digitalizacija-u-zdravstvu-i-medicini/

Dimitropoulos, A. (2014). The influence of environmental concerns on drivers' preferences for electric cars. In *Tinbergen Institute Discussion Papers 14-128/VIII*. Tinbergen Institute.

Dion, K., Berscheid, E., & Walster, E. (1972). What is beautiful is good. *Journal of Personality and Social Psychology*, *24*(3), 285–290. doi:10.1037/h0033731 PMID:4655540

Divert, N. S. (2021). *Fighting Food Waste*. Retrieved from https://divertns.ca/education/fighting%20food-waste

Dixit, N., & Saroj, K. (1970). Acceptance of e-banking among adult customers: An empirical investigation in India. *Journal of Internet Banking and Commerce*, *15*(2), 1–17.

Dixon, R. K., Li, J., & Wang, M. Q. (2016). Progress in hydrogen energy infrastructure development—addressing technical and institutional barriers. In *Compendium of Hydrogen Energy* (pp. 323–343). Woodhead Publishing. doi:10.1016/B978-1-78242-362-1.00013-4

Djukic, N. (2015). *Economic indicators of sustainable tourism in the leading spas in Serbia: master thesis*. Faculty of Mathematics and Natural Sciences, University in Nis, Serbia, p. 3.

Do Paço, F., Arminda, M., Ferreira, J. M., Raposo, M., Gouveia, R., & Dinis, A. (2011). Behaviours and entrepreneurial intention: Empirical findings about secondary students. *Journal of International Entrepreneurship*, *9*(1), 20–38. doi:10.100710843-010-0071-9

Dobos, I., & Vörösmarty, G. (2019). Inventory-related costs in green supplier selection problems with Data Envelopment Analysis (DEA). *International Journal of Production Economics*, *209*, 374–380. doi:10.1016/j.ijpe.2018.03.022

Dorsch, M. T. (2014). Economic development and determinants of environmental concern. *Social Science Quarterly*, *95*(4), 960–977. doi:10.1111squ.12071

Du, L., Dinçer, H., Ersin, İ., & Yüksel, S. (2020). IT2 fuzzy-based multidimensional evaluation of coal energy for sustainable economic development. *Energies*, *13*(10), 2453. doi:10.3390/en13102453

Dunne, L. (2004). *The Design of Wearable Technology: Addressing The Human-Device Interface through Functional Apparel Design* (Unpublished master thesis). https://ecommons.cornell.edu/handle/1813/150

Dunne, L., Ashdown, S., & Smyth, B. (2005). Expanding garment functionality through embedded electronic technology. *Journal of Textile and Apparel Technology and Management*, *4*(3), 1–11.

Duqi, A., & Al-Tamimi, H. (2019). *Factors affecting investors' decision regarding investment in Islamic Sukuk.* Qualitative Research in Financial Markets. doi:10.1108/QRFM-01-2018-0009

Durkalic, D., Furtula, S., & Borisavljevic, K. (2019). Ranking tourism market performance in EMU countries: Results of PROMETHEE-GAIA approach. *Menadžment u hotelijerstvu i turizmu*, *7*(2), 67-76.

Dusoruth, V., & Peterson, H. H. (2020). Food waste tendencies: Behavioral response to cosmetic deterioration of food. *PLoS One*, *15*(5), e0233287. doi:10.1371/journal.pone.0233287 PMID:32469982

Duygun, M., Hashem, S. Q., & Tanda, A. (2020). Financial Intermediation Versus Disintermediation: Opportunities and Challenges in the FinTech Era. *Frontiers in Artificial Intelligence*, 3.

Ecer, F. (2020). Multi-criteria decision making for green supplier selection using interval type-2 fuzzy AHP: A case study of a home appliance manufacturer. *Operations Research*, •••, 1–35. doi:10.100712351-020-00552-y

Egbue, O., & Long, E. (2012). Barriers to widespread adoption of electric vehicles: An analysis of consumer attitudes and perceptions. *Energy Policy*, *48*, 717–729. doi:10.1016/j.enpol.2012.06.009

Ehrlich, P. R. (2008). Key issues for attention from ecological economists. *Environment and Development Economics*, *13*(1), 1–20. doi:10.1017/S1355770X07004019

Elkington, J. (1998). Partnerships from cannibals with forks: The triple bottom line of 21st-century business. *Environmental Quality Management*, *8*(1), 37–51. doi:10.1002/tqem.3310080106

Ellen MacArthur Foundation. (2012). *Towards the Circular Economy: Economic and Business Rationale for an Accelerated Transition. Isle of Wight.* Ellen MacArthur Foundation.

Elliott, R. (2013). The taste for green: The possibilities and dynamics of status differentiation through "green" consumption. *Poetics*, *41*(3), 294–322. doi:10.1016/j.poetic.2013.03.003

Ergen, B., & Ergen, Z. (2011). How does education affect environmental knowledge: A survey in urban and regional planning education? *US-China Education Review*, *7*, 924–931.

Ertugrul, H. M., Gebesoglu, P. F., & Atasoy, B. S. (2018). Mind The Gap: Turkish Case Study of Policy Change in Private Pension Schemes. *Borsa Istanbul Review.*, *18*(2), 140–149. doi:10.1016/j.bir.2017.11.003

Espinilla, M., Halouani, N., & Chabchoub, H. (2015). Pure linguistic PROMETHEE I and II methods for heterogeneous MCGDM problems. *International Journal of Computational Intelligence Systems*, *8*(2), 250–264. doi:10.1080/18756891.2015.1001949

Eti, S., Kalkavan, H., Dinçer, H., & Yüksel, S. (2020). Predicting the role of Islamic banking on sustainable economic development: An analysis for Turkey with ARIMA model. In *Handbook of research on creating sustainable value in the global economy* (pp. 146–164). IGI Global. doi:10.4018/978-1-7998-1196-1.ch009

European Commission. (2002). *Regional Innovation Strategies under the European Regional Development Fund Innovative Actions 2000-2002.* European Commission.

European Commission. (2004). *A Study on the Factors of Regional Competitiveness.* European Commission.

European Commission. (2009). Directive 2009/125/EC of the European Parliament and of the Council of 21 October 2009 Establishing a Framework for the Setting of Ecodesign Requirements for energy-related products (recast).

European Commission. (2010). *SMEs and the Environment in the European Union*. Publications Office of the European Union.

European Commission. (2015). *Internet of Things: Wearable technology*. European Commission.

Eurostat. (2021). *Regions database*. Retrieved from https://ec.europa.eu/eurostat/web/regions/

Evans, D. S., & Leighton, L. S. (1989). Some empirical aspects of entrepreneurship. *The American Economic Review, 79*, 519–535.

Evtodieva, T. E., Chernova, D. V., Voitkevich, N. I., Khramtsova, E. R., & Gorgodze, T. E. (2017). Transformation of logistics organization forms under the conditions of modern economy. In E. G. Popkova (Ed.), *Contributions to Economics*. Springer.

Fahimnia, B., Sarkis, J., & Davarzani, H. (2015). Green supply chain management: A review and bibliometric analysis. *International Journal of Production Economics, 162*, 101–114. doi:10.1016/j.ijpe.2015.01.003

Fallahpour, A., Olugu, E. U., Musa, S. N., Khezrimotlagh, D., & Wong, K. Y. (2016). An integrated model for green supplier selection under fuzzy environment: Application of data envelopment analysis and genetic programming approach. *Neural Computing & Applications, 27*(3), 707–725. doi:10.100700521-015-1890-3

Fallahpour, A., Wong, K. Y., Rajoo, S., & Mardani, A. (2020). An integrated fuzzy carbon management-based model for suppliers' performance evaluation and selection in green supply chain management. *International Journal of Fuzzy Systems, 22*(2), 712–723. doi:10.100740815-019-00759-y

FAO. (2018, March 15). *Beauty (and taste!) are on the inside*. Retrieved from http://www.fao.org/fao-stories/article/en/c/1100391/

FAO. (2021a). *Sustainable Development Goals: 12.3.1 global food losses*. Retrieved from http://www.fao.org/sustainable-development-goals/indicators/12.3.1/en/

FAO. (2021b). *Foof Loss and Food Waste*. Retrieved from http://www.fao.org/food-loss-and-food-waste/flw-data)

Faruque, M. O., Biplob, M. N. K., Al-Amin, M., & Patwary, M. S. H. (2016). Green banking and its potentiality & practice in Bangladesh. *Science Journal of Business and Management, 4*(2), 28–33. doi:10.11648/j.sjbm.20160402.12

Fernández-Villacañas, M.A. (2018). *Las plataformas logísticas 4.0 y la mejora del comercio global: Creando ventaja competitiva logística y desarrollo sostenible*. VI Simposio Internacional Online de Logística y Competitividad, High Logistics Simposios, Medellín, Colombia.

Fernández-Villacañas, M. A. (2019). Desarrollo e Implementación de Plataformas Logísticas de Carga 4.0. *Proceedings of Primer Congreso Internacional de Tecnología e Innovación en Logística 4.0.*

Fernández-Villacañas, M. A. (2020). The New Concept of Logistics Platforms 4.0: Creating Competitiveness Within the Paradigm of Global Sustainable Logistics. In U. Akkucuk (Ed.), *Handbook of Research on Sustainable Supply Chain Management for the Global Economy* (pp. 36–62). IGI Global. doi:10.4018/978-1-7998-4601-7.ch003

FIRA. (2020). *Reporte trimestral de fertilizantes, Dirección de Investigación y Evaluación Económica del Fideicomisos Instituidos en relación con la Agricultura*. FIRA.

Fishbein, M., & Ajzen, I. (1975). *Belief, attitude, intention and behavior: An introduction to theory and research*. Addison-Wesley.

Fishbein, M., & Ajzen, I. (1975). *Belief, attitude, intention, and behavior: An introduction to theory and research.* Addison-Wesley Pub. Co.

Fornell, C., & Larcker, D. F. (1981). Evaluating structural equation models with unobservable variables and measurement error. *JMR, Journal of Marketing Research*, *18*(1), 39–50. doi:10.1177/002224378101800104

Foth, M., Forlano, L., Satchell, C., & Gibbs, M. (Eds.). (2011). *From social butterfly to engaged citizen: Urban informatics, social media, ubiquitous computing, and mobile technology to support citizen engagement*. MIT Press.

Fountas, S., Sorensen, C. G., Tsiropoulos, Z., Cavalaris, C., Liakos, V., & Gemtos, T. (2015). Farm machinery management information system. *Computers and Electronics in Agriculture*, *110*(1), 131–138. doi:10.1016/j.compag.2014.11.011

Freeman, J., & Chen, T. (2015). Green supplier selection using an AHP-Entropy-TOPSIS framework. *Supply Chain Management*, *20*(3), 327–340. doi:10.1108/SCM-04-2014-0142

Fu, J. S., Shumate, M., & Contractor, N. (2020). Organizational and Individual Innovation Decisions in an Interorganizational System: Social Influence and Decision-Making Authority. *Journal of Communication*, *70*(4), 497–521. doi:10.1093/joc/jqaa018

Furfaro, A., Argento, L., Saccá, D., Angiulli, F., & Fassetti, F. (2019). An infrastructure for service accountability based on digital identity and blockchain 3.0. In *IEEE INFOCOM 2019-IEEE Conference on Computer Communications Workshops (INFOCOM WKSHPS)* (pp. 632-637). IEEE.

G20 Sustainable Finance Study Group (SFSG). (2018). *Sustainable Finance Synthesis Report.* Technical report. Available at: http://www.gflp.org.cn/public/ueditor/php/upload/file/20181203 /1543807391177171.pdf

Gajic, T. (2010). Tourism in the Intensification of the Economic Development of Receptive and Emitting Areas - An Example of the South Backa District. *Industry, 38*(3), 139-155.

Galbraith, J. (2012). The evolution of enterprise organization designs. *Journal of Organization Design*, *2*(1), 1–13.

Gallegos Cedillo, G. (2021). *Working paper.* Agronegocios FIRA.

Gamzigradska banja – Specijalna bolnica [Gamzigrad Spa – Special hospital]. (n.d.). https://euroturs.rs/bamjezszrbija-banja/gamzigradska-banja-specijalna-bolnica

Gamzigradska banja [Gamzigrad Spa]. (n.d.). https://www.gamzigradskabanja.org.rs/index.php/vesti

Ganesan, D. R., & Bhuvaneswari, A. (2016). customer perception towards green banking. *IOSR Journal of Economics and Finance*, *7*(5), 5–17. doi:10.9790/5933-0705010517

Gardiner, B. (2003, April). *Regional competitiveness indicators for Europe-audit, database construction and analysis.* Paper presented at the meeting of the Regional Studies Association, Pisa, Italy.

Gardiner, B., Martin, R., & Tyler, P. E. T. E. R. (2006). Competitiveness, productivity and economic growth across the European regions. *Regional Competitiveness, 30*, 55.

Garg, R., Aggarwal, H., Centobelli, P., & Cerchione, R. (2019). Extracting knowledge from big data for sustainability: A comparison of machine learning techniques. *Sustainability*, *11*(23), 6669.

Garmise, S., & Rees, G. (1997). The role of institutional networks in local economic development: A new model of governance? *Local Economy*, *12*(2), 104–118. doi:10.1080/02690949708726379

Gaskin, J., & Lim, J. (2016). *Model fit measures*. AMOS Plugin. Gaskination's StatWiki.

Gegovska, T., Koker, R., & Cakar, T. (2020). Green Supplier Selection Using Fuzzy Multiple-Criteria Decision-Making Methods and Artificial Neural Networks. *Computational Intelligence and Neuroscience*, *8811834*, 1–26. Advance online publication. doi:10.1155/2020/8811834 PMID:33082775

Gelissen, J. (2007). Explaining popular support for environmental protection: A multilevel analysis of 50 nations. *Environment and Behavior*, *39*(3), 392–415. doi:10.1177/0013916506292014

Genç, Ö. (2019). *Notes on Artificial Intelligence, Machine Learning and Deep Learning for curious people*. https://towardsdatascience.com/notes-on-artificial-intelligence-ai-machine-learning-ml-and-deep-learning-dl-for-56e51a2071c2

Genovese, A., Koh, S. L., Bruno, G., & Bruno, P. (2010, October). Green supplier selection: A literature review and a critical perspective. In *2010 8th International Conference on Supply Chain Management and Information* (pp. 1-6). IEEE.

George, G., Merrill, R., & Schillebeeckx, S. (2020). Digital sustainability and entrepreneurship: How digital innovations are helping tackle climate change and sustainable development. *Entrepreneurship Theory and Practice*. Advance online publication. doi:10.1177/1042258719899425

Ghadimi, P., Azadnia, A. H., Heavey, C., Dolgui, A., & Can, B. (2016). A review on the buyer–supplier dyad relationships in sustainable procurement context: Past, present and future. *International Journal of Production Research*, *54*(5), 1443–1462. doi:10.1080/00207543.2015.1079341

Ghasri, M., Ardeshiri, A., & Rashidi, T. (2019). Perception towards electric vehicles and the impact on consumers' preference. *Transportation Research Part D, Transport and Environment*, *77*, 271–291. doi:10.1016/j.trd.2019.11.003

Ghisetti, C., & Pontoni, F. (2015). Investigating policy and R&D effects on environmental innovation: A meta-analysis. *Ecological Economics*, *118*, 57–66. doi:10.1016/j.ecolecon.2015.07.009

Ghorabaee, M. K., Zavadskas, E. K., Amiri, M., & Esmaeili, A. (2016). Multi-criteria evaluation of green suppliers using an extended WASPAS method with interval type-2 fuzzy sets. *Journal of Cleaner Production*, *137*, 213–229. doi:10.1016/j.jclepro.2016.07.031

Gilligan, C. (1982). In a different voice: psychological theory and women's development. Harvard University Press.

Gitinavard, H., Ghaderi, H., & Pishvaee, M. S. (2018). Green supplier evaluation in manufacturing systems: A novel interval-valued hesitant fuzzy group outranking approach. *Soft Computing*, *22*(19), 6441–6460. doi:10.100700500-017-2697-1

Giungato, P., Rana, R., Tarabella, A., & Tricase, C. (2017). Current trends in sustainability of bitcoins and related blockchain technology. *Sustainability*, *9*(12), 2214.

Global Entrepreneurship Monitor Report. (2021). *Entrepreneurial behaviour and attitude*. Retrieved from https://www.gemconsortium.org/data

Goffin, K., & Mitchell, R. (2017). *Innovation Management: Effective strategy and implementation*. Red Globe Press. doi:10.1057/978-1-137-37344-1

Goldenfein, J., & Leiter, A. (2018). Legal engineering on the blockchain: 'Smart contracts' as legal conduct. *Law and Critique*, *29*(2), 141–149.

Gold, S., Hahn, R., & Seuring, S. (2013). Sustainable supply chain management in "Base of the Pyramid" food projects—A path to triple bottom line approaches for multinationals? *International Business Review*, *22*(5), 784–799. doi:10.1016/j.ibusrev.2012.12.006

Goldstein, I., Jiang, W., & Karolyi, G. A. (2019). To FinTech and beyond. *Review of Financial Studies*, *32*(5), 1647–1661.

Goodale, T. (2021). Multivariate analysis of the impact of gender and college major on student levels of environmental concern and knowledge. *International Electronic Journal of Environmental Education*, *11*(1), 1–12.

Goulet, F. (2020). Family Farming and The Emergence of an Alternative Sociotechnical Imaginary in Argentina. *Science, Technology & Society*, *25*(1), 86–105. doi:10.1177/0971721819889920

Govindan, K. (2017). Sustainable consumption and production in the food supply chain: A conceptual framework. *International Journal of Production Economics*, *195*, 419–431. doi:10.1016/j.ijpe.2017.03.003

Govindan, K., Kadziński, M., & Sivakumar, R. (2017). Application of a novel PROMETHEE-based method for construction of a group compromise ranking to prioritization of green suppliers in food supply chain. *Omega*, *71*, 129–145. doi:10.1016/j.omega.2016.10.004

Govindan, K., Rajendran, S., Sarkis, J., & Murugesan, P. (2015). Multi criteria decision making approaches for green supplier evaluation and selection: A literature review. *Journal of Cleaner Production*, *98*, 66–83. doi:10.1016/j.jclepro.2013.06.046

Govindan, K., & Sivakumar, R. (2016). Green supplier selection and order allocation in a low-carbon paper industry: Integrated multi-criteria heterogeneous decision-making and multi-objective linear programming approaches. *Annals of Operations Research*, *238*(1-2), 243–276. doi:10.100710479-015-2004-4

Gray, S. G., Raimi, K. T., Wilson, R., & Árvai, J. (2019). Will Millennials save the world? the effect of age and generational differences on environmental concern. *Journal of Environmental Management*, *242*, 394–402. doi:10.1016/j.jenvman.2019.04.071 PMID:31059952

Greenberg, P., & Bugden, D. (2019). Energy consumption boomtowns in the United States: Community responses to a cryptocurrency boom. *Energy Research & Social Science*, *50*, 162–167.

Growlink. (2021). *The brain of your smart farm.* Retrieved from https://growlink.com/sensors/

Güçlü, F. (2019). *The Rise of Environmental Consciousness in Islamic Finance: Green Sukuk.* Academic Press.

Gujarati, D. (2008). *Basic econometrics.* New York: McGraw-Hill.

Guo, Z., Liu, H., Zhang, D., & Yang, J. (2017). Green supplier evaluation and selection in apparel manufacturing using a fuzzy multi-criteria decision-making approach. *Sustainability*, *9*(4), 650. doi:10.3390u9040650

Gupta, H., & Barua, M. K. (2017). Supplier selection among SMEs on the basis of their green innovation ability using BWM and fuzzy TOPSIS. *Journal of Cleaner Production*, *152*, 242–258. doi:10.1016/j.jclepro.2017.03.125

Gupta, S., Soni, U., & Kumar, G. (2019). Green supplier selection using multi-criterion decision making under fuzzy environment: A case study in automotive industry. *Computers & Industrial Engineering*, *136*, 663–680. doi:10.1016/j.cie.2019.07.038

Gwo-Hshiung, T., & Jih-Jeng, H. (2011). *Multiple attribute decision making: methods and applications.* Academic Press.

Hackbarth, A., & Madlener, R. (2013). Consumer preferences for alternative fuel vehicles: A discrete choice analysis. *Transportation Research Part D, Transport and Environment*, *25*, 5–17. doi:10.1016/j.trd.2013.07.002

Haeri, S. A. S., & Rezaei, J. (2019). A grey-based green supplier selection model for uncertain environments. *Journal of Cleaner Production*, *221*, 768–784. doi:10.1016/j.jclepro.2019.02.193

Hair, J. F., Black, W. C., Babin, B. J., & Anderson, R. E. (2014). *Multivariate data analysis: Pearson new international edition.* Pearson Education Limited.

Hair, J. F. Jr, Hult, G. T. M., Ringle, C. M., & Sarstedt, M. (2017). *A primer on partial least squares structural equation modeling (pls-sem)*. Sage.

Haiyun, C., Zhixiong, H., Yüksel, S., & Dinçer, H. (2021). Analysis of the innovation strategies for green supply chain management in the energy industry using the QFD-based hybrid interval valued intuitionistic fuzzy decision approach. *Renewable & Sustainable Energy Reviews*, *143*, 110844. doi:10.1016/j.rser.2021.110844

Hajkowicz, S. A., Young, M., Wheeler, S., MacDonald, D. H., & Young, D. (2000). *Supportin decisions: understanding natural resource management assessment techniques. A Report to the Land and Water Resources Research and Development Corporation*. CSIRO.

Håkansson, H., & Olsen, P. I. (2015). The roles of money and business deals in network structures. *Industrial Marketing Management*, *45*, 207–217. doi.org/10.1016/j.indmarman.2015.01.009

Halim, Z. A., How, J., & Verhoeven, P. (2017). Agency costs and corporate sukuk issuance. *Pacific-Basin Finance Journal*, *42*, 83–95. doi:10.1016/j.pacfin.2016.05.014

Hall, B. H. (2005). The Financing of Innovation. In S. Shane (Ed.), *The Handbook of Technology and Innovation Management* (pp. 409–430). Wiley.

Hall, B. H., Moncada-Paternó-Castello, P., Montresor, S., & Vezzani, A. (2016). Financing constraints, R&D investments and innovative performances: New empirical evidence at the firm level for Europe. *Economics of Innovation and New Technology*, *25*(3), 183–196. doi:10.1080/10438599.2015.1076194

Hamlin, R., Knight, J., & Cuthbert, R. (2016). Niche marketing and farm diversification processes: Insights from New Zealand and Canada. *Renewable Agriculture and Food Systems*, *31*(1), 86–98. doi:10.1017/S1742170514000489

Hanna, N., & Wozniak, R. (2017). *Consumer behavior an applied approach*. Kendall Hunt Pub Co.

Harring, N., & Jagers, S. C. (2017). Why do people accept environmental policies? the prospects of higher education and changes in norms, beliefs and policy preferences. *Environmental Education Research*, 1–16.

Hart, S. L., & Dowell, G. (2011). Invited editorial: A natural-resource-based view of the firm: Fifteen years after. *Journal of Management*, *37*(5), 1464–1479. doi:10.1177/0149206310390219

Hashemi, S. H., Karimi, A., & Tavana, M. (2015). An integrated green supplier selection approach with analytic network process and improved Grey relational analysis. *International Journal of Production Economics*, *159*, 178–191. doi:10.1016/j.ijpe.2014.09.027

Hayes, A. F. (2012). *Process: A versatile computational tool for observed variable mediation, moderation, and conditional process modeling* [White paper]. http://www.afhayes.com/public/process2012.pdf

Heather, K. L. (2012). *The Environmental Benefits of Urban Agriculture on Unused, Impermeable and Semi-Permeable Spaces in Major Cities With a Focus on Philadelphia*. Retrieved from https://repository.upenn.edu/mes_capstones/46

Hedman, J., & Kalling, T. (2002). *IT and Business Models: Concepts and Theories*. Copenhagen Business School Press.

Henry, M., Bauwens, T., Hekkert, M., & Kirchherr, J. (2020). A typology of circular start-ups: Analysis of 128 circular business models. *Journal of Cleaner Production*, *245*, 118528. doi:10.1016/j.jclepro.2019.118528

He, X., Zhan, W., & Hu, Y. (2018). Consumer purchase intention of electric vehicles in China: The roles of perception and personality. *Journal of Cleaner Production*, *204*, 1060–1069. doi:10.1016/j.jclepro.2018.08.260

Hickey, L. T., Hafeez, A. N., Robinson, H., Jackson, S. A., Leal-Bertioli, S. C. M., & Tester, M. (2019). Breeding crops to feed 10 billion. *Nature Biotechnology*, 37. PMID:31209375

Hidrue, M. K., Parsons, G. R., Kempton, W., & Gardner, M. P. (2011). Willingness to pay for electric vehicles and their attributes. *Resource and Energy Economics*, *33*(3), 686–705. doi:10.1016/j.reseneeco.2011.02.002

Higueras-Castillo, E., Molinillo, S., Coca-Stefaniak, J. A., & Liébana-Cabanillas, F. (2020). Potential early adopters of hybrid and electric vehicles in Spain—Towards a customer profile. *Sustainability*, *12*(11), 4345. doi:10.3390u12114345

Hines, W. W., & Montgomery, D. (1972). *Probability and Statistics in Engineering and Management Science* (3rd ed.). John Wiley and Sons.

Hinson, R., Lensink, R., & Mueller, A. (2019). Transforming agribusiness in developing countries: SDGs and the role of FinTech. *Current Opinion in Environmental Sustainability*, *41*, 1–9.

Hirsh, J. B. (2010). Personality and environmental concern. *Journal of Environmental Psychology*, *30*(2), 245–248. doi:10.1016/j.jenvp.2010.01.004

HLPE High Level Panel of Experts on Food Security and Nutrition. (2014). Food losses and waste in the context of sustainable food systems (No. 8). CFS Committee on World Food Security HLPE, Food and Agriculture Organization of the United Nations (FAO).

Hoang, C. V., Hoang, T. G., & Kane, V. (2020). Technological approaches to sustainability. In D. Crowther & S. Seifi (Eds.), *The Palgrave Handbook of Corporate Social Responsibility*. Palgrave Macmillan. doi:10.1007/978-3-030-22438-7_37-1

Holmes, T. J., & Schmitz, J. A. Jr. (1990). A theory of entrepreneurship and its application to the study of business transfers. *Journal of Political Economy*, *98*(2), 265–294. doi:10.1086/261678

Hong, C.-W. (2012). Using the Taguchi method for effective market segmentation. *Expert Systems with Applications*, *39*(5), 5451–5459. doi:10.1016/j.eswa.2011.11.040

Hoogendoorn, B., Guerra, D., & van der Zwan, P. (2015). What drives environmental practices of SMEs? *Small Business Economics*, *44*(4), 759–781. doi:10.100711187-014-9618-9

Horbach, J. (2016). Empirical determinants of eco-innovation in European countries using the community innovation survey. *Environmental Innovation and Societal Transitions*, *19*, 1–14. doi:10.1016/j.eist.2015.09.005

Horbach, J., Rammer, C., & Rennings, K. (2012). Determinants of eco-innovations by type of environmental impact-the role of regulatory push/pull, technology push and market pull. *Ecological Economics*, *78*, 112–122. doi:10.1016/j.ecolecon.2012.04.005

HORES - Business Association of Hoteliers and Restaurateurs of Serbia. (n.d.). http://www.hores.rs/odrzivost.html

Hossain, M. S., Uddin, M. H., & Kabir, S. H. (2020). Sukuk and bond puzzle: An analysis with characteristics matched portfolios. *Emerging Markets Finance & Trade*, 1–26. doi:10.1080/1540496X.2019.1706478

Hou, Q., & Xie, L. (2019). Research on supplier evaluation in a green supply chain. *Discrete Dynamics in Nature and Society*, *2601301*, 1–14. Advance online publication. doi:10.1155/2019/2601301

Hoyer, W., & MacInnis, D. (2007). Consumer behavior (4th ed.). Boston: Houghton Mifflin Co.

Hribar, J., & DaSilva, L. (2019). Utilising correlated information to improve the sustainability of internet of things devices. In *2019 IEEE 5th World Forum on Internet of Things (WF-IoT)* (pp. 805-808). IEEE.

Huggins, R., & Davies, W. (2006). *European Competitiveness Index 2006-2007*. Retrieved from http://www.cforic.org/pages/european-competitiveness.php

Huggins, R., Izushi, H., Prokop, D., & Thompson, P. (2014). Regional competitiveness, economic growth and stages of development. *Zbornik radova Ekonomskog fakulteta u Rijeci: časopis za ekonomsku teoriju i praksu, 32*(2), 255-283.

Huggins, R. (2003). Creating a UK Competitiveness Index: Regional and Local Benchmarking. *Regional Studies, 37*(1), 89–96. doi.org/10.1080/0034340022000033420

Hui, M. (2012). The Research on Development Trend of Telecommunications Industry. *IERI Procedia, 3*, 232–236. doi:10.1016/j.ieri.2012.09.038

Hu, L., & Bentler, P. M. (1999). Cutoff criteria for fit indexes in covariance structure analysis: Conventional criteria versus new alternatives. *Structural Equation Modeling, 6*(1), 1–55. doi:10.1080/10705519909540118

Humphrey, J. (2014). *Market systems approaches – A literature review*. The BEAM Exchange. Available at: www.beamexchange.org

Huyghebaert, N. (2004). *Comportamiento estratégico de los titulares en los mercados financieros y la salida de los arranques empresariales*. Academic Press.

Huynh, M., Ghimire, P., & Truong, D. (2017). Hybrid app approach: Could it mark the end of native app domination? *Issues in Informing Science and Information Technology Education, 14*, 49–65. doi:10.28945/3723

Hu, Z., Rao, C., Zheng, Y., & Huang, D. (2015). Optimization decision of supplier selection in green procurement under the mode of low carbon economy. *International Journal of Computational Intelligence Systems, 8*(3), 407–421. doi:10.1080/18756891.2015.1017375

Hwang, C. (2014). *Consumers ' acceptance of wearable technology: Examining solar-powered clothing* (Unpublished master thesis). https://lib.dr.iastate.edu/cgi/viewcontent.cgi?article=4957&context=etd

Igarashi, M., de Boer, L., & Fet, A. M. (2013). What is required for greener supplier selection? A literature review and conceptual model development. *Journal of Purchasing and Supply Management, 19*(4), 247–263. doi:10.1016/j.pursup.2013.06.001

Ilić, B. (2019). Social Component of Sustainable Development and Quality of Life: Region of the Balkans, Eastern Serbia: chapter. In Handbook of Research on Creating Sustainable Value in the Global Economy. IGI Global.

Ilic, B., Djukic, G., & Balaban, M. (2020). Sustainable development directions of rural tourism of Timok region. *Economics of Agriculture, 67*(1), 157–174.

Ilic, B., Djukic, G., & Radisavljevic, G. (2020). Understanding ecotourism and protection of natural values - a comparison of America and Serbia. *Conference Proceedings Innovation as a driver of development, Innovation to business success*, 253-262.

Incyt. (2021). *Weather and microclimate mornitoring*. Retrieved from https://incyt.io/pages/weather-monitoring

Institute of Development and Research in Banking Technology. (2013). *Publications: IDRBT*. Retrieved in July 21, 2016 from http://www.idrbt.ac.in/publications/Frameworks/Green%20Banking%20Framework%20

Intermarché. (2021). *Inglorious Fruits and Vegetables*. Retrieved from: https://www.youtube.com/watch?v=p2nSECWq_PE&ab_channel=Marcel

International Institute of Management Development. (2000). *The World Competitiveness Yearbook 2000*. Lausanne, Switzerland: International institute of management development.

İsabet, E. Y., & Kazak, H. (2019). Sukuk As One The Islamic Financial Instrumentes: Investigation of The Murabahah Contracts In The Hybrid Sukuk In Terms Of Compliance With The Islamic Provision. *Journal of Life Economics, 6*(1), 91.

Ishangulyyev, R., Kim, S., & Lee, S. H. (2019). Understanding food loss and waste - Why are we losing and wasting food? *Foods*, *8*(8), 297. doi:10.3390/foods8080297 PMID:31362396

Ishikawa, T., Fujiwara, H., Imai, O., & Okabe, A. (2008). Wayfinding with a GPS-based mobile navigation system: A comparison with maps and direct experience. *Journal of Environmental Psychology*, *28*(1), 74–82. doi:10.1016/j.jenvp.2007.09.002

Islam, M. A., Yousuf, S., Hossain, K. F., & Islam, M. R., (2014). Green financing in Bangaladesh: Challenges and opportunities- a descriptive approach. *International Journal of Green Economics, 8*(1), 74-91.

Jackson, T. (2014). Sustainable consumption. In *Handbook of Sustainable Development* (p. 279). Edward Elgar Publishing. doi:10.4337/9781782544708.00029

Jaeger, S. R., Machín, L., Aschemann-Witzel, J., Antúnez, L., Harker, F. R., & Ares, G. (2018). Buy, eat or discard? A case study with apples to explore fruit quality perception and food waste. *Food Quality and Preference*, *69*, 10–20. doi:10.1016/j.foodqual.2018.05.004

Jahangir, N., & Begum, N. (2008). The role of perceived usefulness, perceived ease of use, security and privacy, and customer attitude to engender customer adaption in the context of electronic banking. *African Journal of Business Management*, *2*(1), 32–40.

Jain, S. K., & Kaur, G. (2006). Role of socio-demographics in segmenting and profiling green consumers. *Journal of International Consumer Marketing*, *18*(3), 107–146. doi:10.1300/J046v18n03_06

Jakobsen, B. J., Tanev, S., & Jensen, S. (2017). The value of business incubation services for early stage start-ups. In *ISPIM Innovation Symposium* (pp. 1-23). The International Society for Professional Innovation Management (ISPIM).

Jalali, A., Seiiedlou, S., Linke, M., & Mahajan, P. (2017). A comprehensive simulation program for modified atmosphere and humidity packaging (MAHP) of fresh fruits and vegetables. *Journal of Food Engineering*, *206*, 88–97. doi:10.1016/j.jfoodeng.2017.03.007

Jani, A. S. (2012). A study of consumer perception on the use of e-technology in the retail banking sector: A comparative study of public sector and private sector banks. *Indian Journal of Marketing*, *42*(1), 46–58.

Janssen, R. (1992). *Multiobjective Decision Support for Environmental Management Dordrecht*. Kluwer Academic Publishers. doi:10.1007/978-94-011-2807-0

Jayabal & Saundarya. (2016). Green Banking: As Banks Initiative for Sustainable Development. *International Journal of Management*, *7*(7), 276–280.

Jeyapaul, R., Shahabudeen, P., & Krishnaiah, K. (2005). Quality management research by considering multi-response problems in the Taguchi method – a review. *International Journal of Advanced Manufacturing Technology*, *26*(11-12), 1331–1337. doi:10.100700170-004-2102-y

Jha, N., & Bhome, S. (2013). A study of green banking trends in India. *International Monthly Referred Journal of Research in Management and Technology*, *2*, 127–132.

Joshi, S., Saxena, S., & Godbole, T. (2016). Developing smart cities: An integrated framework. *Procedia Computer Science*, *93*, 902–909. doi.org/10.1016/j.procs.2016.07.258

Joubert, R.-L., & Jokonya, O. (2021). A systematic literature review of factors affecting the adoption of technologies in food waste management. *Procedia Computer Science*, *181*, 1034–1040. doi:10.1016/j.procs.2021.01.298

Jovčić, M. (2002). *Econometric models*. CID. (in Serbian)

Jové-Llopis, E., & Segarra-Blasco, A. (2018). Eco-innovation strategies: A panel data analysis of Spanish manufacturing firms. *Business Strategy and the Environment, 27*(8), 1209–1220. doi:10.1002/bse.2063

Jovic, M., Jovic, L. J., Milosevic, J., Manic, S., Rajkovic, T., & Djordjevic, P. (2016). Pravni i etički problemi prinudne hospitalizacije – prehospitalni aspekt [Legal and ethical problems of involuntary hospitalization - prehospital aspect]. *HALO, 22*(1), 60–89.

Junquera, B., Moreno, B., & Álvarez, R. (2016). Analyzing consumer attitudes towards electric vehicle purchasing intentions in Spain: Technological limitations and vehicle confidence. *Technological Forecasting and Social Change, 109*, 6–14. doi:10.1016/j.techfore.2016.05.006

Kabongo, J. D., & Boiral, O. (2017). Doing more with less: Building dynamic capabilities for eco-efficiency. *Business Strategy and the Environment, 26*(7), 956–971. doi:10.1002/bse.1958

Kaipia, R., Dukovska-Popovska, I., & Loikkanen, L. (2013). Creating sustainable fresh food supply chains through waste reduction. *International Journal of Physical Distribution & Logistics Management, 43*(3), 262–276. doi:10.1108/IJPDLM-11-2011-0200

Kalita, B. (2017). Marketing Efficiency, Price Spread, Share of Farmers In Case Of Horticultural Markets of Assam. *International Journal of Advance Research and Development, 2*(8), 65–72.

Kamble, S. S., Gunasekaran, A., & Gawankar, S. A. (2020). Achieving sustainable performance in a data-driven agriculture supply chain: A review for research and applications. *International Journal of Production Economics, 219*, 179–194. doi:10.1016/j.ijpe.2019.05.022

Kamilaris, A., Kartakoullis, A., & Prenafeta-Boldù, F. X. (2017). A Review on the Practice of Big Data Analysis in Agriculture. *Computers and Electronics in Agriculture, 143*(1), 23–37. doi:10.1016/j.compag.2017.09.037

Kannan, D., Govindan, K., & Rajendran, S. (2015). Fuzzy axiomatic design approach based green supplier selection: A case study from Singapore. *Journal of Cleaner Production, 96*, 194–208. doi:10.1016/j.jclepro.2013.12.076

Karim, F., Karim, F., & Frihida, A. (2017). Monitoring system using web of things in precision agriculture. *Procedia Computer Science, 110*, 402–409. doi:10.1016/j.procs.2017.06.083

Kates, A., & Galbraith, J. (2007). *Designing your organization using the star model to solve 5 critical design challenges*. Jossey-Bass.

Kawgan-Kagan, I. (2020). Are women greener than men? a preference analysis of women and men from major German cities over sustainable urban mobility. *Transportation Research Interdisciplinary Perspectives*, 8.

Kaya, F., & Kahya, M. (2017). *Sigorta ve Sigortacılık, Beta Yayıncılık, 1*. Baskı.

Kayhan, F. (2020). Türkiye'de Faizsiz Bireysel Emeklilik Sisteminin SWOT Analizi. *Finans Ekonomi ve Sosyal Araştırmalar Dergisi, 5*(4), 745–754.

Kayhan, F., & Islamoglu, M. (2020). A Comprehensive Assessment of Auto-Enrollment System as a Second Pillar of Pension System: The Case of Turkey. *International Journal of Publication and Social Studies, 5*(2), 167–177. doi:10.18488/journal.135.2020.52.167.177

Kayhan, F., Islamoglu, M., & Apan, M. (2020). Portfolio and Benchmark Returns of Private Pension Funds: A Comparative Analysis for Turkey. *Business & Management Studies: An International Journal, 8*(5), 3891–3910.

Kearney, J. (2010). Food consumption trends and drivers. *Philosophical Transactions, 365*(1554), 2793–2807. doi:10.1098/rstb.2010.0149 PMID:20713385

Kedari, P. (2019). *AI, ML and DL — Let's make it more clear!* https://community.letsupgrade.in/post/ai-ml-and-dl----let-s-make-it-more-clear-5eda6110d5a8b90b708491c2

Keshavarz-Ghorabaee, M., Amiri, M., Hashemi-Tabatabaei, M., Zavadskas, E. K., & Kaklauskas, A. (2020). A New Decision-Making Approach Based on Fermatean Fuzzy Sets and WASPAS for Green Construction Supplier Evaluation. *Mathematics*, *8*(12), 2202. doi:10.3390/math8122202

Keshminder, J. S., Singh, G. K. B., Wahid, Z. A., & Abdullah, M. S. (2019). Green sukuk: Malaysia taking the lead. *Malaysian Journal of Consumer and Family Economics, 22*, 1-22.

Khakurel, J., Penzenstadler, B., Porras, J., Knutas, A., & Zhang, W. (2018). The rise of artificial intelligence under the lens of sustainability. *Technologies*, *6*(4), 100.

Khanal, S., Fulton, J., & Shearer, S. (2017). An overview of current and potential applications of thermal remote sensing in precision agriculture. *Computers and Electronics in Agriculture*, *139*, 22–32. doi:10.1016/j.compag.2017.05.001

Khan, S. A. R., Yu, Z., Golpîra, H., Sharif, A., & Mardani, A. (2020). A state-of-the-art review and meta-analysis on sustainable supply chain management: Future research directions. *Journal of Cleaner Production*, 123357.

Khatua, P. K., Ramachandaramurthy, V. K., Kasinathan, P., Yong, J. Y., Pasupuleti, J., & Rajagopalan, A. (2020). Application and assessment of internet of things toward the sustainability of energy systems: Challenges and issues. *Sustainable Cities and Society*, *53*, 101957.

Khorasani, S. T. (2018). Green supplier evaluation by using the integrated fuzzy AHP model and fuzzy copras. *Process Integration and Optimization for Sustainability*, *2*(1), 17–25. doi:10.100741660-017-0027-9

Kim, J., Rasouli, S., & Timmermans, H. (2014). Expanding scope of hybrid choice models allowing for mixture of social influences and latent attitudes: Application to intended purchase of electric cars. *Transportation Research Part A, Policy and Practice*, *69*, 71–85. doi:10.1016/j.tra.2014.08.016

Kim, K., & Shin, D.-H. (2014). An acceptance model for smart watches. *Internet Research*, *25*(4), 527–541. doi:10.1108/IntR-05-2014-0126

Kim, K., & Sundar, S. (2015). Mobile persuasion: Can screen size and presentation mode make a difference to trust? *Human Communication Research*, *42*(1), 45–70. doi:10.1111/hcre.12064

King, A. (2017). Technology: The Future of Agriculture. *Nature*, *544*(7651), S21–S23. doi:10.1038/544S21a PMID:28445450

Kirmani, M. D., & Khan, M. N. (2016). Environmental concern to attitude towards green products: Evidences from India, *Serbian. Journal of Management*, *11*(2), 159–179.

Kirstein, T., Cottet, D., Grzyb, J., & Tröster, G. (2005). Wearable computing systems – electronic textiles. In T. Xiaoming (Ed.), *Wearable electronics and photonics*. Woodhead Publishing.

Kitanović, D., & Krstić, M. (2010). Uticaj deviznog kursa na privredni rast Srbije. *Ekonomske teme, 48*(1), 13-27.

Kitson, M., Martin, R., & Tyler, P. (2004). Regional competitiveness: An elusive yet key concept? *Regional Studies*, *38*(9), 991–999. doi.org/10.1080/0034340042000320816

Klapper, L., & Singer, D. (2017). The opportunities and challenges of digitizing government-to-person payments. *The World Bank Research Observer*, *32*(2), 211–226.

Klein, L. R. (1962). Introduction to econometrics. New York: Prentice Hall.

Klein, A. G., Gerhard, C., Büchner, R. D., Diestel, S., & Schermelleh-Engel, K. (2016). The detection of heteroscedasticity in regression models for psychological data. *Psychological Test and Assessment Modeling, 58*(4), 542–568.

Klein, P. O., & Weill, L. (2016). Why do companies issue sukuk? *Review of Financial Economics, 31*, 26–33. doi:10.1016/j.rfe.2016.05.003

Klein, P. O., Weill, L., & Godlewski, C. J. (2018). How sukuk shapes firm performance. *World Economy, 41*(3), 699–722. doi:10.1111/twec.12509

Klerkx, L., Jakku, E., & Labarthe, P. (2019). A review of social science on digital agriculture, smart farming and agriculture 4.0: New contributions and a future research agenda. *NJAS Wageningen Journal of Life Sciences, 90-91*, 1–16. doi:10.1016/j.njas.2019.100315

Ko, F., El-Aufy, A., Lam, H., & Macdiarmid, A. (2005). Electrostatically generated nanofibres for wearable electronics. In X. Tao (Ed.), *Wearable Electronics and Photonics*. Woodhead Publishing.

Ko, M., Mancha, R., Beebe, N., & Yoon, H. S. (2012). Customers' personality, their perceptions, and green concern on internet banking use. *Journal of Information Technology Management, 23*(4), 21–32.

Konys, A. (2019). Green supplier selection criteria: From a literature review to a comprehensive knowledge base. *Sustainability, 11*(15), 4208. doi:10.3390u11154208

Kosic, K. (2009). *Vojvodina Rural Tourism and Sustainable Development: Ph.D. Thesis.* The University of Novi Sad, Faculty of Science, Department of Geography, Tourism and Hotel Management, Novi Sad.

Krikser, T., Zasada, I., & Piorr, A. (2019). Socio-Economic Viability of Urban Agriculture – A Comparative Analysis of Success Factors in Germany. *Sustainability, 11*(7), 1–12. doi:10.3390u11071999

Krishankumar, R., Gowtham, Y., Ahmed, I., Ravichandran, K. S., & Kar, S. (2020). Solving green supplier selection problem using q-rung orthopair fuzzy-based decision framework with unknown weight information. *Applied Soft Computing, 94*, 106431. doi:10.1016/j.asoc.2020.106431

Krstić, B., Radivojević, V., & Stanišić, T. (2016a). Determinants of CEE countries' tourism competitiveness: A benchmarking study. *Management: Journal of Sustainable Business and Management Solutions in Emerging Economies, 21*(80), 11-22. doi:10.7595/management.fon.2016.0021

Krstić, B., Radivojević, V., & Stanišić, T. (2016b). Measuring and analysis of competition intensity in the sugar market in Serbia. *Economics of Agriculture, 63*(2), 389-406. doi:10.5937/ekoPolj1602389K

Krstić, B., Stanišić, T., & Radivojević, V. (2016c), The impact of innovativeness' factors on the EU countries competitiveness. *Industrija, 44*(2), 101-116. doi:10.5937/industrija41-4000

Krstić, M. (2012a). The role of rules in the evolution of the market system: Hayek's concept of evolutionary epistemology. *Economic Annals, 57*(194), 123-140. doi:10.2298/EKA1294123K

Krstić, M. (2012b). New theory of international trade and new economic geography-Volčkova N.: Novaja teorija meždunarodnoj torgovli i novaja ěkonomičeskaja geografija, Voprosy teorii, 2009. *Geographica Pannonica, 16*(1), 35–37.

Krstić, M. (2021). Higher Education as Determinant of Competitiveness and Sustainable Economic Development. In M. Topić & G. Lodorfos (Eds.), *The Sustainability Debate: policies, gender and the media* (pp. 15–35). Emerald Publishing.

Krstić, M., Filipe, J. A., & Chavaglia, J. (2020b). Higher Education as a Determinant of the Competitiveness and Sustainable Development of an Economy. *Sustainability, 12*(16), 6607. doi.org/10.3390/su12166607

Krstić, M., & Pavlović, N. (2020a). Behavioral Economics: New Dimension in Understanding the Real Economic Behavior. In U. Akkucuk (Ed.), *Handbook of Research on Sustainable Supply Chain Management for the Global Economy* (pp. 281–298). IGI Global.

Krueger, N. F. Jr, Reilly, M. D., & Carsrud, A. L. (2000). Competing models of entrepreneurial intentions. *Journal of Business Venturing*, *15*(5-6), 411–432. doi:10.1016/S0883-9026(98)00033-0

Krueger, R. A., & Casey, M. A. (2014). *Focus Groups: A Practical Guide for Applied Research.* SAGE Publications.

Krugman, P. (1994). Competitiveness: A dangerous obsession. *Foreign Affairs*, *73*(2), 28–44.

Krupa, J. S., Rizzo, D. M., Eppstein, M. J., Lanute, D. B., Gaalema, D. E., Lakkaraju, K., & Warrender, C. E. (2014). Analysis of a consumer survey on plug-in hybrid electric vehicles. *Transportation Research Part A, Policy and Practice*, *64*, 14–31. doi:10.1016/j.tra.2014.02.019

Kudva, S., & Ye, X. (2017). Smart cities, big data, and sustainability union. *Big Data and Cognitive Computing*, *1*(1), 4.

Kulikowski, C. A. (2017). Historical Roots of International Biomedical and Health Informatics: The Road to IFIP-TC4 and IMIA through Cybernetic Medicine and the Elsinore Meetings. *Yearbook of Medical Informatics*, *26*(1), 257–262.

Kumaraswamy, A., Garud, R., & Ansari, S. (2018). Perspectives on Disruptive Innovations. *Journal of Management Studies*, *55*(7), 1025–1042. doi:10.1111/joms.12399

Kumari, R., & Mishra, A. R. (2020). Multi-criteria COPRAS method based on parametric measures for intuitionistic fuzzy sets: Application of green supplier selection. *Iranian Journal of Science and Technology. Transaction of Electrical Engineering*, *44*(4), 1645–1662. doi:10.100740998-020-00312-w

Kumar, P., Singh, R. K., & Vaish, A. (2017). Suppliers' green performance evaluation using fuzzy extended ELECTRE approach. *Clean Technologies and Environmental Policy*, *19*(3), 809–821. doi:10.100710098-016-1268-y

Kumar, R. R., & Alok, K. (2020). Adoption of electric vehicle: A literature review and prospects for sustainability. *Journal of Cleaner Production*, *253*, 1–21. doi:10.1016/j.jclepro.2019.119911

Kuo, Y., Yang, T., & Huang, G. W. (2008). The use of grey relational analysis in solving multiple attribute decision-making problems. *Computers & Industrial Engineering*, *55*(1), 80–93. doi:10.1016/j.cie.2007.12.002

La Trobe, H. L., & Acott, T. G. (2000). A modified NEP/DSP environmental attitudes scale. *The Journal of Environmental Education*, *32*(1), 12–20. doi:10.1080/00958960009598667

Lai, I., Liu, Y., Sun, X., Zhang, H., & Xu, W. (2015). Factors influencing the behavioural intention toward full electric vehicles: An empirical study in Macau. *Sustainability*, *7*(9), 12564–12585. doi:10.3390u70912564

Lantzos, T., Koykoyris, G., & Salampasis, M. (2013). FarmManager: An Android application for the management of small farms. *Procedia Technology*, *8*, 587–592. doi:10.1016/j.protcy.2013.11.084

Lee, A. H., Kang, H. Y., Hsu, C. F., & Hung, H. C. (2009). A green supplier selection model for high-tech industry. *Expert Systems with Applications*, *36*(4), 7917–7927. doi:10.1016/j.eswa.2008.11.052

Lee, J. A., & Holden, S. J. (1999). Understanding the determinants of environmentally conscious behavior. *Psychology and Marketing*, *16*(5), 373–392. doi:10.1002/(SICI)1520-6793(199908)16:5<373::AID-MAR1>3.0.CO;2-S

Lee, J. H., Hardman, S. J., & Tal, G. (2019). Who is buying electric vehicles in California? Characterising early adopter heterogeneity and forecasting market diffusion. *Energy Research & Social Science*, *55*, 218–226. doi:10.1016/j.erss.2019.05.011

Lee, K. H., & Min, B. (2015). Green R&D for eco-innovation and its impact on carbon emissions and firm performance. *Journal of Cleaner Production, 108*, 534–542. doi:10.1016/j.jclepro.2015.05.114

Lengyel, I. (2004). The pyramid model: enhancing regional competitiveness in Hungary. *Acta Oeconomica, 54*(3), 323-342. doi:10.1556/AOecon.54.2004.3.3

Lengyel, I., & Lukovics, M. (2006, August). *An attempt for the measurement of regional competitiveness in Hungary*. Paper presented at the meeting of the European Regional Science Association, Volos, Greece.

Liang, R., & Chong, H. Y. (2019). A hybrid group decision model for green supplier selection: A case study of megaprojects. *Engineering, Construction, and Architectural Management, 26*(8), 1712–1734. doi:10.1108/ECAM-10-2018-0462

Liao, F., Molin, E., & van Wee, B. (2017). Consumer preferences for electric vehicles: A literature review. *Transport Reviews, 37*(3), 252–275. doi:10.1080/01441647.2016.1230794

Libecap, G. D. (2011). Introduction: Entrepreneurship and global competitiveness in regional economies: Determinants and policy implications. In G. D. Libecap (Ed.), *Entrepreneurship and Global Competitiveness in Regional Economies: Determinants and Policy Implications* (pp. 9–12). Emerald Group Publishing Limited.

Liberty, M., & Hongjuan, W. (2010). Greening our future and environmental values: An investigation of awareness, attitudes and awareness of environmental issues in Zambia. *Environmental Values, 19*(4), 485–516. doi:10.3197/096327110X531570

Li, C., & Niu, B. (2020). Design of smart agriculture based on big data and Internet of things. *International Journal of Distributed Sensor Networks, 16*(5), 1–11. doi:10.1177/1550147720917065

Li, J., Li, N., Peng, J., Cui, H., & Wu, Z. (2019). Energy consumption of cryptocurrency mining: A study of electricity consumption in mining cryptocurrencies. *Energy, 168*, 160–168.

Li, J., Zhang, J., Zhang, D., & Ji, Q. (2019). Does gender inequality affect household green consumption behaviour in China? *Energy Policy, 135*, 111071. doi:10.1016/j.enpol.2019.111071

Likas, A., Vlassis, N., & Verbeek, J. J. (2003). The global k-means clustering algorithm. *Pattern Recognition, 36*(2), 451–461. doi:10.1016/S0031-3203(02)00060-2

Lin, B., & Tan, R. (2017). Estimation of the environmental values of electric vehicles in Chinese cities. *Energy Policy, 104*, 221–229. doi:10.1016/j.enpol.2017.01.037

Lin, C. Y., & Syrgabayeva, D. (2016). Mechanism of environmental concern on intention to pay more for renewable energy: Application to a developing country. *Asia Pacific Management Review, 21*(3), 125–134. doi:10.1016/j.apmrv.2016.01.001

Lin, J., Lobo, A., & Leckie, C. (2017). The role of benefits and transparency in shaping consumers' green perceived value, self-brand connection and brand loyalty. *Journal of Retailing and Consumer Services, 35*, 133–141.

Lin, K. P., Hung, K. C., Lin, Y. T., & Hsieh, Y. H. (2018). Green suppliers performance evaluation in Belt and Road using fuzzy weighted average with social media information. *Sustainability, 10*(1), 5. doi:10.3390u10010005

Lin, R. J., Tan, K. H., & Geng, Y. (2013). Market demand, green product innovation, and firm performance: Evidence from Vietnam motorcycle industry. *Journal of Cleaner Production, 40*, 101–107. doi:10.1016/j.jclepro.2012.01.001

Lin, S. T., & Niu, H. J. (2018). Green consumption: Environmental knowledge, environmental consciousness, social norms, and purchasing behavior. *Business Strategy and the Environment, 27*(8), 1679–1688. doi:10.1002/bse.2233

Lins, T., Oliveira, R. A. R., Correia, L. H., & Silva, J. S. (2018, November). Industry 4.0 Retrofitting. In *2018 VIII Brazilian Symposium on Computing Systems Engineering (SBESC)* (pp. 8-15). IEEE. 10.1109/SBESC.2018.00011

Lin, Z., Manser, M. E., & Picot, G. (1998). The Role of Self-Employment in Job Creation in Canada and the U.S. *OECD-CERF-CILN International Conference on Self-Employment*, Burlington, Ontario, Canada.

Liou, J. J., Chuang, Y. C., Zavadskas, E. K., & Tzeng, G. H. (2019). Data-driven hybrid multiple attribute decision-making model for green supplier evaluation and performance improvement. *Journal of Cleaner Production*, *241*, 118321. doi:10.1016/j.jclepro.2019.118321

Liou, J. J., Tamošaitienė, J., Zavadskas, E. K., & Tzeng, G. H. (2016). New hybrid COPRAS-G MADM Model for improving and selecting suppliers in green supply chain management. *International Journal of Production Research*, *54*(1), 114–134. doi:10.1080/00207543.2015.1010747

Liu, H., Sato, H., & Morikawa, T. (2015). Influences of environmental consciousness and attitudes to transportation on electric vehicle purchase intentions. *Asian Transport Studies*, *3*(4), 430–446.

Liu, X., Vedlitz, A., & Shi, L. (2014). Examining the determinants of public environmental concern: Evidence from national public surveys. *Environmental Science & Policy*, *39*, 77–94. doi:10.1016/j.envsci.2014.02.006

Li, W., & Chen, N. (2018). Absolute income, relative income and environmental concern: Evidence from different regions in China. *Journal of Cleaner Production*, *187*, 9–17. doi:10.1016/j.jclepro.2018.03.171

Li, X., Zhu, S., Yüksel, S., Dinçer, H., & Ubay, G. G. (2020). Kano-based mapping of innovation strategies for renewable energy alternatives using hybrid interval type-2 fuzzy decision-making approach. *Energy*, *211*, 118679. doi:10.1016/j.energy.2020.118679

Li, Y. X., Wu, Z. X., Dinçer, H., Kalkavan, H., & Yüksel, S. (2021). Analyzing TRIZ-based strategic priorities of customer expectations for renewable energy investments with interval type-2 fuzzy modeling. *Energy Reports*, *7*, 95–108. doi:10.1016/j.egyr.2020.11.167

Lo, H. W., Liou, J. J., Wang, H. S., & Tsai, Y. S. (2018). An integrated model for solving problems in green supplier selection and order allocation. *Journal of Cleaner Production*, *190*, 339–352. doi:10.1016/j.jclepro.2018.04.105

Lukovics, M. (2007, October). *Statistical Analysis on the Competitiveness of the Hungarian Sub-regions*. Paper presented at the meeting of the European Regional Science Association, Tatras, Slovak Republic.

Lund, E. H., Jaccheri, L., Li, J., Cico, O., & Bai, X. (2019). Blockchain and sustainability: A systematic mapping study. In *2019 IEEE/ACM 2nd International Workshop on Emerging Trends in Software Engineering for Blockchain (WETSEB)* (pp. 16-23). IEEE.

Luo, N., Olsen, T. L., & Liu, Y. (2021). A Conceptual Framework to Analyze Food Loss and Waste within Food Supply Chains: An Operations Management Perspective. *Sustainability*, *13*(2), 927. doi:10.3390u13020927

Luthra, S., Kumar, A., Zavadskas, E. K., Mangla, S. K., & Garza-Reyes, J. A. (2020). Industry 4.0 as an enabler of sustainability diffusion in supply chain: An analysis of influential strength of drivers in an emerging economy. *International Journal of Production Research*, *58*(5), 1505–1521. doi:10.1080/00207543.2019.1660828

Lutovac, D. M. *Informacioni sistem i informacione tehnologije u zdravstvu* [Information system and information technologies in healthcare]. Škola biznisa. http://www.vps.ns.ac.rs/SB/2008/1.15.pdf

Mahfuzur, SM, R., & Barua, S. (2016). The design and adoption of green banking framework for environment protection: Lessons from Bangladesh. *Australian Journal of Sustainable Business and Society*, *2*(1), 1–19.

Mainieri, T., Barnett, E. G., Valdero, T. R., Unipan, J. B., & Oskamp, S. (1997). Green buying: The influence of environmental concern on consumer behavior. *The Journal of Social Psychology*, *137*(2), 189–204. doi:10.1080/00224549709595430

Makhal, A., Robertson, K., Thyne, M., & Mirosa, M. (2020). Normalising the "ugly" to reduce food waste: Exploring the socialisations that form appearance preferences for fresh fruits and vegetables. *Journal of Consumer Behaviour*, 1–15.

MAKStat Database. (2021). *Nadvoresna Trgovija Indikatori*. Retrieved from http://makstat.stat.gov.mk

Maldonado, E. A., Pogrebnyakov, N. A., & van Gorp, A. F. (2006). ICT Policies as a Means to Inhibit Social Exclusion: The South African Case. In E. M. Trauth, D. Howcroft, T. Butler, B. Fitzgerald, & J. I. DeGross (Eds.), *Social Inclusion: Societal and Organizational Implications for Information Systems*. Springer. doi:10.1007/0-387-34588-4_10

Malliga, A.L., & Revathy, K. (2016). *Customers awareness on green banking- an initiative by private sector bank in Theni district*. Academic Press.

Mandi, K., & Patnaik, N. M. (2019). Mobile apps in agriculture and allied sector: An extended arm for farmers. *Agriculture Update*, *14*(4), 334–342. doi:10.15740/HAS/AU/14.4/334-342

Manic, M. (2018). Marketing approach to specific forms of tourism in the context of revitalization of underdeveloped areas of Eastern Serbia. *Marketing*, *49*(4), 305–306.

Manjunatha, S.B., Biradar, D.P. & Aladakatti, R. (2016). Nanotechnology and its applications in agriculture: A review. *Journal of Farm Sciences, 29*(1).

Ma, Q., & Liu, L. (2004). The technology acceptance model: A meta-analysis of empirical findings. *Journal of Organizational and End User Computing*, *16*(1), 59–72. doi:10.4018/joeuc.2004010104

Markus, M. L., & Nan, W. V. (2020). Theorizing the connections between digital innovations and societal transformation: learning from the case of M-Pesa in Kenya. In *Handbook of digital innovation*. Edward Elgar Publishing.

Marquart-Pyatt, S. T. (2012). Contextual influences on environmental concerns cross-nationally: A multilevel investigation. *Social Science Research*, *41*(5), 1085–1099. doi:10.1016/j.ssresearch.2012.04.003 PMID:23017919

Martindale, W., & Schiebel, W. (2017). The impact of food preservation on food waste. *British Food Journal*, *119*(12), 2510–2518. doi:10.1108/BFJ-02-2017-0114 PMID:29720739

Martin, R., Kitson, M., & Tyler, P. (2006). *Regional Competitiveness*. Routledge.

Martyn, Z. (2013). Tourism and local development – the case of Mórahalom. *Geographica Timisiensis*, *22*(2), 49–64.

Masukujjaman, M., & Aktar, S. (2013). Green banking in Bangladesh: A commitment towards the global initiatives. *Journal of Business and Technology (Dhaka)*, *8*(1-2), 17–40. doi:10.3329/jbt.v8i1-2.18284

Masukujjaman, M., Siwar, C., Mahmud, M. R., & Alam, S. S. (2017). Bankers' perception of Green Banking: Learning from the experience of Islamic banks in Bangladesh. *Geografia: Malaysian Journal of Society and Space*, *12*(2), 144–153.

Ma, Y. J., Gam, H. J., & Banning, J. (2017). Perceived ease of use and usefulness of sustainability labels on apparel products: Application of the technology acceptance model. *Fashion and Textiles*, *4*(1), 1–20.

Mazzukato, M. (2020). *Entrepreneurial state - dispelling myths about the conflict between the public and private sectors*. Novi Sad, Serbia: Academic Book.

Mbow, C., Rosenzweig, C. E., Barioni, L. G., Benton, T. G., Herrero, M., . . . Diouf, A. A. (2020). Food security. In *Climate Change and Land: an IPCC special report on climate change, desertification, land degradation, sustainable land management, food security, and greenhouse gas fluxes in terrestrial ecosystems*. Geneva: Intergovernmental Panel on Climate Change (IPCC). Retrieved from https://www.ipcc.ch/site/assets/uploads/sites/4/2019/11/08_Chapter-5.pdf

McAloone, T. C., & Bey, N. (2009). *Environmental Improvement through Product Development: A Guide*. Danish Environmental Protection Agency.

McDonald, S., Oates, C. J., Young, C. W., & Hwang, K. (2006). Toward sustainable consumption: Researching voluntary simplifiers. *Psychology and Marketing*, *23*(6), 515–534. doi:10.1002/mar.20132

McDougall, P. (2000). *International entrepreneurship*. AMJ.

Meadows, D. (2009). *Thinking in systems: A primer*. Earthscan.

Mehta, K., & Sharma, R. (2016). Customers" Persistence for Green Banking in Nepal. *Asian Journal of Research in Banking and Finance*, *6*(10), 30–44. doi:10.5958/2249-7323.2016.00050.X

Meissner, D. (2019). Public-Private Partnership Models for Science, Technology, and Innovation Cooperation. *Journal of the Knowledge Economy*, *10*(4), 1341–1361. doi:10.100713132-015-0310-3

Melander, L. (2018). Customer and supplier collaboration in green product innovation: External and internal capabilities. *Business Strategy and the Environment*, *27*(6), 677–693. doi:10.1002/bse.2024

Mele, C., & Russo-Spena, T. (2015). Eco-innovation practices. *Journal of Organizational Change Management*, *28*(1), 4–25. doi:10.1108/JOCM-08-2013-0146

Migendt, M., Polzin, F., Schock, F., Täube, F. A., & von Flotow, P. (2017). Beyond venture capital: An exploratory study of the finance-innovation-policy nexus in cleantech. *Industrial and Corporate Change*, *26*(6), 973–996. doi:10.1093/icc/dtx014

Mi, L., Xu, T., Sun, Y., Zhao, J., Lv, T., Gan, X., ... Qiao, L. (2021). Playing Ant Forest to promote online green behavior: A new perspective on uses and gratifications. *Journal of Environmental Management*, *278*, 111544. PMID:33129025

Miles, C. (2019). The combine will tell the truth: On precision agriculture and algorithmic rationality. *Big Data & Society*, *6*(1), 1–12. doi:10.1177/2053951719849444

Milicevic, S., Milovanovic, V., & Milutinovic, S. (2012). Wellness tourism in the function of change management in the tourism market. *Economics: Theory and Practice*, *5*(2), 29-41.

Mills, A., Watson, R., Pitt, L., & Kietzmann, J. (2016). Wearing safe: Physical and informational security in the age of the wearable device. *Business Horizons*, *59*(6), 615–622. doi:10.1016/j.bushor.2016.08.003

Mirzaei, O., Micheels, E., & Boecker, A. (2016). Product and Marketing Innovation in Farm-Based Businesses: The Role of Entrepreneurial Orientation and Market Orientation. *The International Food and Agribusiness Management Review*, *19*(2), 99–130.

Mishra, D. K. (2013). Green strategies: Response of Indian banks to climate change. *An International Quarterly Journal of Environmental Sciences*, *3*, 345 – 348.

Mishra, A. (2019). Digitization in Indian Banking Sector. *ITIHAS The Journal of Indian Management*, *9*(4), 73–80.

Mishra, N. (2008). New agro-technology and its impact on rural institutions in Orissa. *Social Change*, *38*(3), 411–433. doi:10.1177/004908570803800302

Mishra, P., & Homa, F. (2019). *Essentials of Statistics in Agricultural Sciences*. Apple Academic Press. doi:10.1201/9780429425769

Mitchell, O. S. (2020). *Building Better Retirement Systems in the Wake of the Global Pandemic*. National Bureau of Economic Research Working Paper 27261.

Moghul, U. F., & Safar-Aly, S. H. (2014). Green Sukuk: The Introduction of Islam's Environmental Ethics to Contemporary Islamic Finance. *Geo. Int'l Envtl. L. Rev.*, *27*, 1.

Mohamed, M., Higgins, C., Ferguson, M., & Kanaroglou, P. (2016). Identifying and characterizing potential electric vehicle adopters in Canada: A two-stage modelling approach. *Transport Policy*, *52*, 100–112. doi:10.1016/j.tranpol.2016.07.006

Mohanty, B. R., & Pawar, B. R. (2019). Cashless (Less Cash) Economy & Digitization of Indian Economy. *Proceedings of 10th International Conference on Digital Strategies for Organizational Success.*

Mohd Suki, N. (2016). Consumer environmental concern and green product purchase in Malaysia: Structural effects of consumption values. *Journal of Cleaner Production*, *132*, 204–214. doi:10.1016/j.jclepro.2015.09.087

Momaya, K. S. (2019). The Past and the Future of Competitiveness Research: A Review in an Emerging Context of Innovation and EMNEs. *International Journal of Global Business and Competitiveness*, *14*(1), 1–10. doi:10.100742943-019-00002-3

Mondal, S. (2015). Financial inclusion: A step towards eradicate poverty. *American Journal of Theoretical and Applied Business*, *1*(1), 21–26.

Montoya, D. (2007). Nuevas necesidades en ingeniería para el desarrollo de la biotecnología. *Revista Colombiana de Biotecnologia*, *9*(2).

Mooij, M., & Hofstede, G. (2002). Convergence and divergence in consumer behavior: Implications for international retailing. *Journal of Retailing*, *78*(1), 61–69. doi:10.1016/S0022-4359(01)00067-7

Morea, D., & Poggi, L. A. (2017). An innovative model for the sustainability of investments in the wind energy sector: The use of green sukuk in an Italian case study. *International Journal of Energy Economics and Policy*, *7*(2).

Morrar, R., Arman, H., & Mousa, S. (2017). The fourth industrial revolution (Industry 4.0): A social innovation perspective. *Technology Innovation Management Review*, *7*(11), 12–20. doi:10.22215/timreview/1117

Morton, C., Anable, J., & Nelson, J. D. (2017). Consumer structure in the emerging market for electric vehicles: Identifying market segments using cluster analysis. *International Journal of Sustainable Transportation*, *11*(6), 443–459. doi:10.1080/15568318.2016.1266533

Mousakhani, S., Nazari-Shirkouhi, S., & Bozorgi-Amiri, A. (2017). A novel interval type-2 fuzzy evaluation model based group decision analysis for green supplier selection problems: A case study of battery industry. *Journal of Cleaner Production*, *168*, 205–218. doi:10.1016/j.jclepro.2017.08.154

Moyo, N. (2018). Environmental concern and purchase intention of electric vehicles in the eastern part of China. *Archives of Business Research*, *6*(8), 297–309.

Mrkajic, B., Murtinu, S., & Scalera, V. G. (2019). Is green the new gold? Venture capital and green entrepreneurship. *Small Business Economics*, *52*(4), 929–950. doi:10.100711187-017-9943-x

Muangprathub, J., Boonnam, N., Kajornkasirat, S., Lekbangponga, N., Wanichsombat, A., & Nillaor, P. (2019). IOT and agriculture data analysis for smart farm. *Computers and Electronics in Agriculture*, *156*, 467–474. doi:10.1016/j.compag.2018.12.011

Mueller, J. T., & Mullenbach, L. E. (2018). Looking for a white male effect in generation Z: Race, gender, and political effects on environmental concern. *Society & Natural Resources*, *31*(8), 925–941. doi:10.1080/08941920.2018.1445331

Muhammad, S., Fathelrahman, E., & Ullah, R. U. T. (2015). Factors affecting consumers' willingness to pay for certified organic food products in United Arab Emirates. *Journal of Food Distribution Research*, *46*(1), 37–45.

Muhmad, S. N., & Muhmad, S. N. (2018). Potential Development of Sukuk in Competitive Market. *International Journal of Business and Management, 2*(2), 26–29.

Munir, S., Masruro, U., Fawaiq, A., & Merlinda, S. (2020). Green Sukuk: Indonesian Youth Investment Prospects For Environmental Sustainability. *Economics Business and Organization Research*, 140-147.

Muñoz, E. (1999). *Biotecnología y desarrollo en distintos contextos culturales. Influencias e impactos*. Consejo Superior de Investigaciones Científicas.

Mwangi, M., & Kariuki, S. (2015). Factors Determining Adoption of New Agricultural Technology by Smallholder Farmers in Developing Countries. *Journal of Economics and Sustainable Development, 6*(5), 208–216.

My Dung, T. T., Luan, N. M., & Quoc, L. H. (2016). The analytic approach applications in green supplier selection: A literature review. *Journal of Engineering and Applied Sciences (Asian Research Publishing Network), 11*(11), 6754–6762.

Nagy, D., Schuessler, J., & Dubinsky, A. (2016). Defining and identifying disruptive innovations. *Industrial Marketing Management, 57*, 119–126. doi:10.1016/j.indmarman.2015.11.017

Nahavandi, S. (2019). Industry 5.0—A human-centric solution. *Sustainability, 11*(16), 4371. doi:10.3390u11164371

Najaf, K., Subramaniam, R. K., & Atayah, O. F. (2021). Understanding the implications of FinTech Peer-to-Peer (P2P) lending during the COVID-19 pandemic. *Journal of Sustainable Finance & Investment*, 1-16.

Nanda, S. (2017). A study on the customer awareness of wearable technology. *International Journal of Multidisciplinary and Current Research, 5*, 293–298.

Närvänen, E., Mesiranta, N., Sutinen, U. M., & Mattila, M. (2018). Creativity, aesthetics and ethics of food waste in social media campaigns. *Journal of Cleaner Production, 195*, 102–110. doi:10.1016/j.jclepro.2018.05.202

Nas, T. F. (2016). *Cost-Benefit Analysis. Theory and Application* (2nd ed.). Lexington Books.

Netzband, M., Stefanov, W. L., & Redman, C. (Eds.). (2007). *Applied remote sensing for urban planning, governance and sustainability*. Springer Science & Business Media.

Neuman, W. R. (2010). *Media, Technology, and Society: Theories of Media Evolution*. Digit Cult Books. doi:10.3998/dcbooks.8232214.0001.001

Newell, S. J., & Green, C. L. (1997). Racial differences in consumer environmental concern. *The Journal of Consumer Affairs, 31*(1), 53–69. doi:10.1111/j.1745-6606.1997.tb00826.x

Newman, T. P., & Fernandes, R. (2016). A re-assessment of factors associated with environmental concern and behavior using the 2010 General Social Survey. *Environmental Education Research, 22*(2), 153–175. doi:10.1080/13504622.2014.999227

Nguyen, H. (2018) *Sustainable food systems. Concept and framework*. Agricultural Development Economics Division (ESA), Food and Agriculture Organization of the United Nations FAO. Last access 22/05/2021 web http://www.fao.org/3/ca2079en/CA2079EN.pdf

Nguyen, H. T., Hoang, T. G., Nguyen, L. Q. T., Le, H. P., & Mai, H. X. V. (2021). Green technology transfer in developing country: Mainstream practitioner views. *The International Journal of Organizational Analysis*. Advance online publication. doi:10.1108/IJOA-11-2019-1941

Nguyen, H., Hoang, T. G., & Luu, H. (2019). Corporate Social Responsibility in Vietnam: Opportunities and Innovation Experienced by Multinational Corporation Subsidiaries. *Social Responsibility Journal, 16*(6), 771–792. doi:10.1108/SRJ-02-2019-0082

Nguyen, T. N., Lobo, A., & Greenland, S. (2017). The influence of cultural values on green purchase behaviour. *Marketing Intelligence & Planning, 35*(3), 377–396. doi:10.1108/MIP-08-2016-0131

Nguyen, T., Li, Z., Spiegler, V., Ieromonachou, P., & Lin, Y. (2018). Big data analytics in supply chain management: A state-of-the-art literature review. *Computers & Operations Research, 98*, 254–264. doi:10.1016/j.cor.2017.07.004

Nicoletti, B., Nicoletti, W., & Weis. (2017). *Future of FinTech.* Basingstoke, UK: Palgrave Macmillan.

Nicoletti, B. (2017). Fintech innovation. In *The Future of FinTech* (pp. 81–159). Palgrave Macmillan.

Nilashi, M., Rupani, P. F., Rupani, M. M., Kamyab, H., Shao, W., Ahmadi, H., ... Aljojo, N. (2019). Measuring sustainability through ecological sustainability and human sustainability: A machine learning approach. *Journal of Cleaner Production, 240*, 118162.

Nishant, R., Kennedy, M., & Corbett, J. (2020). Artificial intelligence for sustainability: Challenges, opportunities, and a research agenda. *International Journal of Information Management, 53*, 102104.

Njoroge, B. M., Fei, T. K., & Thiruchelvam, V. (2018). A Research Review of Precision Farming Techniques and Technology. *Journal of Applied Technology and Innovation, 2*(1), 22–30.

Nowak, B. (2021). Precision Agriculture: Where do We Stand? A Review of the Adoption of Precision Agriculture Technologies on Field Crops Farms in Developed Countries. In *Agricultural Research*. Springer. doi:10.100740003-021-00539-x

Oberhoffer, G. (1962). Report on an international seminar for medical documentation and statistics. *Methods of Information in Medicine, 01*, 27–31.

Ochara, N. M., & Mawela, T. (2015). Enabling social sustainability of e-participation through mobile technology. *Information Technology for Development, 21*(2), 205–228.

OECD. (1994). Tourism strategies and rural development. Organization for Economic Co-Operation and Development.

OECD. (2009). *Sustainable Manufacturing and Eco-innovation: Framework, Practices and Measurement.* Organisation for Economic Cooperation and Development.

OECD. (2020). https://www.oecd.org/south-east-europe/COVID-19-Crisis-in-Serbia.pdf

OECD. (2020). https://www.oecd.org/south-east-europe/COVID-19-Crisis-Response-Western-Balkans.pdf

OECD. (2020a). *Pension Funds in Figures 2020.* OECD.

OECD. (2020b). *Pension Markets in Focus 2020.* Author.

Olds, J., Wei, Sh., Corin, S., Barry, D., Anthony, I., Bharat, S., Suresh, Ch., & Ziaul, H. (2003). *First Year Report.* The Institute For Future Space Transport A University Research Engineering And Technology Institute.

Olmos, L., Ruester, S., & Liong, S.-J. (2012). On the selection of financing instruments to push the development of new technologies: Application to clean energy technologies. *Energy Policy, 43*, 252–266. doi:10.1016/j.enpol.2012.01.001

Ommen, N. O., Blut, M., Backhaus, C., & Woisetschläger, D. M. (2016). Toward a better understanding of stakeholder participation in the service innovation process: More than one path to success. *Journal of Business Research, 69*(7), 2409–2416. doi:10.1016/j.jbusres.2016.01.010

Omoregie, E. M., & Thomson, K. J. (2001). Measuring regional competitiveness in oilseeds production and processing in Nigeria: A spatial equilibrium modelling approach. *Agricultural Economics, 26*(3), 281–294. doi.org/10.1111/j.1574-0862.2001.tb00070.x

Optimizacija zdravstva - Zaječarski okrug [Optimization healthcare, Zajecar district]. (n.d.). https://optimizacijazdravstva.rs/lat/region-istocne-srbije

Organisation for Economic Co-operation and Development (OECD), UN Food and Agriculture Organization (FAO) & UN Capital Development Fund (UNCDF). (2016). Adopting a Territorial Approach to Food Security and Nutrition Policy. OECD Publishing.

Ortega-Anderezand, D., & Lai, E. (2017, February). The Main Factors Influencing Entrepreneurial Activities and Intentions within a Country—A Case Study of Spain. *International Journal of Innovation, Management and Technology*, *8*(1).

Østby, G. (2016). Rural–urban migration, inequality and urban social disorder: Evidence from African and Asian cities. *Conflict Management and Peace Science*, *33*(5), 491–515. doi:10.1177/0738894215581315

Ouafiq, E., Elrharras, A., Mehdary, A., Chehri, A., Saadane, R., & Wahbi, M. (2020). IoT in Smart Farming Analytics, Big Data Based Architecture. In Human Centred Intelligent Systems. Springer.

Ozili, P. K. (2018). Impact of digital finance on financial inclusion and stability. *Borsa Istanbul Review*, *18*(4), 329–340.

Pallant, J. (2011). *SPSS survival manual: a step by step guide to data analysis using SPSS* (4th ed.). Allen & Unwin.

Pang, Q., Yang, T., Li, M., & Shen, Y. (2017). A fuzzy-grey multicriteria decision making approach for green supplier selection in low-carbon supply chain. *Mathematical Problems in Engineering*, *9653261*, 1–9. Advance online publication. doi:10.1155/2017/9653261

Pant, B. (2016). *Financial Inclusion in Nepal: Policy Review and Prescriptions* (No. 34). NRB Working Paper.

Parasuraman, A., Berry, L. L., & Zeithaml, V. A. (1991). Understanding customer expectations of service. *Sloan Management Review*, *32*(3), 39–48.

Parfitt, J., Barthel, M., & Macnaughton, S. (2010). Food waste within food supply chains: Quantification and potential for change to 2050. *Philosophical Transactions of the Royal Society of London. Series B, Biological Sciences*, *365*(1554), 3065–3081. doi:10.1098/rstb.2010.0126 PMID:20713403

Pariag-Maraye, N., Munusami, N., & Ansaram, K. (2017). A customer's perspective of green banking: A case study of commercial banks in Mauritius. *Theoretical Economics Letters*, *7*(07), 19–75. doi:10.4236/tel.2017.77134

Park, D.-H., Lee, J., & Han, I. (2014). The effect of on-line consumer reviews on consumer purchasing intention: The moderating role of involvement. *International Journal of Electronic Commerce*, *11*(4), 125–148. doi:10.2753/JEC1086-4415110405

Patel, R. (2012). The Long Green Revolution. *The Journal of Peasant Studies*, *40*(1), 1–63. Advance online publication. doi:10.1080/03066150.2012.719224

Peksevim, S., & Akgiray, V. (2019). Reforming the Pension System in Turkey: Comparison of Mandatory and Auto-Enrollment Pension Systems in Selected OECD Countries. OECD.

Peng, M. (2012). *Global Strategy*. Thomson South-Western.

Perera, C., & Zaslavsky, A. (2014). Improve the sustainability of internet of things through trading-based value creation. In *2014 IEEE World Forum on Internet of Things (WF-IoT)* (pp. 135-140). IEEE.

Peres, R., Muller, E., & Mahajan, V. (2010). Innovation diffusion and new product growth models: A critical review and research directions. *International Journal of Research in Marketing*, *27*(2), 91–106. doi:10.1016/j.ijresmar.2009.12.012

Peters, A., & Dütschke, E. (2014). How do consumers perceive electric vehicles? a comparison of German consumer groups. *Journal of Environmental Policy and Planning*, *16*(3), 359–377. doi:10.1080/1523908X.2013.879037

Petkovic, M.& Lazarevic, S. (2012). Upravljanje interorganizacionim odnosima: dizajn centra zajedničkih usluga [Inter-organizational relations management: the design of a shared services center]. *Management, 64*.

Petkovic, M., & Lukic, J. (2013). Uticaj informacione tehnologije na dizajn organizacije: Primer organizacije u zdravstvu [The influence of information technology on the design of the organization: the example of the organization in health care]. *Sociologija*, *55*(3), 439–460.

Petrović, P., & Nikolić, G. (2018). Schumpeterian Growth Theory: empirical testing of barriers to competition effect. *Economic Annals, 63*(217), 7-37. doi:10.2298/EKA1817007P

Pfeiffer, F., & Reize, F. (2000). Business Start-ups by the Unemployed - an Econometric Analysis Based on Firm Data. *Labour Economics*, *7*(5), 629–663. doi:10.1016/S0927-5371(00)00016-6

Piliyanti, I. (2019). Fintech achieving sustainable development: The side perspective of crowdfunding platform. *Shirkah: Journal of Economics and Business*, *3*(2).

Plötz, P., Schneider, U., Globisch, J., & Dütschke, E. (2014). Who will buy electric vehicles? Identifying early adopters in Germany. *Transportation Research Part A, Policy and Practice*, *67*, 96–109. doi:10.1016/j.tra.2014.06.006

Plouffe, S., Lanoie, P., Berneman, C., & Vernier, M.-F. (2011). Economic benefits tied to ecodesign. *Journal of Cleaner Production*, *19*(6), 573–579. doi:10.1016/j.jclepro.2010.12.003

Poonia, A. S., Banerjee, C., Banerjee, A., & Sharma, S. K. (2020). Comparative Analysis of Internet of Things Monitoring Devices in Smart Agriculture Using Deep-Learning Techniques. In Soft Computing: Theories and Applications. Springer. doi:10.1007/978-981-15-0751-9_129

Porter, M. (1980). *Competitive Strategy*. Free Press.

Porter, M. E. (1998). Location, clusters and the new economics of competition. *Business Economics (Cleveland, Ohio)*, *33*(1), 7–17.

Porter, M. E. (2001). Regions and the New Economics of Competition. In A. J. Scott (Ed.), *Global City Regions: Trends, Theory, Policy* (pp. 139–156). Oxford University Press.

Potena, C., Khanna, R., Nieto, J., Siegwart, R., Nardi, D. & Pretto, A. (2020). AgriColMap: Aerial-Ground Collaborative 3D Mapping for Precision Farming. *IEEE Robotics and Automation Letters, 4*(2), 1085-1092.

Powell, T. C. (2001). Competitive advantage: Logical and philosophical considerations. *Strategic Management Journal*, *22*(9), 875–888. doi.org/10.1002/smj.173

Prasad, V. M. (2017). Green Banking- Is The Concept Known By The Bank Customers? *Journal of Advances in Business Management.*, *3*(1), 23–27.

Przychodzen, J., & Przychodzen, W. (2015). Relationships between eco-innovation and financial performance–evidence from publicly traded companies in Poland and Hungary. *Journal of Cleaner Production*, *90*, 253–263.

Purvis, B., Mao, Y., & Robinson, D. (2019). Three pillars of sustainability: In search of conceptual origins. *Sustainability Science*, *14*(3), 681–695.

Puschmann, T. (2017). Fintech. *Business & Information Systems Engineering*, *59*(1), 69–76.

Qing-tao, Z., Qing, X., Liu, C., & Geng, S. (2013). Technologies for Efficient Use of Irrigation Water and Energy in China. *Journal of Integrative Agriculture*, *12*(8), 1363–1370. doi:10.1016/S2095-3119(13)60544-4

Qin, J., Liu, X., & Pedrycz, W. (2017). An extended TODIM multi-criteria group decision making method for green supplier selection in interval type-2 fuzzy environment. *European Journal of Operational Research*, *258*(2), 626–638. doi:10.1016/j.ejor.2016.09.059

Qiu, D., Dinçer, H., Yüksel, S., & Ubay, G. G. (2020). Multi-faceted analysis of systematic risk-based wind energy investment decisions in E7 economies using modified hybrid modeling with IT2 fuzzy sets. *Energies*, *13*(6), 1423. doi:10.3390/en13061423

Queirós, A., Faria, D., & Almeida, F. (2017). Strengths and Limitation of Qualitative and Quantitative Research Methods. *European Journal of Education Studies*, *3*(9), 369–387.

Qu, W., Sun, H., & Ge, Y. (2020). The effects of trait anxiety and the big five personality traits on self-driving car acceptance. *Transportation*, 1–17.

Radivojević, V., Kahrović, E., & Krstić, M. (2019a). Population skills as an indicator of European countries' competitiveness in the modern economy. *Vojno delo, 71*(5), 105-116. doi:10.5937/vojdelo1905105R

Radivojević, V., Krstić, B., Krstić, M., & Petković, M. (2019b). Benchmarking agricultural and other environmental performances of Central and East European countries. *Economics of Agriculture, 66*(2), 471-484. doi:10.5937/ekoPolj1902471R

Radivojevic, A. (2020). *The Tourism sector in the Republic of Serbia during the pandemic. Quarterly Monitor, 61.*

Radjenovic, Z., Milovanovic, S., & Milovanovic. (2017). Heliant Health Information System as a Support to Electronic Business of Healthcare Organizations in Serbia. *Facta Universitatis Series: Economics and Organization*, *14*(3), 267. doi:10.22190/FUEO1703265R

Rahimi, I., Gandomi, A. H., Fong, S. J., & Ülkü, M. A. (Eds.). (2020). *Big Data Analytics in Supply Chain Management: Theory and Applications*. CRC Press. doi:10.1201/9780367816384

Rahim, S. R. M., & Mohamad, Z. Z. (2018). Green Sukuk for Financing Renewable Energy Projects. *TUJISE*, *5*(2), 129–144. doi:10.26414/m031

Rahul, D., & Pretesh, J. (2018). Application of Taguchi-Based Design of Experiments for Industrial Chemical Processes. *Intech Open*, 137-155.

Raj, D. G. P., & Rajan, D. A. P. (2017). A Study on the customer awareness on green banking initiatives. *Intercontinental Journal of Finance Research Review*, *5*(7), 54–65.

Rajper, S. Z., & Albrecht, J. (2020). Prospects of electric vehicles in the developing countries: A literature review. *Sustainability*, *12*(5), 1–19. doi:10.3390u12051906

Ramadhan, I. A. (2020, March). Green Sukuk Issuance as an Investment Instrument for Sustainable Development. In *International Conference on Law, Governance and Islamic Society (ICOLGIS 2019)* (pp. 95-98). Atlantis Press. 10.2991/assehr.k.200306.189

Ramos-Soler, I., Martínez-Sala, A. M., & Campillo-Alhama, C. (2019). ICT and the sustainability of World Heritage Sites. Analysis of senior citizens' use of tourism Apps. *Sustainability*, *11*(11), 3203.

Ra, S., Ahmed, M., & Teng, P. S. (2019). Creating high-tech 'agropreneurs' through education and skills development. *International Journal of Training Research*, *17*(sup1, s1), 41–53. doi:10.1080/14480220.2019.1629736

Rashid, A., Asif, F. M. A., Krajnik, P., & Nicolescu, C. M. (2013). Resource Conservative Manufacturing: An essential change in business and technology paradigm for sustainable manufacturing. *Journal of Cleaner Production, 57*, 166–177. doi:10.1016/j.jclepro.2013.06.012

Rastogi, C. (2018). M-Kopa solar: Lighting up the dark continent. *South Asian Journal of Business and Management Cases, 7*(2), 93–103.

Rauschnabel, P., Brem, A., & Ro, Y. (2015). *Augmented Reality Smart Glasses: Definition, Conceptual Insights, and Managerial Importance*. Retrieved from Semantic Scholar: https://pdfs.semanticscholar.org/b2d8/bbbf1c654f8b-b157ad83ffdc9543598c6724.pdf

Razak, S. S., Saiti, B., & Dinç, Y. (2019). The contracts, structures and pricing mechanisms of sukuk: A critical assessment. *Borsa Istanbul Review, 19*, S21-S33.

Republic of Serbia. (2018). *Spa Law*. Official Gazette of RS, 95/2018, Belgrade, the Republic of Serbia.

Republic of Serbia. (2018). *Spa Law*. Official Gazette of RS, 95/2018. https://www.paragraf.rs/propisi/zakon-o-banjama-republike-srbije.html

Republic Statistical Office (2019). *Municipalities and regions in the Republic of Serbia*. Belgrade, Serbia: Republic Statistical Office. (in Serbian)

Revell, A., Stokes, D., & Chen, H. (2010). Small businesses and the environment: Turning over a new leaf? *Business Strategy and the Environment, 19*(5), 273–288.

Revest, V., & Sapio, A. (2012). Financing technology-based small firms in Europe: What do we know? *Small Business Economics, 39*(1), 179–205. doi:10.100711187-010-9291-6

Risal, N., & Joshi, S. K. (2018). Measuring Green Banking Practices on Bank's Environmental Performance: Empirical Evidence from Kathmandu valley. *Journal of Business and Social Sciences, 2*(1), 44–56. doi:10.3126/jbss.v2i1.22827

Ritchie, H., & Roser, M. (2021). *Emissions by sector*. https://ourworldindata.org/emissions-by-sector

Roberts, J. A. (1996). Green consumers in the 1990s: Profile and implications for advertising. *Journal of Business Research, 36*(3), 217–231. doi:10.1016/0148-2963(95)00150-6

Roberts, J. A., & Bacon, D. R. (1997). Exploring the subtle relationships between environmental concern and ecologically conscious consumer behavior. *Journal of Business Research, 40*(1), 79–89. doi:10.1016/S0148-2963(96)00280-9

Robson, K., Kietzmann, J., & Pitt, L. (2016). APC forum: Extending business values through wearables. *MIS Quarterly Executive, 15*(2), 169–179.

Rogers, E. (1962). *Diffusion of innovations*. The Free Press.

Rohm, H., Oostindjer, M., Aschemann-Witzel, J., Symmank, C., & Almli, V., De Hooge, I. E., ... Karantininis, K. (2017). Consumers in a sustainable food supply chain (COSUS): Understanding consumer behavior to encourage food waste reduction. *Foods, 6*(12), 104. doi:10.3390/foods6120104 PMID:29186883

Rosati, F., & Diniz, L. G. (2019). Addressing the Sustainable Development Goals in sustainability reports: The relationship with institutional factors. *Journal of Cleaner Production, 215*, 1. doi:10.1016/j.jclepro.2018.12.107

Rose, D. C., Wheeler, R., Winter, M., Lobley, M., & Chivers, C. A. (2021). Agriculture 4.0: Making it work for people, production, and the planet. *Land Use Policy, 100*(January), 104933. doi:10.1016/j.landusepol.2020.104933

Rose, D., & Bruce, T. (2018). Finding the right connection: What makes a successful decision support system? *Food and Energy Security*, *7*(1), 1–2. doi:10.1002/fes3.123 PMID:29938109

Rose, D., & Chilvers, J. (2018). Agriculture 4.0: Broadening Responsible Innovation in an Era of Smart Farming. *Frontiers in Sustainable Food Systems*, *2*(87), 1–7. doi:10.3389/fsufs.2018.00087

Rosemary, I., Kraybill, D., & Erbaugh, M. (2014). Impact of integrated pest management technologies on farm revenues of rural households: The case of smallholder Arabica coffee farmers. *African Journal of Agricultural and Resource Economics*, *9*(2), 1–13.

Rosencrance, L. (2017). *Blockchain technology will help the world go green*. Bitcoin Magazine.

Roszkowska, E. (2011). Multi-Criteria Decision- Making Models by Applying the TOPSIS Method to Crisp and Interval Data. Multiple Criteria Decision Making.

Rouyendegh, B. D., Yildizbasi, A., & Üstünyer, P. (2020). Intuitionistic fuzzy TOPSIS method for green supplier selection problem. *Soft Computing*, *24*(3), 2215–2228. doi:10.100700500-019-04054-8

Roy, B., & Vanderpooten, D. (1996). The European School of MCDA, Emergence, Basic Features, and Current Works. *Journal of Multi-Criteria Decision Analysis*, *5*(1), 22–37.

Roy, R. K. (1990). *A Primer on The Taguchi Method*. Van Nostrand Reinhold.

Ruiz-Martínez, I., & Esparcia, J. (2020). Internet Access in Rural Areas: Brake or Stimulus as Post-Covid-19 Opportunity? *Sustainability*, *12*(22), 1–17. doi:10.3390u12229619

Rupnik, R., Kukar, M., Vracar, P., Kosir, D., Pevec, D., & Bosnic, Z. (2019). AgroDSS: A decision support system for agriculture and farming. *Computers and Electronics in Agriculture*, *161*, 260–271. doi:10.1016/j.compag.2018.04.001

Ryu, H. S., & Ko, K. S. (2020). Sustainable Development of Fintech: Focused on Uncertainty and Perceived Quality Issues. *Sustainability*, *12*(18), 7669.

Saad, M. M., Ibrahim, M. A., & Sayad, Z. M. (2017). Eco-City as Approach for Sustainable Development. *American Scientific Research Journal for Engineering, Technology, and Sciences*, *28*(1), 54–74.

Saaty, T. L. (1990). How to make a decision: The analytic hierarchy process. *European Journal of Operational Research*, *48*(1).

Saberi, S., Kouhizadeh, M., Sarkis, J., & Shen, L. (2019). Blockchain technology and its relationships to sustainable supply chain management. *International Journal of Production Research*, *57*(7), 2117–2135. doi:10.1080/00207543.2018.1533261

Sabharwal, M. (2013). The Use of Eco-Friendly Technology and Gree Methods to Bring Down the Carbon Footprint by Indian Banks. *Masters International Journal of Management Research and Development*, *1*(1), 76–85.

Sachs, J., & Ki-moon, B. (2015). *The Age of Sustainable Development*. Columbia University Press. doi:10.7312ach17314

Sahay, M. R., von Allmen, M. U. E., Lahreche, M. A., Khera, P., Ogawa, M. S., Bazarbash, M., & Beaton, M. K. (2020). *The promise of fintech: Financial inclusion in the post COVID-19 era*. International Monetary Fund.

Sahin, S., Ozdemir, Z. A., & Onal, Y. B. (2019). Türkiye'de Bireysel Emeklilik Sisteminin Sermaye Piyasasının Gelişimi Üzerine Etkisi. *Dokuz Eylul Universitesi Iktisadi ve Idari Bilimler Dergisi*, *33*(2), 517–541.

Sahni, D., & Dhamija, A. (2018). Green Banking: Perception and Willingness of Customer to Adapt Green Banking. *International Journal of Financial Management*, *7*(2), 1–8.

Saini, S., Sharma, S., Gulati, A., Hussain, S., & von Braun, J. (2017). Indian food and welfare schemes: Scope for digitization towards cash transfers. *ZEF Discussion Papers on Development Policy*, 241.

Sáiz, V., & Rovira, F. (2020). From Smart Farming towards Agriculture 5.0: A Review on Crop Data Management. *Agronomy (Basel)*, *10*(2), 1–21. doi:10.3390/agronomy10020207

Sakaguchi, L., Pak, N., & Potts, M. D. (2018). Tackling the issue of food waste in restaurants: Options for measurement method, reduction and behavioral change. *Journal of Cleaner Production*, *180*, 430–436. doi:10.1016/j.jclepro.2017.12.136

Salam, A. (2020). Internet of things for sustainability: perspectives in privacy, cybersecurity, and future trends. In *Internet of Things for Sustainable Community Development* (pp. 299–327). Springer.

Salvia, A. L., & Brandli, L. L. (2020). Energy sustainability at universities and its contribution to SDG 7: a systematic literature review. *Universities as Living Labs for Sustainable Development*, 29-45.

Samdahl, D. M., & Robertson, R. (1989). Social determinants of environmental concern: Specification and test of the model. *Environment and Behavior*, *21*(1), 57–81. doi:10.1177/0013916589211004

Samuel, K. S. (2014). The Role of Technology in Enhancing Corporate Performance. *International Journal of Research in Management & Business Studies*, *1*(1), 116–123.

Sands, D. C., Morris, C. E., Dratz, E. A., & Pilgeram, A. L. (2009). Elevating optimal human nutrition to a central goal of plant breeding and production of plant-based foods. *Plant Science*, *177*(5), 377–389. Advance online publication. doi:10.1016/j.plantsci.2009.07.011 PMID:20467463

Santos-Valle, S., & Kienzle, J. (2020). *Agriculture 4.0 – Agricultural robotics and automated equipment for sustainable crop production*. FAO.

Sarkis, J., Zhu, Q., & Lai, K. H. (2011). An organizational theoretic review of green supply chain management literature. *International Journal of Production Economics*, *130*(1), 1–15. doi:10.1016/j.ijpe.2010.11.010

Sathye, M. (1999). Adoption of Internet banking by Australian consumers: An empirical investigation. *International Journal of Bank Marketing*, *17*(7), 324–334. doi:10.1108/02652329910305689

Scheyvens, R., Banks, G., & Hughes, E. (2016). The private sector and the SDGs: The need to move beyond 'business as usual'. *Sustainable Development*, *24*(6), 371–382.

Schiffman, L., & Kanuk, L. (1978). *Consumer Behavior*. Prentice Hill.

Schuitema, G., Anable, J., Skippon, S., & Kinnear, N. (2013). The role of instrumental, hedonic and symbolic attributes in the intention to adopt electric vehicles. *Transportation Research Part A, Policy and Practice*, *48*, 39–49. doi:10.1016/j.tra.2012.10.004

Schwab, K., & Porter, M. E. (2018). *The Global Competitiveness Report 2018-2019*. Geneva, Switzerland: World Economic Forum.

Schwertner, J., Juray, T., & Martyn, Z. (2014). *Integrated spa development strategy for spas situated in the area along the main highway traffic route crossing the Hungarian-Serbian border region. In Cooling Cubes Project*. The University of Szeged, Department of Economic and Social Geography.

Sedlmeir, J., Buhl, H. U., Fridgen, G., & Keller, R. (2020). The energy consumption of blockchain technology: Beyond myth. *Business & Information Systems Engineering*, *62*(6), 599–608.

Semios. (2021). *Water management*. Retrieved from https://semios.com/solutions/water-management/

Sepic, D. (2005). *The regional competitiveness: some notions*. Russian-European Centre for Economic Policy.

Sergey, S., & Kweku, M. (2013). Using Data Envelopment Analysis (DEA) for monitoring efficiency-based performance of productivity- driven organizations. Design and implementation of a decision support system. *Omega*, *41*(1), 131–142. doi:10.1016/j.omega.2011.02.010

Seuring, S., & Müller, M. (2008). From a literature review to a conceptual framework for sustainable supply chain management. *Journal of Cleaner Production*, *16*(15), 1699–1710. doi:10.1016/j.jclepro.2008.04.020

Shackelford, S. J. (2020). *Governing New Frontiers in the Information Age: Toward Cyber Peace*. Cambridge University Press. doi:10.1017/9781108604000

Shaikh, S., & Saeed, S. (2010). *Sukuk bond: The global Islamic financial instrument*. Academic Press.

Shampa, T. S., & Jobaid, M. I. (2017). Factors Influencing Customers' Expectation Towards Green Banking Practices in Bangladesh. *European Journal of Business and Management*, *9*(12), 140–152.

Shapero, A., & Sokol, L. (1982). The social dimensions of entrepreneurship. In C. A. Kent, D. L. Sexton, & K. H. Vesper (Eds.), *Encyclopedia of Entrepreneurship* (pp. 72–90). Prentice Hall.

Sharma, N., Sarika, K., & Gopal, R. (2014). A Study on customer awareness on green banking initiatives in selected public and private sector banks with special reference to Mumbai. *IOSR Journal of Economics and Finance.*, *3*(7), 28–35.

Shaumya, K., & Anton Arulrajah, A. (2017). The Impact of Green Banking Practices on Bank's Environmental Performance: Evidence from Sri Lanka. *Journal of Finance and Bank Management*, *5*(1), 77–90.

Shaumya, K., & Arulrajah, A. (2016). Measuring green banking practices: evidence from Sri Lanka. *University of Sri Jayewardenepura, Sri Lanka, 13th International Conference on Business Management (ICBM)*, 999-1023.

Shen, L., Kyllo, J. M., & Guo, X. (2013). An integrated model based on a hierarchical indices system for monitoring and evaluating urban sustainability. *Sustainability*, *5*(2), 524–559.

Shen, Z.-J. M., Feng, B., Mao, C., & Ran, L. (2019). Optimization models for electric vehicle service operations: A literature review. *Transportation Research Part B: Methodological*, *128*, 462–477. doi:10.1016/j.trb.2019.08.006

Shojaei, P., & bolvardizadeh, A. (2020). Rough MCDM model for green supplier selection in Iran: A case of university construction project. *Built Environment Project and Asset Management*, *10*(3), 437–452. doi:10.1108/BEPAM-11-2019-0117

Shook, C. L., Priem, R. L., & McGee, J. E. (2003). Venture creation and the enterprising individual: A review and synthesis. *Journal of Management*, *29*(3), 379–399. doi:10.1016/S0149-2063(03)00016-3

Sierzchula, W., Bakker, S., Maat, K., & Wee, B. (2014). The influence of financial incentives and other socio-economic factors on electric vehicle adoption. *Energy Policy*, *68*, 183–194. doi:10.1016/j.enpol.2014.01.043

Simpson, T. W., Peplinski, J., Koch, P. N., & Allen, J. K. (1997). On The Use Of Statistics In Design And The Implications For Deterministic Computer Experiments. *Proceedings of DETC'97, 1997 ASME Design EngineeringTechnical Conferences.*

Singh, A., Parizi, R. M., Zhang, Q., Choo, K. K. R., & Dehghantanha, A. (2020). Blockchain smart contracts formalization: Approaches and challenges to address vulnerabilities. *Computers & Security*, *88*, 101654.

Singhal, A., & Malik, G. (2018). The attitude and purchasing of female consumers towards green marketing related to cosmetic industry. *Journal of Science and Technology Policy Management*, 1-18.

Singh, H., & Singh, B. P. (2012). An Effective and Resourceful Contribution of Green Banking towards Sustainability. *International Journal of Advances in Engineering Science and Technology, 1*(2), 3.

Siswantoro, D. (2018, November). Performance of Indonesian green sukuk (islamic bond): A sovereign bond comparison analysis, climate change concerns? *IOP Conference Series. Earth and Environmental Science, 200*(1), 012056. doi:10.1088/1755-1315/200/1/012056

Smaoui, H., & Khawaja, M. (2017). The determinants of Sukuk market development. *Emerging Markets Finance & Trade, 53*(7), 1501–1518. doi:10.1080/1540496X.2016.1224175

Smaoui, H., & Nechi, S. (2017). Does sukuk market development spur economic growth? *Research in International Business and Finance, 41*, 136–147. doi:10.1016/j.ribaf.2017.04.018

Soda, S., Sachdeva, A., & Garg, R. K. (2016). Literature review of multi-aspect research works carried out on the concept and implementation of GSCM. *International Journal of Industrial and Systems Engineering, 23*(2), 223–253. doi:10.1504/IJISE.2016.076404

Solomon, M. (2009). Consumer Behavior: Buying, Having, and Being (8th ed.). Prentice Hall.

Sommerville, I. (2015). *Software Engineering*. Pearson Education.

Sorin, N., Simona, D., & Ioan, D. (2016). Fuzzy TOPSIS: A General View. *Information Technology and Quantitative Management*, 823-831.

Souitaris, V., Zerbinatib, S., & Al-Lahamc, A. (2007). Do entrepreneurship programmes raise entrepreneurial intention of science and engineering students? The effect of learning, inspiration and resources. *Journal of Business Venturing, 22*(4), 566–591. doi:10.1016/j.jbusvent.2006.05.002

Sousa, F., Gian, N. & Home, R. (2016). Information Technologies as a Tool for Agricultural Extension and Farmer-to-Farmer Exchange: Mobile-Phone Video Use in Mali and Burkina Faso. *International Journal of Education and Development using Information and Communication Technology, 12*(3), 19-36.

Sovacool, B. K., Kester, J., Noel, L., & de Rubens, G. Z. (2018). The demographics of decarbonizing transport: The influence of gender, education, occupation, age, and household size on electric mobility preferences in the Nordic region. *Global Environmental Change, 52*, 86–100. doi:10.1016/j.gloenvcha.2018.06.008

Spanaki, K., Sivarajah, U., Fakhimi, M., Despoudi, S., & Irani, Z. (2021). (in press). Disruptive technologies in agricultural operations: A systematic review of AI-driven AgriTech research. *Annals of Operations Research*. Advance online publication. doi:10.100710479-020-03922-z

Sreen, N., Purbey, S., & Sadarangani, P. (2018). Impact of culture, behavior and gender on green purchase intention. *Journal of Retailing and Consumer Services, 41*, 177–189. doi:10.1016/j.jretconser.2017.12.002

Srivastava, S. K. (2007). Green supply-chain management: A state-of-the-art literature review. *International Journal of Management Reviews, 9*(1), 53–80. doi:10.1111/j.1468-2370.2007.00202.x

Statistical Office of Montenegro – MONSTAT. (2019). *External trade in goods of Montenegro January - December 2019*. Retrieved from https://www.monstat.org/cg/novosti.php?id=3195

Statistical Office of the Republic of Serbia. (n.d.). https://publikacije.stat.gov.rs/ G2020 / Pdf / G20201205.pdf

Statistical Office of the Republic of Slovenia. (2021). *Exports, imports, trade balance and coverage of imports by exports, Slovenia, annually (cummulative data)*. Retrieved from https://pxweb.stat.si/SiStatData/pxweb/en/Data/-/2490002S.px/

Steiber, A., & Alänge, S. (2015). Organizational innovation: A comprehensive model for catalyzing organizational development and change in a rapidly changing world. *Triple Helix (Heidelberg)*, *2*(9), 1–25. doi:10.118640604-015-0021-6

Stelmach, B. L. (2011). A Synthesis of International Rural Education Issues and Responses. *Rural Educator*, *32*(2), 32–42.

Stinson, B. (2013). *Google Glass and its rival smart glasses: everything you need to know*. https://www.stuff.tv/features/google-glass-and-its-rival-smart-glasses-everything-you-need-know

Stone, M. (2015). The evolution of the telecommunications industry - What can we learn from it? *Journal of Direct, Data and Digital Marketing Practice*, *16*(3), 157–165. doi:10.1057/dddmp.2014.80

Straughan, R. D., & Roberts, J. A. (1999). Environmental segmentation alternatives: A look at green consumer behavior in the new millennium. *Journal of Consumer Marketing*, *16*(6), 558–575. doi:10.1108/07363769910297506

Stuart, D., Schewe, R., & McDermott, M. (2014). Reducing nitrogen fertilizer application as a climate change mitigation strategy: Understanding farmer decision-making and potential barriers to change in the US. *Land Use Policy*, *36*(1), 210–218. doi:10.1016/j.landusepol.2013.08.011

Suherman, S. (2019). Identifikasi Potensi Pasar Green Sukuk Republik Indonesia. *Human Falah: Jurnal Ekonomi dan Bisnis Islam*, *1*(6).

Sun, Y., Liu, N., & Zhao, M. (2019). Factors and mechanisms affecting green consumption in China: A multilevel analysis. *Journal of Cleaner Production*, *209*, 481–493. doi:10.1016/j.jclepro.2018.10.241

Suroso, D. S., Setiawan, B., Pradono, S. P., & Fitriyanto, M. S. (2020). *Cross-Country Study of Indonesia Green Sukuk and REDD*. Academic Press.

Sutton, P. C. (2003). An empirical environmental sustainability index derived solely from nighttime satellite imagery and ecosystem service valuation. *Population and Environment*, *24*(4), 293–311.

Swaminathan, M. (2009). Norman E. Borlaug (1914–2009). *Nature*, *461*(894). doi:10.1038/461894a

Tabachnick, B. G., & Fidell, L. S. (2007). *Using multivariate statistics*. Pearson Education.

Tafidis, P., Sdoukopoulos, A., & Pitsiava-Latinopoulou, M. (2017). Sustainable urban mobility indicators: Policy versus practice in the case of Greek cities. *Transportation Research Procedia*, *24*, 304–312. doi:10.1016/j.trpro.2017.05.122

Tara, K., Singh, S., & Kumar, R. (2015). Green banking for environmental management: A paradigm shift. *Current World Environment*, *10*(3), 1029–1038. doi:10.12944/CWE.10.3.36

TEHAD. (n.d.). http://tehad.org/2021/01/16/2020-yili-elektrikli-ve-hibrid-otomobil-satis-rakamlari-belli-oldu/

Teller, C., Holweg, C., Reiner, G., & Kotzab, H. (2018). Retail store operations and food waste. *Journal of Cleaner Production*, *185*, 981–997. doi:10.1016/j.jclepro.2018.02.280

Thananusak, T., Rakthin, S., Tavewatanaphan, T., & Punnakitikashem, P. (2017). Factors affecting the intention to buy electric vehicles: Empirical evidence from Thailand. *International Journal of Electric and Hybrid Vehicles*, *9*(4), 361–381. doi:10.1504/IJEHV.2017.089875

The Joint Research Centre. (2021). *Smart Specialization Platform*. Retrieved from https://s3platform.jrc.ec.europa.eu/s3-trade-tool

The National Institute of Statistics. (2019). *Monthly statistical bulletin of counties no 11/2019*. The National Institute of Statistics.

Thombre, K. A. (2011). The new face of banking: Green banking. *Indian Streams Research Journal*, *1*(2), 1–4.

Tian, H., Grover, V., Zhao, J., & Jiang, Y. (2020). The differential impact of types of app innovation on customer evaluation. *Information & Management*, *57*(7), 1–17. doi:10.1016/j.im.2020.103358

Tian, Z. P., Zhang, H. Y., Wang, J. Q., & Wang, T. L. (2018). Green supplier selection using improved TOPSIS and best-worst method under intuitionistic fuzzy environment. *Informatica (Vilnius)*, *29*(4), 773–800. doi:10.15388/Informatica.2018.192

Tijanić, L. (2010). Regional (non) competitiveness in the Republic of Croatia [in Croatian]. *Economic Review (Kansas City, Mo.)*, *61*(7-8), 419–454.

Tilkov, S., & Vinoski, S. (2010). Node.js: Using JavaScript to Build High-Performance Network Programs. *IEEE Internet Computing*, *14*(6), 80–83. doi:10.1109/MIC.2010.145

Tiller, L. L. C. (2014). *Nationwide Survey Finds: Americans Believe Environmental Issues Pose Greater Long-Term Threat to Their Health and Well-Being Than Terrorism or Global Epidemics Like Ebola.* https://www.prnewswire.com/news-releases/nationwide-survey-finds-americans-believe-environmental-issues-pose-greater-long-term-threat-to-their-health-and-well-being-than-terrorism-or-global-epidemics-like-ebola-300014118.html

Togan-Egrican, A., & Kayhan, F. (2020). *Voluntary Private Pension Funds and Capital Market Development.* Working paper.

Topalovic, S. (2013). *Status and perspectives of spa tourism in Serbia. Globe, No. 37-38.* Serbian Geographical Society. http://sgd.org.rs/publikacije/globus/3738/7-banjski%20turizam.pdf

Topić, M., & Lodorfos, G. (2021). Introduction. In M. Topić & G. Lodorfos (Eds.), *The Sustainability Debate: policies, gender and the media* (pp. 15–30). Emerald Publishing.

Tran, M., Banister, D., Bishop, J. D., & McCulloch, M. D. (2013). Simulating early adoption of alternative fuel vehicles for sustainability. *Technological Forecasting and Social Change*, *80*(5), 865–875. doi:10.1016/j.techfore.2012.09.009

Triantaphyllou, E. (2000). Multi-criteria decision making methods. In *Multi-criteria decision making methods: A comparative study* (pp. 5–21). Springer. doi:10.1007/978-1-4757-3157-6_2

Trindade, E. P., Hinnig, M. P., Costa, E., Marques, J. S., Bastos, R. C., & Vigitcanlar, T. (2017). Sustainable development of smart cities: A systematic review of the literature. *Journal of Open Innovation*, *3*(11), 1–14. doi:10.118640852-017-0063-2

Trommer, S., Jarass, J., & Kolarova, V. (2015). Early adopters of electric vehicles in Germany unveiled. *World Electric Vehicle Journal*, *7*(4), 722–732. doi:10.3390/wevj7040722

Troshani, I., & Doolin, B. (2007). Innovation diffusion: A stakeholder and social network view. *European Journal of Innovation Management*, *10*(2), 176–200. doi:10.1108/14601060710745242

Tsai, S. B., Wei, Y. M., Chen, K. Y., Xu, L., Du, P., & Lee, H. C. (2016). Evaluating green suppliers from a green environmental perspective. *Environment and Planning. B, Planning & Design*, *43*(5), 941–959. doi:10.1177/0265813515600897

Tsouros, D. C., Bibi, S., & Sarigiannidis, P. G. (2019). A Review on UAV-Based Applications for Precision Agriculture. *Information (Basel)*, *10*(11), 349. doi:10.3390/info10110349

Tsui, C. W., & Wen, U. P. (2014). A hybrid multiple criteria group decision-making approach for green supplier selection in the TFT-LCD industry. *Mathematical Problems in Engineering*, *709872*, 1–13. Advance online publication. doi:10.1155/2014/709872

TUIK. Gelir ve Yaşam Koşulları Araştırması, 2019. (2020). https://tuikweb.tuik.gov.tr/PreHaberBultenleri.do?id=33820

Tu, J. C., & Yang, C. (2019). Key factors influencing consumers' purchase of electric vehicles. *Sustainability, 11*(14), 3863. doi:10.3390u11143863

Turkey's Savings Trends Research. (n.d.). *4Q2020 Report.* ING & Ipsos & Bilkent University.

Ubay, G. G., & Karakuş, H. (2020). The Role of Technology-Based Renewable Energy Investments on Macroeconomic Stability. In *Strategic Outlook for Innovative Work Behaviours* (pp. 119–130). Springer. doi:10.1007/978-3-030-50131-0_6

Ugochukwu, A. I., & Phillips, P. W. B. (2018). Technology Adoption by Agricultural Producers: A Review of the Literature. In N. Kalaitzandonakes, E. Carayannis, E. Grigoroudis, & S. Rozakis (Eds.), *From Agriscience to Agribusiness. Innovation, Technology, and Knowledge Management.* Springer. doi:10.1007/978-3-319-67958-7_17

Ülkü, M. A., Skinner, D. M., & Yıldırım, G. (forthcoming). Towards Sustainability: A Review of Analytical Models for Circular Supply Chains. In Circular Economy Supply Chains - From Chains to Systems. Emerald.

Ülkü, M. A., & Engau, A. (2021). Sustainable Supply Chain Analytics. In W. Leal Filho, A. M. Azul, L. Brandli, A. Lange Salvia, & T. Wall (Eds.), *Industry, Innovation and Infrastructure. Encyclopedia of the UN Sustainable Development Goals.* Springer. doi:10.1007/978-3-319-71059-4_117-1

Ulutaş, A., Topal, A., & Bakhat, R. (2019). An application of fuzzy integrated model in green supplier selection. *Mathematical Problems in Engineering, 2019*(4256359), 1–11. doi:10.1155/2019/4256359

United Nations FAO. (2020). *The State of Food and Agriculture.* FAO.

United Nations FAO. (2021) Developing sustainable food systems and value chains for climate-smart agriculture (B10 2 Food systems and value chains: definitions and characteristics) (Update). Food and Agriculture Organization of the United Nations (FAO).

United Nations. (2021). *Peace, dignity, and equality on a healthy planet: Population.* Last access 22/05/2021 web https://www.un.org/en/global-issues/population

United Nations. (2021a). *Take Action for the Sustainable Development Goals.* Retrieved from https://www.un.org/sustainabledevelopment/poverty/hunger/

United Nations. (2021b). *United Nations Goal 12: Ensure Sustainable Consumption and Production Patterns.* Retrieved from https://www.un.org/sustainabledevelopment/sustainable-consumption-production

University of South Australia. (2021, March 15). Standard digital camera and AI to monitor soil moisture for affordable smart irrigation. *ScienceDaily.* Retrieved from www.sciencedaily.com/releases/2021/03/210315110210.htm

Unterhalter, E. (2019). The many meanings of quality education: Politics of targets and indicators in SDG 4. *Global Policy, 10,* 39–51.

Urban, J., & Ščasný, M. (2012). Exploring domestic energy-saving: The role of environmental concern and background variables. *Energy Policy, 47,* 69–80. doi:10.1016/j.enpol.2012.04.018

Urquhart, C., Liyanage, S., & Kah, M. (2008). ICTs and Poverty Reduction: A Social Capital and Knowledge Perspective. *Journal of Information Technology, 24*(3), 203–213. doi:10.1057/palgrave.jit.2000121

Ursache, M. (2015). *Niche Tourism Markets – Means of Enhancing Sustainable Economic Development in EU's Eastern Periphery.* CES Working Papers, ISSN 2067-7693. Centre for European Studies, Alexandru Ioan Cuza University, 7(2a), p. 654.

Uslu, T. (2007). Faktöriyel Tasarım ve Süreç Optimizasyonu. *Dokuz Eylül Üniversitesi Sosyal Bilimler Enstitüsü Hakemli Dergisi, 9*(1), 535–551.

Uygun, Ö., & Dede, A. (2016). Performance evaluation of green supply chain management using integrated fuzzy multi-criteria decision making techniques. *Computers & Industrial Engineering*, *102*, 502–511. doi:10.1016/j.cie.2016.02.020

Van Hove, L., & Dubus, A. (2019). M-PESA and financial inclusion in Kenya: Of paying comes saving? *Sustainability*, *11*(3), 568.

Van Stel, A., & Suddle, K. (2007). The Impact of New Firm Formation on Regional Development in the Netherlands. *Small Business Economics*.

Vassileva, I., & Campillo, J. (2017). Adoption barriers for electric vehicles: Experiences from early adopters in Sweden. *Energy*, *120*, 632–641. doi:10.1016/j.energy.2016.11.119

Veciana, J. M., Aponte, M., & Urbano, D. (2005). University students' attitudes towards entrepreneurship: A two countries comparison. *The International Entrepreneurship and Management Journal*, *1*(2), 165–182. doi:10.100711365-005-1127-5

Venkatesh, V., & Bala, H. (2008). Technology acceptance model 3 and a research agenda on interventions. *Journal of the Decision Science Institute*, *39*(2), 273–315. doi:10.1111/j.1540-5915.2008.00192.x

Venkatesh, V., & Davis, F. (2000). A theoretical extension of the technology acceptance model: Four longitudinal field studies. *Management Science*, *46*(2), 169–332. doi:10.1287/mnsc.46.2.186.11926

Venkatesh, V., Morris, M., Davis, G., & Davis, F. (2003). User acceptance of information technology: Toward a unified view. *Management Information Systems Quarterly*, *27*(3), 425–478. doi:10.2307/30036540

Venkatesh, V., Thong, J., & Xu, X. (2012). Consumer acceptance and use of information technology: Extending the unified theory of acceptance and use of technology. *Management Information Systems Quarterly*, *36*(1), 157–178. doi:10.2307/41410412

Verhoef, P., Stephen, A., Kannan, P., Luo, X., Abhishek, V., Andrews, M., Bart, Y., Datta, H., Fong, N., Hoffman, D. L., Hu, M. M., Novak, T., Rand, W., & Zhang, Y. (2017). Consumer connectivity in a complex, technology-enables, and mobile-oriented world with smart products. *Journal of Interactive Marketing*, *40*, 1–72. doi:10.1016/j.intmar.2017.06.001

Verma, V. K., Chandra, B., & Kumar, S. (2019). Values and ascribed responsibility to predict consumers' attitude and concern toward green hotel visit intention. *Journal of Business Research*, *96*, 206–216. doi:10.1016/j.jbusres.2018.11.021

Veselinović, P., & Knežević, M. (2015). Competitiveness enhancing as the incentive for the investment in the city of Kragujevac. *Ekonomski pogledi*, *17*(3), 139-163. (in Serbian)

Vicente-Molina, M. A., Fernandez-Sainz, A., & Izagirre-Olaizola, J. (2013). Environmental knowledge and other variables affecting pro-environmental behavior: Comparison of university students from emerging and advanced countries. *Journal of Cleaner Production*, *61*, 130–138. doi:10.1016/j.jclepro.2013.05.015

Vijay & Natarajan. (2015). Customers' Awareness towards Green Banking Products of the Select Commercial Banks in Cuddalore District. *An Empirical Assessment*, *3*(10), 1–5.

Villanueva-Ponce, R., Avelar-Sosa, L., Alvarado-Iniesta, A., & Cruz-Sánchez, V. G. (2015). The green supplier selection as a key element in a supply chain: A review of cases studies. *Dyna*, *82*(194), 36–45. doi:10.15446/dyna.v82n194.54466

Vincent, O., & Evans, O. (2019). Can cryptocurrency, mobile phones, and internet herald sustainable financial sector development in emerging markets? *Journal of Transnational Management*, *24*(3), 259–279.

Vincke, P. (1992). *Multicriteria Decision - Aid*. John Wiley and Sons.

Wang, Q., Wang, C., Feng, Z. Y., & Ye, J. F. (2012). Review of K-means clustering algorithm. *Electronic Design Engineering*, *20*(7), 21-24.

Wang, Y.-M., & Elhag, T. (2006). Fuzzy TOPSIS method based on alpha level sets with an application to bridge risk assessment. *Expert Systems with Applications*, *31*(2), 309–319. doi:10.1016/j.eswa.2005.09.040

Wan, S. P., & Li, D. F. (2015). Fuzzy mathematical programming approach to heterogeneous multiattribute decision-making with interval-valued intuitionistic fuzzy truth degrees. *Information Sciences*, *325*, 484–503. doi:10.1016/j.ins.2015.07.014

Wan, S. P., Lin, L. L., & Dong, J. Y. (2017). MAGDM based on triangular Atanassov's intuitionistic fuzzy information aggregation. *Neural Computing & Applications*, *28*(9), 2687–2702. doi:10.100700521-016-2196-9

Wan, S. P., Zou, W. C., Zhong, L. G., & Dong, J. Y. (2019). Some new information measures for hesitant fuzzy PROMETHEE method and application to green supplier selection. *Soft Computing*, *24*(12), 9179–9203. doi:10.100700500-019-04446-w

WB6 CIF (Western Balkans 6). (2020). Prospects for Travelling and Tourism Sector in the Western Balkans in 2020: Study no. 1. *Studies on obstacles and opportunities for doing business in the region: study no. 1*, 1-52.

We are Social. Digital 2021: Turkey. (n.d.). https://datareportal.com/reports/digital-2021-turkey

WE Outlook International Energy Agency (2020). https://www.iea.org/reports/world-energy-outlook-2020

Weiner, N. (1948). *Cybernetics or Control and Communication in the Animal and the Machine*. Wiley.

Wennekers, A. R. M., & Thurik, A. R. (1999). Linking entrepreneurship and economic growth. *Small Business Economics*, *13*(1), 27–55. doi:10.1023/A:1008063200484

Wessels, L., & Drennan, J. (2010). An investigation of consumer acceptance of M-banking. *International Journal of Bank Marketing*, *28*(7), 547–568. doi:10.1108/02652321011085194

WFPUSA. (2021). World Food Program USA-*Food Waste vs. Food Loss: Know the Difference and Help #StopTheWaste Today*. Retrieved from https://www.wfpusa.org/articles/food-loss-vs-food-waste-primer/

Wheeler, D., McKague, K., Thomson, J., Davies, R., Medalye, J., & Prada, M. (2005). Creating sustainable local enterprise networks. *MIT Sloan Management Review*, *47*(1), 33–41.

Wilson, N. L., Miao, R., & Weis, C. S. (2019). When in Doubt, Throw It Out! The Complicated Decision to Consume (or Waste) Food by Date Labels. *Choices*, *34*(1), 1–7.

Witkowski, K. (2017). Internet of things, big data, industry 4.0–innovative solutions in logistics and supply chains management. *Procedia Engineering*, *182*, 763–769. doi:10.1016/j.proeng.2017.03.197

Wong & Hazley (2020). The Future of Health Tourism in the Industrial Revolution 4.0. Era. *Journal of Tourism Futures*. DOI doi:10.1108/JTF-01-2020-0006

Wong, C. C., & Hiew, P. L. (2005). Diffusion of mobile entertainment in Malaysia: Drivers and barriers. *Proceedings of World Academy of Science. Engineering and Technology*, *5*, 263–266.

World Bank. (2012). *Mobile applications for agriculture and rural development*. Retrieved from https://documents.worldbank.org/curated/en/167301467999716265/Mobile-applications-for-agriculture-and-rural-development/

World Economic Forum. (2017). *Shaping the future of global food systems: A scenarios analysis*. WEF.

World Health Organization. (2020). *Climate change and human health. Risks and responses*. UN.

WRAP. (2021). *Consumer Attitudes to Food Waste and Food Packaging*. Retrieved from https://www.wrap.org.uk/sites/files/wrap/Report%20%20Consumer%20attitudes%20to%20food%20waste%20and%20packaging_0.pdf

WTO (World Tourism Organization) &European Travel Commission (ETC). (2018). *Exploring Health Tourism – Executive Summary*. UNWTO. doi:10.18111/978928442030.8

Wu, H. H. (2002). A comparative study of using grey relational analysis in multiple attribute decision making problems. *Quality Engineering*, *15*(2), 209–217. doi:10.1081/QEN-120015853

Wu, J., Liao, H., Wang, J. W., & Chen, T. (2019). The role of environmental concern in the public acceptance of autonomous electric vehicles: A survey from China. *Transportation Research Part F: Traffic Psychology and Behaviour*, *60*, 37–46. doi:10.1016/j.trf.2018.09.029

Wu, Q., Zhou, L., Chen, Y., & Chen, H. (2019). An integrated approach to green supplier selection based on the interval type-2 fuzzy best-worst and extended VIKOR methods. *Information Sciences*, *502*, 394–417. doi:10.1016/j.ins.2019.06.049

Wyman. (2020). *The Future of Travel &Tourism in the Wake of COVID-19*. World Travel & Tourism Council.

Xiao, C., & Dunlap, R. E. (2007). Validating a comprehensive model of environmental concern cross-nationally: A US-Canadian comparison. *Social Science Quarterly*, *88*(2), 471–493. doi:10.1111/j.1540-6237.2007.00467.x

Xiao, C., & McCright, A. M. (2015). Gender differences in environmental concern: Revisiting the institutional trust hypothesis in the USA. *Environment and Behavior*, *47*(1), 17–37. doi:10.1177/0013916513491571

Xie, Y., Zhou, Y., Peng, Y., Dinçer, H., Yüksel, S., & an Xiang, P. (2021). An Extended Pythagorean Fuzzy Approach to Group Decision-Making With Incomplete Preferences for Analyzing Balanced Scorecard-Based Renewable Energy Investments. *IEEE Access: Practical Innovations, Open Solutions*, *9*, 43020–43035. doi:10.1109/ACCESS.2021.3065294

Yadav, P., Heynen, A. P., & Palit, D. (2019). Pay-As-You-Go financing: A model for viable and widespread deployment of solar home systems in rural India. *Energy for Sustainable Development*, *48*, 139–153.

Yadav, R., & Pathak, G. (2013). Environmental sustainability through green banking: A study on private and public sector banks in India. *OIDA International Journal of Sustainable Development*, *6*(08), 37–48.

Yakinni, O. T., Ladigbolu, T. A., Adeniyi, R. T., & Adebisi, G. L. (2019). Benefits Derived from the Use of Information and Communication Technologies among Rural Farmers in Northeast Nigeria. *Journal of Agricultural Extension*, *23*(3), 117–125. doi:10.4314/jae.v23i3.10

Yang, C., Tu, J. C., & Jiang, Q. (2020). The influential factors of consumers' sustainable consumption: A case on electric vehicles in China. *Sustainability*, *12*(8), 3496. doi:10.3390u12083496

Yao, A., & Sio. (2013). Analysis and Design of a Taguchi–Grey Based Electricity Demand Predictor for Energy Management Systems. *Energy Conversion and Management*, *45*(7–8), 1205–1217.

Yao, X., Zhu, D., Yun, W., Peng, F., & Li, A. (2017). A WebGIS-based decision support system for locust prevention and control in China. *Computers and Electronics in Agriculture*, *140*, 148–158. doi:10.1016/j.compag.2017.06.001

Yazdani, M. (2014). An integrated MCDM approach to green supplier selection. *International Journal of Industrial Engineering Computations*, *5*(3), 443–458. doi:10.5267/j.ijiec.2014.3.003

Yazdani, M., Chatterjee, P., Pamucar, D., & Abad, M. D. (2019). A risk-based integrated decision-making model for green supplier selection: A case study of a construction company in Spain. *Kybernetes*, *49*(4), 1229–1252. doi:10.1108/K-09-2018-0509

Yue, C., Alfnes, F., & Jensen, H.H. (2009). Discounting Spotted Apples: Investigating Consumers' Willingness to Accept Cosmetic Damage in an Organic Product. *Journal of Agricultural and Applied Economics, 41*, 29-46.

Yüksel, S., Dinçer, H., Karakuş, H., & Ubay, G. G. (2020). The Negative Effects of Carbon Emission on FDI: A Comparative Analysis Between E7 and G7 Countries. In Handbook of Research on Sustainable Supply Chain Management for the Global Economy (pp. 20-35). IGI Global.

Yu, Q., & Hou, F. (2016). An approach for green supplier selection in the automobile manufacturing industry. *Kybernetes*, *45*(4), 571–588. doi:10.1108/K-01-2015-0034

Yusuf, S. E., Ge, M., Hong, J. B., Kim, H. K., Kim, P., & Kim, D. S. (2016, December). Security modelling and analysis of dynamic enterprise networks. Paper presented at the IEEE International Conference on Computer and Information Technology, Nadi, Turkey.

Zain, N. R. M., Hasan, A., Yusof, S. A. M., & Ali, E. R. A. E. (2021). Combating Climate Change in Malaysia: Green Sukuk and Its Potential. In Handbook of Research on Islamic Social Finance and Economic Recovery After a Global Health Crisis (pp. 264-279). IGI Global.

Zarrouk, H., Ghak, T. E., & Bakhouche, A. (2021). Exploring Economic and Technological Determinants of FinTech Startups' Success and Growth in the United Arab Emirates. *Journal of Open Innovation*, *7*(1), 50.

Zeithaml, V. A., Bitner, M. J., Gremler, D. D., & Pandit, A. (2013). *Services Marketing: Integrating customer focus across the firms* (6th ed.). Tata McGraw Hill Education Private Limited.

Zelezny, L. C., Chua, P., & Aldrich, C. (2000). Elaborating on gender differences in environmentalism-statistical data included. *The Journal of Social Issues*, *56*(3), 443–445. doi:10.1111/0022-4537.00177

Zhang, M., Guo, S., Bai, C., & Wang, W. (2019). Study on the impact of haze pollution on residents' green consumption behavior: The case of Shandong Province. *Journal of Cleaner Production*, *219*, 11–19. doi:10.1016/j.jclepro.2019.02.077

Zhang, S., Wei, G., Wang, R., Wu, J., Wei, C., Guo, Y., & Wei, Y. (2021). Improved CODAS Method Under Picture 2-Tuple Linguistic Environment and Its Application for a Green Supplier Selection. *Informatica (Vilnius)*, *32*(1), 195–216. doi:10.15388/20-INFOR414

Zhang, Y., Chen, J., Han, Y., Qian, M., Guo, X., Chen, R., ... Chen, Y. (2021). The contribution of Fintech to sustainable development in the digital age: Ant forest and land restoration in China. *Land Use Policy*, *103*, 105306.

Zhao, H. H., Gao, Q., Wu, Y. P., Wang, Y., & Zhu, X. D. (2014). What affects green consumer behavior in China? a case study from Qingdao. *Journal of Cleaner Production*, *63*, 143–151. doi:10.1016/j.jclepro.2013.05.021

Zhao, H., & Guo, S. (2014). Selecting green supplier of thermal power equipment by using a hybrid MCDM method for sustainability. *Sustainability*, *6*(1), 217–235. doi:10.3390u6010217

Zhao, H., Zhao, Q. H., & Ślusarczyk, B. (2019). Sustainability and digitalization of corporate management based on augmented/virtual reality tools usage: China and other world IT companies' experience. *Sustainability*, *11*(17), 4717.

Zhao, R., Geng, Y., Liu, Y., Tao, X., & Xue, B. (2018). Consumers' perception, purchase intention, and willingness to pay for carbon-labeled products: A case study of Chengdu in China. *Journal of Cleaner Production*, *171*, 1664–1671. doi:10.1016/j.jclepro.2017.10.143

Zhao, X., Fan, H., Zhu, H., Fu, Z., & Fu, H. (2015, April). The design of the internet of things solution for food supply chain. In *2015 International Conference on Education, Management, Information and Medicine*. Atlantis Press. 10.2991/emim-15.2015.61

Zhao, Y., Xu, Y., Yüksel, S., Dinçer, H., & Ubay, G. G. (2021). Hybrid IT2 fuzzy modelling with alpha cuts for hydrogen energy investments. *International Journal of Hydrogen Energy*, *46*(13), 8835–8851. doi:10.1016/j.ijhydene.2020.12.211

Zheng, X., & Meng, Q. (2018). Business Model of Green Finance under the Background of "Internet Plus": A Case Study of Ant Forest. *Open Journal of Social Sciences*, *6*(8), 127–134.

Zhong, J., Hu, X., Yüksel, S., Dinçer, H., & Ubay, G. G. (2020). Analyzing the investments strategies for renewable energies based on multi-criteria decision model. *IEEE Access: Practical Innovations, Open Solutions*, *8*, 118818–118840. doi:10.1109/ACCESS.2020.3005064

Zhou, P., Luo, J., Cheng, F., Yüksel, S., & Dinçer, H. (2021). Analysis of risk priorities for renewable energy investment projects using a hybrid IT2 hesitant fuzzy decision-making approach with alpha cuts. *Energy*, *224*, 120184. doi:10.1016/j.energy.2021.120184

Zhou, Q., Chen, X., & Li, S. (2018). Innovative financial approach for agricultural sustainability: A case study of Alibaba. *Sustainability*, *10*(3), 891.

Zhu, L., Hu, L., Yüksel, S., Dinçer, H., Karakuş, H., & Ubay, G. G. (2020). Analysis of strategic directions in sustainable hydrogen investment decisions. *Sustainability*, *12*(11), 4581. doi:10.3390u12114581

Zolfaghari, P. (2017). *An introduction to Islamic securities (sukuk)*. Academic Press.

About the Contributors

Ulas Akkucuk is an Associate Professor of Quantitative Methods at Bogazici University Department of Management, Istanbul, Turkey. Dr. Akkucuk has a BS in Industrial Engineering from Bilkent University (1997), an MBA from Bogazici University (2000), and a PhD in Management Science and Information Systems from Rutgers University (2004). He has taught Statistics, CRM and Data Mining, and Global Aspects of Management courses, among others, at both graduate and undergraduate levels. He is associate editor and editorial board member of International Journal of Research in Business and Social Science (IJRBS) and has been an ad hoc reviewer for journals such as Journal of Multivariate Analysis, Bogazici Journal, International Journal of Business Economics and Management, Journal of Cleaner Production and Journal of Computational and Graphical Statistics. He is currently the editor of International Journal of Environmental Sustainability and Green Technologies. He has edited eight books so far with IGI Global, two of them have been indexed by WoS. He has published in journal such as Journal of Classification, Fresenius Environmental Bulletin, Sustainability, and European Journal of Health Economics.

* * *

Serap Akcan is an Assoc. Prof. Dr. in Industrial Engineering Department at Tarsus University. Her interesting areas are multi-criteria decision making, environmental sustainability, healthcare, operations research, and optimization.

Tamer Aksoy is a professor of accounting and finance at School of Business, Ibn Haldun University. He obtained his BA in Economics at Ankara University, Faculty of Political Sciences. He received his MA in Economics at Hacettepe University and PhD in international banking at Gazi University. He has a wide experience in research and teaching at undergraduate and postgraduate levels. Strategic auditing, internal control, financial accounting, financial management, corporate risk management, international banking, digitalization, professional ethics, artificial intelligence, blockchain ecosystem, sustainability and corporate governance are the main pillars of his interdisciplinary research. He authored 7 books and many articles. Dr. Aksoy has various editorial roles in international journals. In addition to his academic career, he has also a long and extensive work experience spaning over 33 years. He worked for Citicorp Investment Bank, Zurich, Switzerland for a short time. He served as an internal auditor, inspector (for 10 years) and as the Chief Audit Executive (CAE) of Group's internal audit function (for 23 years) for OYAK which is the second largest leading conglomerate group with 90 subsidiaries in different sectors in Turkey and all over the world. Furthermore, he worked for OYAK Maritime and Port Management

Inc. of the Group as Advisor to CEO. Besides, He also acted as a member of board of directors of various leading commercial companies in different sectors such as Adana Cement, Bolu Cement, Goodyear Tyres-Turkey, Omsan Logistics, Oysa Cement, and Oyak Bank. Furthermore, including the Global Institute of Internal Auditors (IIA) in USA, he undertook many duties on voluntary-basis for local/international professional foundations as committee member, head of committee, member of BODs/advisory boards. Moreover, he has also international professional licenses of CPA, CFE, Independent Auditor (BD), CRMA.

Fernando Luís Almeida has a Ph.D. in Computer Science Engineering from the Faculty of Engineering of University of Porto (FEUP). He also holds an MSc in Innovation and Entrepreneurship and in Informatics Engineering from FEUP. He has around 10 years of teaching experience at higher education levels in the field of computer science and management. He has also worked for 15 years in several positions as a software engineer and project manager for large organizations and research centers like Critical Software, CICA/SEF, INESC TEC, and ISR Porto. He is a founder member of the International Association of Innovation Professionals and he is involved in the development of the US TAG group for ISO 56000 (Innovation Management). His current research areas include innovation policies, entrepreneurship, software development, and decision support systems.

U. Zeynep Ata is an Assistant Professor of Operations and Supply Chain Management at Boğaziçi University Department of International Trade, Istanbul, Turkey. Dr. Ata has a degree in Industrial Engineering from Istanbul Technical University. She completed her MBA degree at METU and PhD in Management at Boğaziçi University. Dr. Ata teaches undergraduate and graduate level courses in International Logistics, Supply Chain Management, Strategic Sourcing Management and Statistics. Her current research interests include Supply Chain Management, Strategic Sourcing, Decision Processes and Models in Logistics and Sustainable Supply Chains.

S. Su Baysal is currently completing her education in Dalhousie University's Bachelor of Commerce Program with a major in Supply Chain and Logistics Management.

Bengi Meriç Benderlioğlu studied Business Administration in Marmara University with a major in Marketing. After graduating with high honors, she went on to her Master's Degree in International Trade Management at Bogazici University. During both her educational cycles, Bachelor's and Master's Degrees, she has participated in International Exchange Programs in two different universities in Germany, Berlin School of Economics & Law and Free University of Berlin. Currently she is working in the marketing area of the software industry for the global market and is resided in Berlin/Germany.

Esna Betül Buğday is an Assistant Professor at Hacettepe University, Department of Family and Consumer Sciences. She completed her master's study at Gazi University and received her Ph.D. at Hacettepe University, Department of Family and Consumer Sciences. After receiving her Ph.D., she has been to the University of Autonoma as a visiting scholar. She has attended several international conferences and published numerous papers in national and international scientific journals. Her research interests include Consumer Behavior, Conscious Consumption, Sustainable Consumption, and Consumer Education. Also, since 2016, she has been serving as a deputy manager of the Center of Consumer and Market Research, Consulting, Test, and Education.

Gamze Ozturk Danisman is an Assistant Professor of Finance at Kadir Has University, Istanbul, Turkey. She holds a BS degree in Mathematics and a minor degree in Economics from Middle East Technical University (2004), and an MA degree in Mathematics from the University of Kansas, USA (2008). She completed her Ph.D. in Finance at Boğaziçi University in 2018. With seven years of experience in private sector prior to her doctoral studies, she has worked as a risk manager and financial controller in both financial and non-financial institutions. Dr. Danisman's research interests include bank stability, financial institutions and corporate finance. She has published her work in top-ranked journals such as Finance Research Letters, Journal of International Financial Markets, Institutions & Money, The European Journal of Finance, Business Strategy and the Environment, Journal of Multinational Financial Management, North American Journal of Economics and Finance, Applied Economics, International Review of Economics & Finance, etc.

Niranjan Devkota is an economist with the expertise on climate change related issues focusing adaptation. He has over 10 years of experience in the field of economics related research with varietal dynamics. He has experience in impact analysis and model building. He received his PhD degree from Tribhuvan University Nepal in 2018.

Hasan Dinçer is a Professor of finance at Istanbul Medipol University, Faculty of Economics and Administrative Sciences, Istanbul-Turkey. Dr. Dinçer has BAs in Financial Markets and Investment Management at Marmara University. He received PhD in Finance and Banking with his thesis entitled "The Effect of Changes on the Competitive Strategies of New Service Development in the Banking Sector". He has work experience in finance sector as portfolio specialist and his major academic studies focusing on financial instruments, performance evaluation, and economics. He is the executive editor of the International Journal of Finance and Banking Studies (IJFBS) and the founder member of the Society for the Study of Business and Finance (SSBF).

Gordana P. Djukic graduated from the University of Belgrade, Faculty of Economics, Serbia, 2006. She defended her master's thesis in 2010 at the same faculty. She defended her doctoral dissertation in 2016 and received her PhD in economics from John Naisbitt University in Belgrade, Faculty of Management in Zajecar. She has acquired the scientific research title of Research Associate in 2017 at the Institute of Economic Sciences in Belgrade.

Manuel Antonio Fernández-Villacañas Marín is a PhD of Economic and Business Sciences, PhD of Administration and Logistics of S&D Systems, Master's Degree of Security & Defence, Master's Degree of Business Studies, Executive Master of Public Management, Consultant in Logistics & Management at M&M Planning and Project Management, Colonel at Spanish Air Force (R), Associate Professor at the School of Aeronautical and Space Engineering and member of the Research Group in Organization and Logistics Engineering, Technical University of Madrid (Spain). With a long professional career of over 34 years, I have a great multidisciplinary experience in the development of logistics, administration and technology projects.

Thinh Gia Hoang is senior research fellow at the Centre for Applied Economics and Business Research, Hanoi, Vietnam and lecturer at RMIT University Vietnam. His research interest focus on sustainability, corporate disclosure, corporate social responsibility, technology and innovation management. His work has been published widely in international journals, books and international conference including Technological Forecasting & Social Change, Service Business, Social Responsibility Journal, The Routledge Companion to Accounting Information Systems, The Palgrave Handbook of Corporate Social Responsibility, and Academy of Management Proceedings. He also serves in editorial board for the SAGE Handbook of Qualitative Research in Asian Context.

Biljana Ilic, PhD, was born on 10/01/1970, in Zajecar, Serbia. She graduated on Business Higher School of Belgrade University, in 1996. She continued further education on the Faculty of Management Zajecar, Megatrend University of Belgrade, in 2007. She Graduated on the Faculty of Management in Zajecar, in 2009. She received the degree of Master of Academic Studies in Management on the same faculty, defended master thesis with mark 10.00, and acquired the title of master manager. She defended her doctoral thesis titled "Strategic directions of regional economic and environmental development of the tourist potential of Gamzigrad Spa", in April, 2016, mark 10.00. She was elected as assistant professor at the Faculty of Management in Zajecar on July, 2016.

Gencay Karakaya is an Assoc. Prof. at Istanbul Commerce University Department of Accounting and Auditing, Istanbul-Turkey. Karakaya has a BS in Business Administration from Sakarya University (2011) and MBA from Istanbul Commerce University (2014). Karakaya has PhD (2019) in Accounting and Auditing from Istanbul Commerce University. His research interests; international trade, financial accounting system, auditing, internal control, international financial report system and financial risk.

Hüsne Karakuş is a graduate student of Banking and Insurance department in Istanbul Medipol University. Her research interests are energy economics, wind energy and renewable energy projects.

Miloš Stanko Krstić received his PhD in Faculty of Economics from the University of Niš (Serbia) in 2014. In 2014 he was hired as the research associete by the Faculty of Sciences and Mathematics at The University of Niš and is currently Guest Editor of Special Issue "Business Games and Numeric Simulations in Economics and Management" of Journal "Mathematics" (MDPI). He writes and presents widely on issues of contemporary economic paradigms (rational choice theory, game theory, social choice theory, rational expectations theory), and is the author of Rational Choice Theory in the Social Sciences (2020, Niš).

Dat Anh Le currently works as a senior data scientist specialized in Human Resources Planning and Budgeting at the Asia Commercial Bank, Vietnam. He earned the Master of Science in Econometrics and Informatics from The University of Zielona Góra, Poland. His research interest focuses on econometric, behavioral economic, innovation, and sustainability. His previous research work has been published in The Palgrave Handbook of Corporate Social Responsibility and Academy of Management Proceedings.

Ignacio Marcos is a strategic consultant for Banco Santander, a technology consultant at Planning and Project Management consulting, and a PhD candidate of Economics and Business Management at University of Alcala. Bachelor's degree of Aerospace engineering from Technical University of Madrid, Master's Degree of Big Data & Business Analytics from University of Alcala; and Master in Science of Thermal Power from Cranfield University.

Goran R. Milovanovic is a full professor at the Faculty of Economics, University of Nis, Serbia. He was born on January 27, 1963 in Sokobanja, Serbia. He graduated from the Faculty of Economics in Nis in 1988. He received his master's degree from the Faculty of Economics in Belgrade in 1993, and then received his doctorate from the same faculty and obtained a doctorate in economics in 1997. He was an assistant trainee from 1989-1993, assistant from 1993-1998, assistant professor from 1998-2003, associate professor from 2003- 2008. He acquired the title of full professor in 2008, since when he was employed at the Faculty of Economics in Nis. Possesses skills: MS Word; MS Exel; MS Photo Paint; Corel Draw; Adobe Photoshop Macromedia Flash; Web Design Scientific and professional training abroad: - Otto-von-Guericke University, Magdeburg (Lehrstuhl Logistik, IFSL) - Technical University, Dresden (Faculty of Mechanical Engineering; Institut für Fördertechnik; Baumaschinen und Logistik) - University of Ljubljana, University of Ljubljana, Slovenia).

Nuno Miranda has a BSc. in Computer Science Engineering from Higher Polytechnic Institute of Gaya (ISPGaya). He works in the computer industry field, where he is responsible for the development of new applications for Web and mobile platforms. His current research areas include software engineering, human interaction, and Web applications.

Giang N. T. Nguyen is a PhD Candidate in Leisure and Tourism in the Newcastle Business School, the University of Newcastle, Australia. She is a young and dynamic lecturer with a particular teaching and research interests in sustainable supply chain management, responsible leadership, corporate social responsibility, sustainable development and emerging technologies in operations management. She has contributed to several research articles in referred journals including International Journal of Corporate Social Responsibility, Social Responsibility Journal, and leading conferences like ANZAM, ANZIBA, EURAM, etc. She was also a researcher of USAID Green Invest Asia (GIA) and BINUS University, Indonesia to join research projects related to Woman in Sustainable Coffee. Giang has also been the recipient of the 2014 Full-fee waiver Transform Together Scholarship of Sheffield Hallam University, UK for her master degree and the 2020 Vice Chancellor's Higher Degree by Research Training Program Scholarship of the University of Newcastle, Australia for her doctoral degree.

Ihtsham Ul Haq Padda is currently serving as chairman at Department of Economics in Federal Urdu University of Arts, Science and Technology, Islamabad. He has more than 10 years of research and teaching experience. He is development and environment economist. He has more than fifty articles and one book. He is also member of many environment and development organizations.

Seeprata Parajuli is currently engaged as a research assistant at Quest Research Management Cell, Quest International College, Pokhara University, Nepal. She has been assisting research work of graduate and undergraduate management students at Quest. Her area of interest is behavioral studies. Her publication is related to perception and determinants of the customers in various field of management. She has presented her research ideas in different conferences.

Vladimir Radivojević received his PhD in Economics from the University of Niš – Faculty of Economics in 2014. In 2017 he was promoted to research associate from the Ministry of Education, Science and Technological Development of the Republic of Serbia. He was hired as an assistant professor at the University of Priština in Kosovska Mitrovica – Faculty of Economics in 2020. He is a member of the editorial board of the scientific journals Ekonomika and Economics of Sustainable Development. He writes and presents widely on issues of functioning and competitiveness of national economies, competition policy, and market concentration, and is the author of one book and more than 40 scientific papers published in domestic and international scientific journals, thematic collections, and collections from scientific conferences.

Mehmet Ali Taş completed his undergraduate education in Industrial Engineering at Yıldız Technical University and master's degree at Süleyman Demirel University. He continues PhD program at Galatasaray University. His research interests include fuzzy logic, multi-criteria decision making, and sustainability.

Niray Tunçel is working as an Assistant Professor at Hacettepe University, Department of Business Administration. She earned a master's degree in Marketing at Bahçeşehir University. Then, she received her Ph.D. in Business Administration from Hacettepe University. In her doctoral period, she spent a semester at the University of Primorska conducting a cross-cultural study. She also researched at the University of Mannheim as a visiting fellow after her Ph.D. Besides, she has participated in various international conferences, been involved in different research projects, and published several papers in international scientific journals. Her research interests include Consumer Behavior, Consumer Ethics, Digital Marketing, Sharing Economy, and Sustainable Consumption.

M. Ali Ülkü is a Full Professor and the Director of the Centre for Research in Sustainable Supply Chain Analytics (CRSSCA), in the Rowe School of Business at Dalhousie University, Canada. Dr. Ülkü's writings include the theoretical modelling of sustainable supply chain and logistics systems, operations-marketing interface, and mathematical modelling of consumer behaviour and societal problems. He published in such leading scholarly journals as European Journal of Operational Research, International Journal of Production Economics, Journal of Business Research, Journal of Cleaner Production, and Service Science. He taught at various universities in Canada, Turkey, and the USA. Dr. Ülkü is honoured to have received the university-wide 2007 Exceptional Teaching Award at the University of Waterloo and the 2019 Distinguished Professor Award from the International Society of Industrial Engineering and Operations Management. His co-edited book Big Data Analytics in Supply Chain Management: Theory and Applications is recently published by CRC Press-Taylor & Francis Group.

José G. Vargas-Hernández, M.B.A., Ph.D., Member of the National System of Researchers of Mexico and a research professor at University Center for Economic and Managerial Sciences, University of Guadalajara. Professor Vargas-Hernández has a Ph. D. in Public Administration and a Ph.D. in Organizational Economics. He has undertaken studies in Organisational Behaviour and has a Master of Business Administration, published four books and more than 200 papers in international journals and reviews (some translated to English, French, German, Portuguese, Farsi, Chinese, etc.) and more than 300 essays in national journals and reviews. He has obtained several international Awards and recognition.

Bruno Vieira has a BSc. in Computer Science Engineering from Higher Polytechnic Institute of Gaya (ISPGaya). He works in the computer industry field, where he is responsible for the development of new applications for Web and mobile platforms. His current research areas include software engineering and development.

Serhat Yüksel is Associate Professor of finance in İstanbul Medipol University. Before this position, he worked as a senior internal auditor for seven years in Finansbank, Istanbul-Turkey and 1 year in Konya Food and Agriculture University as an assistant professor. Dr. Yüksel has a BS in Business Administration (in English) from Yeditepe University (2006) with full scholarship. He got his master degree from the economics in Boğaziçi University (2008). He also has a PhD in Banking from Marmara University (2015). His research interests lie in banking, finance, and financial crisis. He has more than 220 publications (books, book chapters, scientific articles, etc.).

Index

G

H

I

K

M

O

P

Q

R

S

T

U

W

X

www.ingramcontent.com/pod-product-compliance
Lightning Source LLC
LaVergne TN
LVHW062136300926
849161LV00010B/2547